T0275533

# Optimization Models in Electricity Markets

Get up-to-speed with the fundamentals of how electricity markets are structured and operated with this comprehensive textbook.

**Key features**

- Coverage of key topics in electricity markets, including power system and power market operations, transmission, unit commitment, reserves, demand response, hydrothermal planning, investment in generation capacity and risk management.
- Over 140 practical examples, inspired by industry applications, connecting key theoretical concepts to practical scenarios in electricity market design.
- Over 100 coding-based examples and exercises with mathematical programming models, and selected solutions for readers.

Without requiring an advanced background in power systems or energy economics, this is the ideal introduction to electricity markets for senior undergraduate and graduate students in electrical engineering, economics, and operations research, and a robust introduction to the field for professionals in utilities, energy system operations, and energy regulation. Accompanied online by datasets, AMPL code, supporting videos, and full solutions and lecture slides for instructors.

**Anthony Papavasiliou** is an Assistant Professor of Electrical and Computer Engineering at the National Technical University of Athens, and a former Associate Professor and holder of the ENGIE Chair at UCLouvain. He obtained his PhD and conducted post-doctoral research at the department of Industrial Engineering and Operations Research at UC Berkeley. He is the recipient of the Francqui Foundation research professorship, an ERC Starting Grant, and the Bodossaki Foundation Distinguished Young Scientist award. He is a former Associate Editor for Operations Research and IEEE Transactions on Power Systems.

"Strong interactions in electricity networks require coordination to support competition. The key to successful electricity market design exploits the convex case of equivalence between economic dispatch and market equilibrium. This places optimization theory and models as essential ingredients for the great variety of applications explored in this vital book."

William Hogan, *Harvard University*

"This book highlights the important role of optimization models in modern power systems planning, operations and markets. It provides an excellent introduction to the application of optimization models for power systems students who do not have an operations research background, and introduces operations research students to the variety of problems arising in the context of power systems."

Shmuel S. Oren, *University of California, Berkeley*

# Optimization Models in Electricity Markets

**Anthony Papavasiliou**
National Technical University of Athens

CAMBRIDGE
UNIVERSITY PRESS

Shaftesbury Road, Cambridge CB2 8EA, United Kingdom

One Liberty Plaza, 20th Floor, New York, NY 10006, USA

477 Williamstown Road, Port Melbourne, VIC 3207, Australia

314–321, 3rd Floor, Plot 3, Splendor Forum, Jasola District Centre, New Delhi – 110025, India

103 Penang Road, #05–06/07, Visioncrest Commercial, Singapore 238467

Cambridge University Press is part of Cambridge University Press & Assessment, a department of the University of Cambridge.

We share the University's mission to contribute to society through the pursuit of education, learning and research at the highest international levels of excellence.

www.cambridge.org
Information on this title: www.cambridge.org/highereducation/isbn/9781009416610

DOI: 10.1017/9781009416627

© Anthony Papavasiliou 2024

This publication is in copyright. Subject to statutory exception and to the provisions of relevant collective licensing agreements, no reproduction of any part may take place without the written permission of Cambridge University Press & Assessment.

First published 2024

*A catalogue record for this publication is available from the British Library*

*A Cataloging-in-Publication data record for this book is available from the Library of Congress*

ISBN 978-1-009-41661-0 Hardback

Additional resources for this publication at www.cambridge.org/Papavasiliou

Cambridge University Press & Assessment has no responsibility for the persistence or accuracy of URLs for external or third-party internet websites referred to in this publication and does not guarantee that any content on such websites is, or will remain, accurate or appropriate.

# Contents

# Preface

**Intended readers** This book is intended to be used by students, researchers, and industry professionals with a background in operations research or power systems who are interested in learning about or conducting research in electricity markets. Readers would benefit from an elementary background in operations research. A background in economics or power systems engineering is not required. The learning of the material is enhanced by solving problems on a computer solver using a mathematical programming language (e.g. AMPL, GAMS, JuMP, Pyomo). Basic familiarity with a mathematical programming language is therefore especially useful.

**Summary of content** Chapter 1 provides an overview of mathematical programming models that are used in the energy industry. The chapter then continues with an introductory discussion about the economic interpretation of mathematical programming models for electricity markets in the specific context of the capacity expansion planning problem. The link is made between the capacity expansion model and the missing money problem in order to motivate the relevance of modeling for policy analysis.

Chapter 2 covers introductory concepts in duality theory, Karush–Kuhn–Tucker (KKT) conditions, and sensitivity. Supplementary introductory material on linear programming is provided in appendix A. The material on duality is developed as a self-contained learning unit. KKT conditions, which are used extensively throughout the text, are derived from duality theory.

Chapter 3 provides an overview of power system operations and power market operations. A minimum engineering background is provided regarding the operation of electric power systems, ranging from production, to transmission and distribution, down to the consumption of electricity. The timing of power system operations is classified between short-term, medium-term, and long-term operations. This classification guides the organization of the material in subsequent chapters. Electricity markets, which are tightly linked with power system operations, are introduced. Important design dilemmas related to electricity markets are discussed, including the exchange/pool dilemma and the design of auctions. A generic blueprint of a typical electricity market is then outlined, which is analyzed in increasing detail in subsequent chapters.

The next four chapters cover short-term electricity market operations, ranging from real-time to hour-ahead operations. Chapter 4 introduces the simplest possible model of an electricity market, where only energy is traded. After defining competitive market equilibrium, the interpretation of the economic dispatch model as a competitive market equilibrium is discussed in detail. This interpretation is generalized in the last section of the chapter, where the equivalence between competitive market equilibrium and optimization is established. This result is invoked repeatedly throughout the text for providing economic interpretations to models of increasing complexity.

Chapter 5 adds one layer of detail to the economic dispatch model by introducing the transmission network, leading to the optimal power flow model. This chapter presents various linear approximations of the physical equations that govern the flow of electric power. The economic interpretation of the optimal power flow problem is discussed and is linked to the so-called locational marginal pricing market design, which is compared to the zonal pricing market design.

Chapter 6 further refines the economic dispatch model by introducing reserves, which are used for operating the system securely under uncertain conditions. The simultaneous and sequential procurement of reserves are discussed and compared. The activation of reserves is discussed in the context of balancing markets.

Chapter 7 concludes the modeling of short-term operations by introducing unit commitment, the task of scheduling generators in the day-ahead and intra-day time frame. The binary (on/off) nature of this scheduling task presents challenges in the economic interpretation of the model and the design of mechanisms for rewarding units that incur fixed scheduling costs. The pool-versus-exchange dilemma is discussed on the basis of the unit commitment model.

The next two chapters focus on medium-term (month-ahead and year-ahead) aspects of power market operations, specifically hydrothermal planning and financial hedging. Chapter 8 focuses on the hydrothermal planning problem. The chapter commences with a discussion of modeling decision making under uncertainty in two and in multiple periods, as well as various ways for representing uncertainty that include scenario trees, lattices, and stagewise independence. These ideas are then applied to the hydrothermal planning problem, and the concept of dynamic programming is introduced, along with a discussion of the value of water. The chapter concludes with metrics for evaluating the performance of a stochastic programming model relative to alternatives such as perfect foresight or deterministic policies.

Various instruments that are widely used in electricity markets are defined in chapter 9, including forward contracts, futures contracts, financial transmission rights, call options, and bundles of the above. The use of these instruments for hedging risk in the production, transmission, and consumption of electricity are demonstrated through examples.

The next two chapters focus on long-term (multi-year ahead) decision problems. Chapter 10 focuses on the design of mechanisms which can be used for mobilizing demand response. This is classified as a long-term topic because the focus is on long-term (e.g. multi-year) contracts that can be used for enlisting flexible consumers to participate actively in supporting system operation. Two specific approaches are discussed, time of use pricing and priority service pricing.

Chapter 11 focuses on the capacity expansion planning model and its economic interpretation in terms of how energy markets can be used for rewarding investment costs. This chapter then discusses various alternative designs that can be used for remunerating investment cost, including capacity mechanisms and adaptations to ancillary services markets.

Chapter 12 expands the analytical methodology of the text to models beyond electricity markets, with a specific focus on hydrocarbons (oil and natural gas), biofuels, and nonrenewable resources. Rather than focusing on a detailed analysis of the energy sectors themselves, the goal of this chapter is to illuminate how KKT analysis, which is an overarching theme in the textbook, can be used for the analysis of diverse phenomena in energy sectors beyond electricity markets.

Background knowledge and exercise solutions are provided in the appendices. Appendix A covers basic knowledge in linear programming, and appendix B introduces the direct current power flow. Appendix C provides brief solutions to exercises.

**Use of the material in courses**    I have used the material in this book for a number of courses spanning various subjects (power system economics, energy economics, mathematical programming), background training (operations research, energy engineering), and levels of expertise (undergraduate, graduate).

The core material was developed for a graduate course in quantitative models for electricity markets at UCLouvain, which targets students in mathematical engineering with an optimization background but no exposure to power systems. For this purpose, I followed the same structure as of the book by selecting topics across chapters 1 to 11.

The material has also been used for undergraduate and graduate courses in power system economics at the department of electrical and computer engineering at the National Technical University of Athens. The main difference here is that chapter 2 is replaced by appendix A, and certain parts of the main content that rely more on duality (for instance, discussions on pricing in unit commitment models, chapter 7) are dropped from the lectures. The remainder of the structure can be followed straightforwardly (chapter 1 and choices from chapters 3–11), with less of an emphasis on background engineering knowledge in chapter 3.

A course on power system economics such as the one described in the previous paragraph can be supplemented in a subsequent semester by topics related to reliability and specialized topics in electricity market design. For instance, an undergraduate course on economic and reliable system operations can cover topics related to reserves, unit commitment, demand response, and expansion planning. An indicative sequence of lectures includes a review of chapter 1, appendix A, and chapters, 3, and 4, followed by a detailed treatment of chapters 6, 7, 10, and 11.

The material has further been used for undergraduate and graduate courses in energy economics at the National Technical University of Athens, where students are not assumed to have prior exposure to operations research. Apart from replacing chapter 2 with appendix A, topics that are more specific to the electricity sector can be skipped. An indicative sequence of lectures includes chapter 1, appendix A, and chapters 3, 4, 9, 11, and 12.

**Table 0.1 Possible use of the book material in courses.**

| Student background | Course topic | Chapters/Appendices |
|---|---|---|
| Operations research graduate | Electricity markets | 1–11 |
| Energy undergraduate/graduate | Electricity markets | 1, A, 3–11 |
| Energy undergraduate | Reliable system operations | 1, A, 3, 4, 6, 7, 10, 11 |
| Energy graduate/undergraduate | Energy markets | 1, A, 3 4, 9, 11, 12 |
| Operations research/energy graduate/undergraduate | Math programming models | A, 2, 4.1, 7.1, B, 5.1, 6.2, 11.1 |

Finally, parts of the textbook have been used in an undergraduate course on mathematical programming models. Here, students are not assumed to have any exposure to energy or power systems, and the emphasis is rather on implementing mathematical programming models with industry relevance. An indicative sequence of lectures that can be used for this purpose is appendix A, chapter 2, section 4.1, section 7.1, sections 5.1 and 6.2, chapter 8, and section 11.1.

This information is summarized in table 0.1.

**Exercises**   Exercises are provided at the end of each chapter. Terse solutions are provided in appendix C. Additional resources are available to instructors, including a detailed solutions manual for the textbook exercises, as well as detailed solutions to additional exercises beyond the textbook. Certain exercises involve programming, and the available data and codes are available on the textbook website: www.cambridge.org/Papavasiliou. Exercises include theoretical proofs of results that are presented or alluded to in the textbook (with varying difficulty), the formulation of mathematical programs on paper and their analysis or implementation in code, simple applications of concepts presented in the text on numerical examples, or more elaborate numerical examples that require code. There is no specific indication of the level of difficulty of any given exercise.

**Notation and terminology**   Mathematical notation in the textbook varies from chapter to chapter, nevertheless I do try to follow some standard notation that is used in the scientific literature, and notation is anyway explained every time a new model is introduced. English lowercase letters are often used for decision variables, English uppercase letters are often used for parameters, and Greek lowercase letters are often used for dual variables. Market clearing prices are typically denoted as $\lambda$ or $\rho$, quantities produced are often denoted as $p$, and quantities consumed are often denoted as $d$. Binary commitment decisions are often denoted as $u$. There is no bold, uppercase, or otherwise specific notation for vectors. Whenever introduced, vectors typically correspond to column vectors, and their transpose is typically denoted with a T superscript.

Part of what makes the energy industry and research area fascinating, but also challenging, is the speed and uncertainty of its evolution. The book uses numerous examples throughout, and refers to technologies such as "gas," "coal," "oil," or "nuclear." The cost figures that are used for these technologies, and their presence in the fuel mix, are representative of their cost and role at the time of writing of this book. Nevertheless, and as recent evolutions in the petroleum and gas markets suggest, things can change very rapidly in energy markets, in terms of both costs and the role of various technologies in future energy mixes. For this reason, these technology names should be understood as labels rather than literal references to a specific technology.

# Acknowledgments

I am grateful to numerous people for the development and evolution of this book. Firstly, I wish to thank the numerous students who have approached me with suggestions for and corrections to content. In addition, I wish to single out certain people who have influenced my thinking about our scientific area and therefore the content of this book. Shmuel Oren, my PhD advisor, has been a constant source of inspiration. And Yves Smeers, my informal advisor during my time at UCLouvain, has greatly shaped my understanding of electricity market models. Special thanks to William Hogan for making himself available during my visits to Harvard University during my sabbatical under the Francqui Foundation professorship. I am also especially grateful to my students for constant corrections and enrichments of the text, with special thanks to Ignacio Aravena for kicking off the solutions manual, and Quentin Lété for spotting important corrections in the text. Many of the examples and remarks in the book have been inspired by discussions with various colleagues and former students, including (in alphabetical order) Gilles Bertrand, Mette Bjørndal, Pantelis Capros, Jacques Cartuyvels, Philippe Chevalier, Jehum Cho, Athanasios Dagoumas, Gauthier de Maere d'Aertrycke, Gerard Doorman, Andreas Ehrenmann, Céline Gerard, Kory Hedman, Yves Langer, Mehdi Madani, Marijn Maenhoudt, Alain Marien, Yuting Mou, Alex Papalexopoulos, Nicolas Stevens, Mathieu Van Vyve. Special thanks to Stephen Boyd for granting permission for the notation and organization of the material in chapter 2, which is inspired by his own material. Also special thanks to Christos Karydas, whose teaching material largely inspired the content of chapter 12. I am especially grateful to the Francqui Foundation and the European Research Council for important financial support, which created resources (including time) that made this project possible.

# 1    Introduction

The study of electricity markets through quantitative models is challenging due to the interdisciplinary nature of the task. A background in power systems engineering, economics, and operations research is useful, if not necessary, for developing meaningful quantitative models. For people entering a career in electricity markets, the requirement for such a broad background may appear daunting.

During my graduate studies, I recall collecting information from journals and textbooks that did not necessarily convey as strong a message when read as stand-alone information, but was far more interesting when (i) assembled and compared, and (ii) related to the institutional and physical constraints that govern the operation of electric power systems and electricity markets. This textbook is an attempt to assemble all the information I wish I had access to before kicking off my graduate studies, not so much as stand-alone models, but as an ensemble. The principal goal of the text is to draw the big picture for the reader who is not an expert in power systems engineering, economics, or operations research, and provide an understanding that is sufficient for kicking off analysis and research in the quantitative modeling of electricity markets.

This chapter introduces the reader to the study of electricity markets through optimization models. Section 1.1 motivates the study of electricity markets through an optimization lens. Section 1.2 provides a motivating example by discussing the missing money problem. The example is intended to highlight that even relatively simple models can become puzzling unless a holistic view towards electricity market modeling is adopted.

## 1.1    Optimization meets energy

The energy industry has been a major domain for the application of operations research since the birth of the field of operations research. Numerous applications in electricity markets require advanced decision making and economic modeling due to the large scale and complex operations of electric power systems. Such applications include (i) tactical operations such as the optimal scheduling of power production, asset management, and the integration of renewable energy sources, (ii) strategic operations such as risk management and the optimal sizing and expansion of the

1

system, and (iii) policy analysis such as price and welfare analysis, competition analysis, and the analysis of environmental policies, to mention a few examples. The ability of the energy industry to adopt advanced optimization algorithms such as branch-and-bound Lagrange relaxation, and dynamic programming testifies that the energy sector is thirsty for models that are not only capable of capturing the complexity of electric power systems, but also scalable and computationally tractable.

Operations research is poised to remain relevant for electric power system operations for years to come. The deregulation of electricity markets, the large-scale integration of renewable energy sources, the coordination of market operations, the expansion of environmental policies, and the advent of smart grids and demand response are a number of policy-driven changes that inspire questions that can be addressed meaningfully through operations research. Moreover, the remarkable evolution of optimization software and the advent of parallel computing suggests that the research community will be capable of formulating models of increasing scale and addressing questions of increasing complexity.

As the title of the textbook suggests, the material focuses on models, not algorithms. Moreover, the text is focused on optimization models, not equilibrium models which cannot be cast in equivalent optimization form. Therefore, the assumptions about the economic behavior of agents are fairly simple. In particular, questions of strategic behavior are not addressed. Although optimization models may oversimplify important behavioral aspects of market agents, they remain scalable to the size of realistic systems, can be used to address an enormous array of interesting questions, and serve as a very useful benchmark for comparing more complex models involving strategic agent interaction.

In the remainder of this section we discuss certain classes of mathematical programming models, as well as corresponding applications in the energy sector, algorithms that can be used for tackling these optimization models, and relevant software packages. The goal is not to be exhaustive, but rather to allow the reader to place some of the terms that may be encountered in the literature as part of the broader landscape. The classification of problems is between optimization models (which this textbook is concerned with) and models beyond optimization (which extend beyond the scope of the textbook). The discussion about algorithms is limited to naming them, with no attempt whatsoever to expand on how or why they work, since this goes well beyond the scope of the book. The reference to software is also merely indicative, without attempting to be exhaustive.

### 1.1.1  Optimization

Linear programming is the simplest and most broadly used class of mathematical programs. These are models with a linear objective function and linear constraints.[1]

---

[1] For readers who lack a background in mathematical programming, linear programs are briefly introduced in appendix A, while certain optimization notions that are specifically pertinent to this textbook are developed in chapter 2.

Linear programs are applied extensively in the energy sector. The economic dispatch model that we develop in chapter 4 is an example of a linear program. Economic dispatch is a real-time application. By the time economic dispatch is solved, operators have largely positioned the system, and the only things left to decide are adjustments to power output that can balance the system. This decision problem can be expressed as a linear program. Typical algorithms that one may use for tackling linear programs include the classic simplex algorithm and interior point methods (a specific case of which is the barrier method). Indicative commercial software for this class of problems includes CPLEX, Gurobi, and Xpress, as well as open-source software that is available online, e.g. in the COIN-OR library (such as Cbc).

Large-scale linear programming includes two-stage and multi-stage stochastic linear programming. Stochastic programs attempt to represent uncertainty endogenously within the optimization model by adapting decisions to the uncertainty that unfolds throughout the decision-making horizon. A famous application of multi-stage stochastic linear programming in the energy sector is hydrothermal planning in medium-term operations, where the goal is to regulate the water levels of hydro reservoirs in hydrothermal systems with river networks subject to monthly uncertainty about rainfall. Another application is multi-year generation capacity expansion under uncertainty about climate conditions and component outages. The generation capacity expansion problem in deterministic form is formulated in chapter 11. Certain algorithms in this space include stochastic dual dynamic programming, progressive hedging, and other special-purpose decomposition algorithms. Such tailored algorithms often use the aforementioned commercial solvers as modules. Certain commercial packages for analyzing power systems (such as PSR-SDDP and Plexos) implement such algorithms internally.

Mixed integer linear programming is a generalization of linear programs, where the objective function and constraints of the model remain linear, but where decisions are mixed integer. This means that, although some decisions can be continuous, others should be integer. A famous application of mixed integer linear programming in power systems is day-ahead unit commitment, which is covered in chapter 7. The goal here is to schedule the on–off status of units (which are integer decisions) as well as their power output and the amount of reserve that these units offer to the system (which are continuous decisions). The backbone algorithm for mixed integer linear programming is the branch-and-bound algorithm, which uses linear programming solvers as modules. Commercial software that can handle linear programs can often handle mixed integer linear programs, as is the case for CPLEX, Gurobi, and Xpress.

Nonlinear convex optimization is a class of models where the objective or constraints may be nonlinear, but where the problem is convex (convex mathematical programs are defined in chapter 2). An example of a nonlinear convex objective function is the quadratic fuel cost of certain units. Interior point methods can be used for tackling this class of problems, and certain commercial software that can handle linear programs can also be used for this class of problems.

Nonlinear non-convex optimization problems are entirely general models, which have an objective function or constraints that are not convex. The optimal power flow problem (which is discussed in chapter 5) is originally a nonlinear non-convex optimization model. The non-convexity arises from the fact that the power flow equations (formulated in appendix B), which describe how the flow of power on transmission lines depends on the injection of power in different parts of the power network, result in a non-convex mapping. Non-convex models also emerge in the gas sector due to the complex physical relations that govern the flow of gas on pipelines. Nonlinear non-convex optimization problems pose significant computational difficulties, and models of this form are typically not as scalable as their linear or convex counterparts. The economic interpretation of these models is also challenging because the connection between optimization and decentralized coordination is broken. We note that numerous other problems (such as certain formulations of optimization under uncertainty, e.g. robust optimization, or problems with complementarity constraints) can be cast as nonlinear non-convex optimization problems. Nonlinear optimization solvers include ipopt, baron, and knitro.

## 1.1.2  Beyond optimization

A result that is invoked repeatedly in this textbook is the equivalence between certain classes of convex optimization models and decentralized decision making, as stated in proposition 4.11. This result allows us to attach economic interpretations to many of the models presented in this textbook. However, the assumptions that are needed for arriving to this equivalence include the notion of perfect competition, i.e. the fact that market agents do not explicitly account for the actions of other market agents (but rather implicitly through a fixed market price) when deciding on their own operations.

This perfect competition assumption is often not fulfilled in practice. Instead, strategic interactions are a common phenomenon in energy markets, whereby agents explicitly account for the strategies of their competitors when deciding how to proceed with their own actions. Certain types of strategic interactions, such as strategic competition of agents in quantities, are captured through classes of mathematical programs beyond optimization problems. These include mixed complementarity problems (MCPs), mathematical programs with equilibrium constraints (MPECs), and equilibrium problems with equilibrium constraints (EPECs). For instance, the Cournot competition of a small number of producers who decide how much power to produce while attempting to balance out loss of market share with sustained higher prices is an example of a model that can be expressed as an MCP.

The present textbook is largely not concerned with such models. The premise is that even the simplest perfect competition models include significant insights that are valuable to understand before advancing to the more complex phenomena that are captured by MCPs, MPECs, and EPECs. One notable exception is the priority service model for demand response in chapter 10. In this model, there is a hierarchical interaction between the designer of a price menu for demand service and a population of retail customers who react by selecting options among the menu as a function of how the menu designer sets prices in the first place. Strategic models are also discussed in chapter 12 in the context of applications in the petroleum and natural gas industry.

## 1.2    An example: missing money

In this section, a simple capacity expansion planning model is analyzed. This simple model not only highlights a major policy challenge faced in electricity markets, but also raises interesting questions in terms of modeling. The development purposefully avoids any math, but rather attempts to present notions (such as the optimal expansion plan of a system or competitive equilibrium) in the abstract. One of the intentions in doing so is to trigger the appetite of the reader for formal and unambiguous definitions, which follow in the subsequent chapters.

**Screening curves**   The chronological evolution of system load and wind power production in Belgium in 2013 is shown in figure 1.1. The net load of the country is obtained by subtracting total load from wind power production. Sorting the time series of net load in descending order yields the **load duration curve** of the system, which is shown in figure 1.2. This curve describes the number of hours in the year that the load was greater than or equal to a given level. For example, net load was at or above 10000 MW for 2000 hours during the year.

Using the load duration curve, load can be stratified horizontally, rather than chronologically. Figure 1.2 presents a step-wise approximation of the load duration curve. According to this stepwise approximation, base load corresponds to the lower horizontal block in the figure and amounts to 7086 MW. This corresponds to loads that last for the entire year. The medium horizontal block corresponds to loads between 7086 and 9004 MW. These loads last for 7000 hours. Peak load corresponds

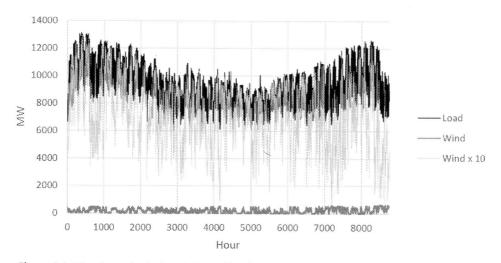

**Figure 1.1**  The chronological evolution of load and wind power in Belgium in 2013.

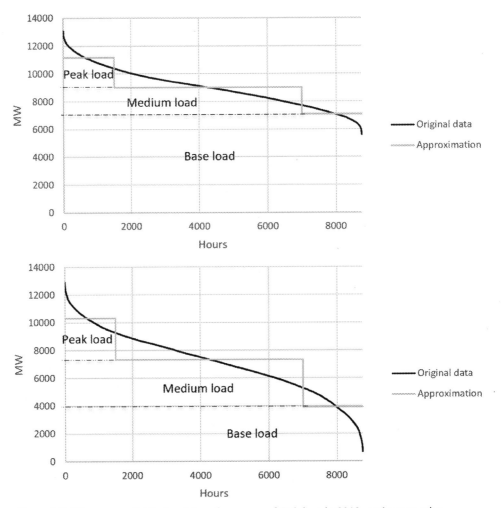

**Figure 1.2** (Upper panel) The load duration curve of Belgium in 2013, and a step-wise approximation. (Lower panel) The load duration curve if wind production were 10 times greater.

to the upper horizontal block and ranges between 9004 MW and 11169 MW. These loads last for 1500 hours.

The set of technologies that can be used for serving demand is shown in table 1.1. Fuel cost corresponds to the variable cost of each plant and is proportional to the amount of energy produced. Investment cost is independent of output and proportional to the total capacity built. The investment cost has been amortized in the table in order to reflect the *hourly* cash flow that is required in order to support an investment in 1 MW of the given technology. Overnight and amortized investment cost are discussed more extensively in chapter 3.

The optimal investment problem can be posed as follows: find the mix of technologies that should be built in order to minimize the total fixed and variable cost of serving demand. A graphical solution to the problem can be derived from the

**Table 1.1** The set of options for serving demand in the example of section 1.2.

| Technology | Fuel cost ($/MWh) | Inv cost ($/MWh) |
|---|---|---|
| Coal | 25 | 16 |
| Gas | 80 | 5 |
| Nuclear | 6.5 | 32 |
| Oil | 160 | 2 |

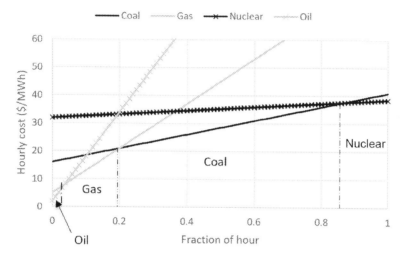

**Figure 1.3** A screening curve describes the total hourly cost of each technology as a function of the fraction of time that it is producing.

**screening curve** of figure 1.3. The total cost of using 1 MW of each technology depends on the amount of time that the technology is used. In order to determine optimal investment, it is therefore useful to view load as a collection of horizontal slices. The only difference among slices is their duration, and this determines the technology that should be used for serving each slice. Base load should be served by a technology with low variable cost (even if fixed cost may be relatively high), whereas peak load should be served by a technology with low fixed cost (even if variable cost may be relatively high).

The exact fraction of time for which each technology should be functioning can be found from the lower envelope of the total cost function:

- For oil: $2 + 160 \cdot f \leq 5 + 80 \cdot f \Leftrightarrow f \leq 0.0375$. This corresponds to 0–328 hours.
- For gas: $f > 0.0375$ and $5 + 80 \cdot f \leq 16 + 25 \cdot f \Leftrightarrow f \leq 0.2$. This corresponds to 328–1752 hours.
- For coal: $f > 0.2$ and $16 + 25 \cdot f \leq 32 + 6.5 \cdot f \Leftrightarrow f \leq 0.8649$. This corresponds to 1752–7576 hours.
- For nuclear: $0.8649 \leq f \leq 1$. This corresponds to 7576–8760 hours.

It follows from the load duration curve of the system that if the goal is to always satisfy load, then the least cost capacity investment plan is the following (and the reader is asked to confirm this in problem 1.3 using a linear programming model):

- Base load is assigned to nuclear: 7086 MW
- Medium load is assigned to coal: 1918 MW
- Peak load is assigned to gas: 2165 MW
- No load is assigned to oil: 0 MW

The results correspond to the step-wise approximation of the load duration curve. If the actual load duration curve had been used instead, there would have also been an investment in peaking oil units.

**Short-run versus long-run equilibrium**   Consider how the energy price will evolve in this system. Suppose that suppliers are price takers, meaning that they do not account for the impact of their decisions on market prices. The **competitive market equilibrium**, which is formally introduced in chapter 4, is a combination of market prices and production quantities such that (i) no producer can benefit from changing their production quantity and (ii) supply satisfies demand.

The competitive equilibrium price is not necessarily unique. Assuming the configuration of plants has been decided already, *one* possible price in the competitive equilibrium is equal to the fuel cost of the **marginal unit**, i.e. the most expensive unit that is producing energy. Each hour of the year corresponds to a different marginal unit. To see that this is a competitive equilibrium price, note that at this price, for any hour of the year, (i) the marginal unit is serving the horizontal segment corresponding to its capacity, (ii) more expensive units are off, and (iii) cheaper units are serving the horizontal segment corresponding to their capacity. No unit can profitably deviate from its production schedule, and supply exactly equals demand.

Suppose that the optimal mix of generators determined previously (7086 MW of nuclear, 1918 MW of coal, 2165 MW of gas) has *already* been decided. Then, the competitive equilibrium price that is determined from the marginal cost of the marginal unit[2] is equal to 6.5 $/MWh during hours of base load, 25 $/MWh during hours of medium load, and 80 $/MWh during hours of peak load.

The profit of each technology can now be computed. Nuclear units earn zero profits when they are marginal, they earn a profit margin of $25 - 6.5 = 18.5$ $/MWh when coal is marginal (this lasts $(7000 - 1500)/8760 = 62.8\%$ of the year), and they earn a profit margin of $80 - 6.5 = 73.5$ $/MWh when gas is marginal (this lasts $1500/8760 = 17.1\%$ of the year). This corresponds to an average hourly profit of $0.628 \cdot 18.5 + 0.171 \cdot 73.5 = 24.2$ $/MWh. Similarly, coal earns a profit of $80 - 55 = 25$

---

[2] Returning to the point that the equilibrium price is not necessarily unique, it is left to the reader to verify that *given* that the configuration of plants has already been decided to be the optimal mix (7086 MW of nuclear, 1918 MW of coal, 2165 MW of gas), the following prices also result in an equilibrium: (i) any price in the interval from 6.5 $/MWh to 25 $/MWh for the hours of base load, (ii) any price in the interval from 25 $/MWh to 80 $/MWh for the hours of medium load, and (iii) any price at or above 80 $/MWh for the hours of peak load.

$/MWh during peak periods, and thus an average hourly profit of $0.1 / 1 \cdot 55 = 9.4$ $/MWh. Gas earns an average hourly profit of 0 $/MWh.

This analysis leads to an apparent contradiction: it seems that what has been defined (informally) as a competitive equilibrium results in a situation where no technology can survive in the market, in the sense of recovering its investment costs! More specifically, comparing the average profitability of each technology with its investment cost, it can be concluded that nuclear units are losing $32 - 24.2 = \$7.8$ per hour for each MW of investment, coal units are losing $16 - 9.4 = 6.6$ $/MWh, and gas units are losing $5.0 - 0 = 5.0$ $/MWh. None of these units are earning enough profit in the energy market to survive. Notably, the capacity of a typical nuclear plant is no smaller than 1000 MW, which would imply that such an investment would be suffering damages of at least $7800 for every hour that goes by during the lifetime of the plant! The results are not coincidental, in the sense that they do not happen to be the result of the specific choice of numbers for the example. Any choice of load duration curve, fuel costs, and investment costs produces the same result (some technologies do not earn sufficient profit to cover their fixed investment cost) if electricity is priced at the short-term marginal cost of the marginal unit. The reader can verify this immediately for the peaking technology: if a price cap is imposed that is equal to the marginal cost of the most expensive technology, the *peaking units* (units that serve peak demand) are never able to earn profits that cover their investment cost.

In order to resolve the apparent contradiction that arises in the example, it is important to distinguish a short-run equilibrium from a long-run equilibrium. In a short-run equilibrium, the capacities of plants are assumed fixed, in the sense that they have already been decided. For a long-run equilibrium, the capacities of the plants are also part of the decision of the generators. In such a case, the price determined previously as the marginal cost of the marginal unit is not an equilibrium price, because under such pricing generators would not be willing to invest in the optimal capacity in the first place. Instead (and this is shown formally in chapter 11), long-run equilibrium prices will have to rise above short-term marginal costs for certain hours of the year in order to convince investors to enter this market in the first place.

This example is intended to motivate the reader to reflect on the notion of economic equilibrium. The concept is defined precisely in chapter 4. At this stage of the development, it suffices to note that the concept can be powerful in terms of interpreting agent interactions, and that it needs to be defined precisely in order to avoid ambiguous results.

**Offer caps and the missing money problem**   Electricity markets are vulnerable to the strategic behavior of agents. **Market power** is exercised by producers who withhold capacity from the market in order to *profitably* increase market prices above competitive levels. The detection of market power is difficult because it can be challenging to distinguish the rise of prices as a result of strategic behavior from the rise of prices due to true scarcity in generating capacity. In order to prevent the gaming of the market, regulators impose *offer price caps* that limit the price at which generators can offer electricity to the market. There are also market price caps that limit the price at which the market clears.

The rise of price above marginal cost could (sometimes wrongly) be perceived as an indication of market power. For the case of the example, consider a price cap of 80 $/MWh. Note that this would be a fatal market design move: investors would never invest in gas generators if the market price is not allowed to hike above 80 $/MWh, and it has been previously shown that gas generators *should* be part of the optimal long-run fuel mix in this example. This example highlights the **missing money** problem, which occurs when energy prices are depressed due to price caps, leading to losses for generators and under-investment.

Suppose, now, that the actual wind production in the system is 10 times greater than the production assumed previously, as shown in figure 1.1. The load duration curve of the system is shown in the lower panel of figure 1.2. It can be noted that the load duration curve has now become less flat, and there are hours when the net load of the system is near zero. With the new load duration curve, the required amount of coal and gas capacity also increases.[3] It has been argued that a price cap of 80 $/MWh implies that in a competitive economic environment there will be no voluntary build-out of peaking gas units. Notice the tension: the increased integration of renewable resources increases the amount of natural gas capacity that is required in the optimal mix, whereas the price cap guarantees that under competitive economic conditions this capacity will not be built out voluntarily. In reality, the need for peaking units is in fact even greater, since these units typically have more flexible operating characteristics. These characteristics have been ignored here, and are considered in more detail when the unit commitment problem is developed in chapter 7.

The missing money persists even when the average cost of interrupting service to customers is factored into the analysis. If the load duration curve has a peak that is too short in duration, it is optimal to leave this peak demand unserved in order to avoid investing in generators that are underutilized. The average cost of service interruption can be incorporated into the screening curve by adding a line that crosses the origin of the screening curve,[4] the slope of which is equal to the average cost of interruption, as shown in figure 1.4. It can be observed that, for a small number of hours, load curtailment will be the "peaking unit." The price for those hours will be the "marginal cost" of load interruption. Since the marginal cost of service interruption is typically 10–100 times greater than the marginal cost of the most expensive unit in the system, the price cap is, in practice, likely to be lower than the marginal cost of service interruption. As is shown in chapter 11, such a price cap may interfere with the optimal economic outcome of a long-run competitive market equilibrium.

---

[3] Note that the reference to coal and gas is purely indicative, since our analysis does not factor in environmental externalities. You can think of coal and gas as labels for the example. What ultimately matters in the discussion is the fuel and investment costs of these technologies.

[4] It is assumed that load interruption requires zero investment cost. The cost of smart meters, load controllers, and so on, can be incorporated in the analysis as an investment cost, but is ignored in this numerical example.

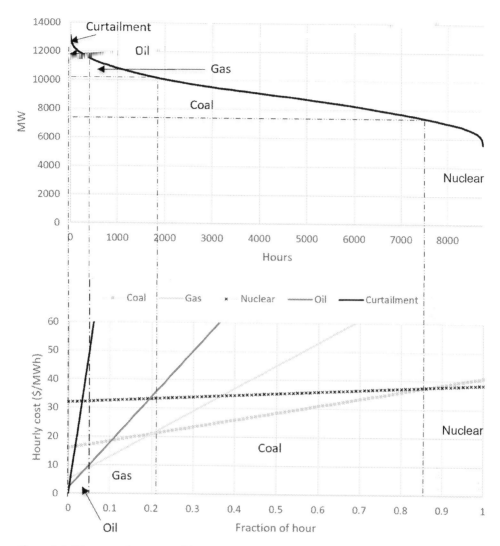

**Figure 1.4** The screening curve with demand response valued at 1000 $/MWh.

## Problems

**1.1** *Capacity expansion with demand curtailment using screening curves.* Consider a capacity expansion problem with the technology options of table 1.1, as well as a demand response "technology" with a "fuel cost" of 1000 $/MWh, as indicated in figure 1.4. The load duration curve of the upper panel of figure 1.2 is available on the textbook website. Solve the problem using screening curves.

**1.2** *Capacity expansion with demand curtailment using linear programming.* Solve problem 1.1 using a mathematical programming language and the model that is presented in section 11.1, which relies on a vertical representation of the load duration curve. Confirm that the results are (almost) identical in both problems, and explain any possible differences.

**1.3** *Optimal expansion for a three-step approximation of the load duration curve.* Consider the three-step approximation of the load duration curve that is presented in the upper panel of figure 1.2. It is argued in the text, using screening curve analysis, that the optimal investment in the system corresponds to 7086 MW of nuclear, 1918 MW of coal, and 2165 MW of gas. Confirm the results using the same linear programming model as that of problem 1.2.

**1.4** *Reading a load duration curve.* Denote a screening curve as $L(t)$, where $t$ is expressed in hours and $L$ is expressed in MW. Then, $L(2000) = 4000$ is interpreted as follows (select only one answer):
1. The load in the system is 4000 MW at the 2000th hour of the year.
2. The load of the system is at most 4000 MW for no more than 2000 hours in the year.
3. The load of the system is at least 4000 MW for 2000 hours in the year.

**1.5** *Relation between load duration curve and a distribution function.* Suppose that we sample load from a continuous cumulative distribution function $G$. Then, given enough samples, what would be the mathematical expression of the load duration curve $L(f)$, where $0 \leq f \leq 1$ corresponds to a fraction of time (as opposed to hours)?

**1.6** *Limitations of screening curve analysis.* True or false: we can use screening curves to determine the optimal technology mix in a system where one of the candidate technologies is storage.

# Bibliography

**Section 1.1**  The potential of using equilibrium models in order to analyze deregulated electricity and gas markets was developed by Smeers (1997). The institutional challenges that are posed from the integration of demand response are analyzed by Borenstein, Jaske, and Rosenfeld (2002). The application of Lagrange relaxation in unit commitment was pioneered by Muckstadt and Koenig (1977). The evolution from Lagrange relaxation to branch-and-bound for solving unit commitment in operations is discussed by Carrion and Arroyo (2006). The role of mixed complementarity problems, mathematical programs with equilibrium constraints, and equilibrium problems with equilibrium constraints in energy systems modeling is discussed in Gabriel et al. (2012).

**Section 1.2**  The relation between equilibrium market prices, peak load pricing, and decisions of investment in power generation capacity is investigated in early research by Boiteux (1960). Shanker (2003) coined the term "missing money" that was to anchor later discussions on insufficient investment in generation: missing money appears when the market insufficiently remunerates a service or a commodity. Hogan (2005) developed the concept and explains that the missing money is a market imperfection resulting from regulatory intervention capping energy prices in a restructured power market. Fabra (2018) provides an interesting discussion about the relationship between market price caps, market power, and capacity markets.

# 2    Mathematical background

This chapter covers mathematical background that leads to notions that are used extensively in the text. The chapter uses the notation of and closely follows Boyd and Vandenberghe (2008). Section 2.1 defines convex optimization problems. Sections 2.2 and 2.3 develop duality and the KKT conditions as self-contained units. Section 2.4 provides a sensitivity interpretation of dual multipliers that is used repeatedly in the text in order to gain intuition about the models that are developed in the textbook. Before developing the duality results, a number of definitions need to be recalled. For readers who have no prior experience in linear and, more generally, mathematical programming, an introduction is provided in appendix A.

## 2.1    Convex optimization problems

Consider a set of points $x_i \in \mathbb{R}^n, i = 1, \ldots, n$. A **convex combination** of these points is a point $\sum_{i=1}^{n} \lambda_i x_i$, such that $\sum_{i=1}^{n} \lambda_i = 1$ and $\lambda_i \geq 0, i = 1, \ldots, n$. $X$ is a **convex set** if it contains any convex combination of points $x_i \in X$. Geometrically, a convex combination of two points is a line connecting these points. A set is non-convex if it does not contain all the convex combinations of its points. These concepts are illustrated graphically in figure 2.1.

We say that $f$ is a **convex function** if for all $0 \leq \lambda \leq 1$ and any $x_1, x_2$ we have $f(\lambda x_1 + (1-\lambda)x_2) \leq \lambda f(x_1) + (1-\lambda)f(x_2)$. The function $f$ is **concave** if $-f$ is convex. The definition of a convex function is illustrated graphically in figure 2.2.

An **optimization problem** is the problem of finding the minimum of a function $f$ over a set $X \subset \mathbb{R}^n$:

$$\min_x f(x)$$
$$\text{subject to } x \in X.$$

We say that $X$ is the **feasible set** of the problem and the function $f$ is the **objective function** of the problem. Any $x \in X$ is a **feasible solution**. Any $x^\star \in X$ such that $f(x^\star) \leq f(x)$ for any $x \in X$ is an **optimal solution**. "Subject to" is abbreviated as "s.t.," or omitted, in what follows. A **convex optimization problem** is an optimization problem with a convex objective function and a convex set of feasible solutions.

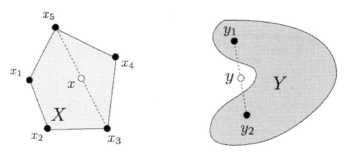

**Figure 2.1** Consider points $x_1, \ldots x_5 \in \mathbb{R}^2$. The point $x$ is a convex combination of $x_3$ and $x_5$ because it can be expressed as $x = 0.5x_3 + 0.5x_5$. The set $X \subset \mathbb{R}^2$ is a convex set because any point that can be expressed as a convex combination of distinct points in $X$ also belongs to $X$. The set $Y \subset \mathbb{R}^2$ is not convex because there are convex combinations of points in $Y$ that do not belong in $Y$, for example $y = 0.5y_1 + 0.5y_2$ does not belong in $Y$.

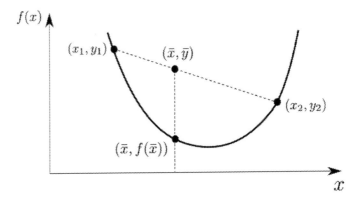

**Figure 2.2** A graphical illustration of a convex function with a domain in $\mathbb{R}$. Consider points $x_1, x_2 \in \mathbb{R}$ with $y_1 = f(x_1)$ and $y_2 = f(x_2)$. Here, $f$ is a convex function because, for any points $x_1, x_2$, and for any choice of $0 \leq \lambda \leq 1$, the value of the function at $\bar{x} = \lambda x_1 + (1 - \lambda)x_2, f(\bar{x})$, is no greater than the convex combination $\bar{y} = \lambda y_1 + (1 - \lambda)y_2$.

## 2.2  Duality

Duality theory is based on the notion of allowing certain constraints of an optimization problem to be violated at a penalty. This gives rise to a new optimization problem, called the dual problem, the optimal solution of which admits important economic interpretations, including the price of a market that is in equilibrium (a precise definition of equilibrium will be provided in the next chapter), agent profits, and so on. In order to define the dual problem, it is necessary to first define a Lagrangian function.

Consider the following optimization problem:

$$p^\star = \min_x f_0(x)$$
$$\text{s.t.} \, f_i(x) \le 0, i = 1, \ldots, m$$
$$h_i(x) = 0, i = 1, \ldots, p$$
$$x \in \text{dom} f_0 \subset \mathbb{R}^n,$$

where the **domain** of a function $f$, dom $f$, is the set where $f$ is defined.

Clearly, any vector $x$ that has a chance at being optimal should lie within dom $f$. The $m$ inequality constraints, $f_i(x) \le 0$, and $p$ equality constraints, $h_i(x) = 0$, further restrict the feasible set of vectors that should be considered.

Consider now the possibility of *relaxing* the inequality and equality constraints: allowing them to be violated, but penalizing their violation. In particular, consider the following **Lagrangian function**, the domain of which is a subset of $\mathbb{R}^n \times \mathbb{R}^m \times \mathbb{R}^p$:

$$L(x, \lambda, \nu) = f_0(x) + \sum_{i=1}^m \lambda_i f_i(x) + \sum_{i=1}^p \nu_i h_i(x).$$

The Lagrangian function can be interpreted as a new objective function that adds to the objective function $f_0$ a weighted sum of the constraint functions that are relaxed. The weights $\lambda_i$ (for inequality constraints) and $\mu_i$ (for equality constraints) are referred to as **Lagrange multipliers** and can be interpreted as the penalty coefficients by which the violation of the relaxed constraints is penalized.

In much of the remaining text, the multipliers that correspond to a certain set of constraints are placed to the left of the corresponding constraints in the formulation of the problem, as in the following notation:

$$\min_x f_0(x)$$
$$\text{s.t.} \, x \in \text{dom} f_0 \subset \mathbb{R}^n$$
$$(\lambda_i): \, f_i(x) \le 0, i = 1, \ldots, m$$
$$(\nu_i): \, h_i(x) = 0, i = 1, \ldots, p.$$

Now consider fixing the Lagrange multipliers $\lambda$ and $\nu$ to a certain value. The following **Lagrange dual function** is defined as:

$$g(\lambda, \nu) = \min_{x \in \text{dom} f_0} L(x, \lambda, \nu)$$
$$= \min_{x \in \text{dom} f_0} \left( f_0(x) + \sum_{i=1}^m \lambda_i f_i(x) + \sum_{i=1}^p \nu_i h_i(x) \right).$$

Note that the dual function $g(\lambda, \nu)$ can be $-\infty$ for some $\lambda, \nu$. The values of $(\lambda, \nu)$ for which the dual function is not $-\infty$ correspond to the domain of the dual function.

It is easy to check that the dual function provides a lower bound to the original optimization problem if $\lambda \ge 0$. To see this, note that the minimization of the Lagrangian function (which defines the dual function) permits a violation of the relaxed constraints, but if they are respected then any feasible solution $x$ receives a "bonus" in the Lagrangian function whenever $\lambda \geq 0$, in the sense that $L$ evaluates to something less than $f_0(x)$. This argument is developed formally in proposition 2.1.

**Proposition 2.1**  *If $\lambda \geq 0$ then $g(\lambda, \nu) \leq p^\star$.*

*Proof*  Consider an $\bar{x}$ such that $f_i(\bar{x}) \leq 0$ for all $i = 1, \ldots, m$ and $h_i(\bar{x}) = 0$ for all $i = 1, \ldots, p$. For $\lambda \geq 0$,

$$f_0(\bar{x}) \geq L(\bar{x}, \lambda, \nu) \geq \min_{x \in \text{dom} f_0} L(x, \lambda, \nu) = g(\lambda, \nu).$$

Minimizing over all $\bar{x}$ such that $f_i(\bar{x}) \leq 0$ and $h_i(\bar{x}) = 0$ results in the desired inequality, $p^\star \geq g(\lambda, \nu)$.  □

---

**Example 2.1**  *Coordination of multiple agents.* We say that $f$ is an **additively separable function** if it can be written as $f(x_1, x_2, \ldots, x_n) = f_1(x_1) + f_2(x_2) + \cdots + f_n(x_n)$. Consider a set of agents $G$ with cost functions $f_g(x_g), g \in G$, where $x_g$ is the decision vector of each agent. As a group, agents need to decide on activities $x_g$ so as to respect a set of additively separable coordinating constraints, $\sum_{g \in G} h1_g(x_g) = 0$. Moreover, each agent is required to adhere to private operating constraints $h2_g(x_g) \leq 0$. Consider the following optimization problem:

$$\min_x \sum_{g \in G} f_g(x_g)$$
$$\text{s.t. } h2_g(x_g) \leq 0, g \in G$$
$$(\lambda): \sum_{g \in G} h1_g(x_g) = 0. \tag{2.1}$$

Relaxing the coordination constraints (2.1), the following Lagrangian function is obtained:

$$L(x, \lambda) = \sum_{g \in G} \left( f_g(x_g) + \lambda^T h1_g(x_g) \right),$$

where $\lambda$ is the vector of Lagrange multipliers that corresponds to the relaxed constraints. The dual function is given by

$$g(\lambda) = \sum_{g \in G} \min_{x_g : h2_g(x_g) \leq 0} \left( f_g(x_g) + \lambda^T h1(x_g) \right).$$

Note that the computation of $g(\lambda)$ for a given $\lambda$ can be accomplished by solving $|G|$ *independent* optimization problems.

---

The dual function $g(\lambda, \nu)$ is *always* a concave function, even if $f_0(x), f_i(x),$ or $h_i(x)$ are not necessarily convex functions.

**Proposition 2.2**  $g(\lambda, v)$ *is a concave function.*

*Proof*   Consider any $(\lambda_1, v_1)$, $(\lambda_2, v_2)$. Consider $\alpha \in [0, 1]$. Then,

$$g(\alpha\lambda_1 + (1 - \alpha)\lambda_2, \alpha v_1 + (1 - \alpha)v_2)$$

$$= \min_{x \in \mathrm{dom}\, f_0} \left( f_0(x) + \sum_{i=1}^{m} (\alpha\lambda_{1,i} f_i(x) + (1 - \alpha)\lambda_{2,i} f_i(x)) \right.$$

$$\left. + \sum_{i=1}^{p} \left( \alpha v_{1,i} h_i(x) + (1 - \alpha) v_{2,i} h_i(x) \right) \right)$$

$$\geq \alpha \min_{x \in \mathrm{dom}\, f_0} \left( f_0(x) + \sum_{i=1}^{m} \lambda_{1,i} f_i(x) + \sum_{i=1}^{p} v_{1,i} h_i(x) \right)$$

$$+ (1 - \alpha) \min_{x \in \mathrm{dom}\, f_0} \left( f_0(x) + \sum_{i=1}^{m} \lambda_{2,i} f_i(x) + \sum_{i=1}^{p} v_{2,i} h_i(x) \right)$$

$$= \alpha g(\lambda_1, v_1) + (1 - \alpha) g(\lambda_2, v_2). \qquad \square$$

It is natural to maximize the dual function in order to obtain the tightest possible lower bound from the relaxation of the problem. This gives rise to the **Lagrange dual problem**:

$$d^\star = \max_{\lambda, v} g(\lambda, v)$$

$$\text{s.t. } \lambda \geq 0.$$

This problem can be easier to solve compared to the original optimization problem because the dual function is concave and there exists a large family of efficient algorithms for minimizing convex functions (or maximizing concave functions) over convex sets. More importantly from a modeling point of view, the optimal solution of the dual problem often contains interesting economic interpretations.

Since $g(\lambda, v) \leq p^\star$ for any $\lambda, v$, it follows that $p^\star \geq d^\star$. This relationship is referred to as **weak duality** and holds for any optimization problem, whether it is convex or not. In fact, $d^\star$ often yields nontrivial bounds to difficult problems. **Strong duality** refers to the case where $p^\star = d^\star$ and does not hold in general. If the relationship holds, then the original optimization problem is equivalent to solving the maximization of the dual. Strong duality usually holds for convex problems. The conditions that guarantee strong duality in convex problems are called constraint qualifications. The difference $p^* - d^*$, whether zero or not, is defined as the **duality gap** of a relaxation.

---

**Example 2.2** *Linear programming duality.* The dual problem of a linear program can be written out automatically by following the mnemonic rule of table 2.1. Prove the mnemonic rules of table 2.1.

**Solution**

First note that every linear program can be written in the following form:

$$\min_{x} c_1^T x_1 + c_2^T x_2 + c_3^T x_3,$$

$$(\lambda_1): \quad A_{11} x_1 + A_{12} x_2 + A_{13} x_3 \geq b_1$$

**Table 2.1 Linear programming duality mnemonic table.**

| Primal | Minimize | Maximize | Dual |
|---|---|---|---|
| Constraints | $\geq b_i$ | $\geq 0$ | Variables |
|  | $\leq b_i$ | $\leq 0$ |  |
|  | $= b_i$ | Free |  |
| Variables | $\geq 0$ | $\leq c_j$ | Constraints |
|  | $\leq 0$ | $\geq c_j$ |  |
|  | Free | $= c_j$ |  |

$$(\lambda_2):\ A_{21}x_1 + A_{22}x_2 + A_{23}x_3 \leq b_2$$
$$(\lambda_3):\ A_{31}x_1 + A_{32}x_2 + A_{33}x_3 = b_3$$
$$(\sigma_1):\ x_1 \geq 0$$
$$(\sigma_2):\ x_2 \leq 0.$$

This results in the following Lagrangian function:

$$L(x, \lambda, \sigma) = \sum_{j=1}^{3} c_j^T x_j - \sum_{i=1}^{3} \lambda_i^T \left( \sum_{j=1}^{3} A_{ij}x_j - b_i \right) - \sum_{j=1}^{2} \sigma_j^T x_j.$$

For $L(x, \lambda, \sigma)$ to be a lower bound to the original problem, it must be that $\lambda_1 \geq 0$, $\lambda_2 \leq 0$, $\sigma_1 \geq 0$, and $\sigma_2 \leq 0$. Furthermore, as $x_j$ are free variables in the relaxation (since the last two constraints of the original problem have been relaxed), the dual function $g(\lambda, \sigma)$ will be finite if and only if

$$c_j^T - \sum_{i=1}^{3} \lambda_i^T A_{ij} - \sigma_j^T = 0, \quad j = 1, 2,$$

$$c_3^T - \sum_{i=1}^{3} \lambda_i^T A_{i3} = 0.$$

In this domain, the dual function becomes $g(\lambda, \sigma) = \sum_{i=1}^{3} b_i^T \lambda_i$. Taking into account the sign of $\sigma$, the dual problem can be stated as

$$\max_{\lambda}\ b_1^T \lambda_1 + b_2^T \lambda_2 + b_3^T \lambda_3$$
$$\text{s.t.}\ A_{11}^T \lambda_1 + A_{21}^T \lambda_2 + A_{31}^T \lambda_3 \leq c_1$$
$$A_{12}^T \lambda_1 + A_{22}^T \lambda_2 + A_{32}^T \lambda_3 \geq c_2$$
$$A_{13}^T \lambda_1 + A_{23}^T \lambda_2 + A_{33}^T \lambda_3 = c_3$$
$$\lambda_1 \geq 0$$
$$\lambda_2 \leq 0,$$

which proves the mnemonic rules of table 2.1.

**Example 2.3** *Dual function of a unit commitment problem.* Consider three generators with the technical and economic specifications of table 2.2. The goal is to satisfy a

**Table 2.2 List of generators for example 2.3.**

| Generator | Activation cost ($) | Marg. cost ($/MWh) | Capacity (MW) |
|---|---|---|---|
| Cheap | 500 | 0 | 20 |
| Moderate | 1000 | 10 | 100 |
| Expensive | 2000 | 80 | 100 |

demand of 200 MW at minimum cost. Each generator can be activated at a certain cost, and it is necessary to activate a generator in order to produce power. The power produced by a generator cannot exceed its capacity. Denote $p_i$ to be the power produced by generator $i$ and $u_i$ as a binary variable that indicates whether generator $i$ has been activated or not. The overall optimization problem can be expressed as follows:

$$\min_{p,u} 500 \cdot u_1 + 1000 \cdot u_2 + 10 \cdot p_2 + 2000 \cdot u_3 + 80 \cdot p_3$$
$$\text{s.t. } 0 \leq p_1 \leq 20 \cdot u_1$$
$$0 \leq p_2 \leq 100 \cdot u_2$$
$$0 \leq p_3 \leq 100 \cdot u_3$$
$$(\lambda): \ p_1 + p_2 + p_3 = 200 \tag{2.2}$$
$$u_i \in \{0,1\}.$$

The dual function that is obtained by relaxing constraint (2.2) can be expressed as follows:

$$g(\lambda) = \min_{p,u} 500 \cdot u_1 + 1000 \cdot u_2 + 10 \cdot p_2 + 2000 \cdot u_3 + 80 \cdot p_3$$
$$-\lambda \cdot (p_1 + p_2 + p_3 - 200)$$
$$\text{s.t. } p_1 \leq 20 \cdot u_1, p_2 \leq 100 \cdot u_2, p_3 \leq 100 \cdot u_3$$
$$p_i \geq 0, u_i \in \{0,1\}.$$

Having relaxed the complicating constraint, the dual function computation can be decomposed into one subproblem per generator:

$$g(\lambda) = g_1(\lambda) + g_2(\lambda) + g_3(\lambda) + 200 \cdot \lambda,$$

where

$$g_1(\lambda) = \min_{p,u}\{500 \cdot u_1 - \lambda \cdot p_1, 0 \leq p_1 \leq 20 \cdot u_1, u_1 \in \{0,1\}\},$$
$$g_2(\lambda) = \min_{p,u}\{1000 \cdot u_2 + (10 - \lambda) \cdot p_2, 0 \leq p_2 \leq 100 \cdot u_2, u_2 \in \{0,1\}\},$$
$$g_3(\lambda) = \min_{p,u}\{2000 \cdot u_3 + (80 - \lambda) \cdot p_3, 0 \leq p_3 \leq 100 \cdot u_3, u_3 \in \{0,1\}\}.$$

Consider the evaluation of $g_1(\lambda)$. For $\lambda \geq 25$, the optimal solution is to turn the generator on and produce at the technical limit ($u_1^\star = 1, p_1^\star = 20$). For $\lambda < 25$ the generator should be kept off ($u_1^\star = 0, p_1^\star = 0$):

$$g_1(\lambda) = \begin{cases} 0, & \lambda \leq 25, \\ 500 - 20 \cdot \lambda, & \lambda > 25. \end{cases}$$

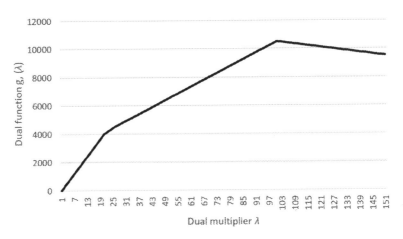

Figure 2.3 The dual function $g(\lambda)$ of example 2.3.

Note that $g_1$ is a concave function of $\lambda$. Following a similar procedure for $g_2$ and $g_3$, the following closed form expression can be obtained for $g$:

$$g(\lambda) = \begin{cases} 200 \cdot \lambda, & \lambda \leq 20, \\ 2000 + 100 \cdot \lambda, & 20 < \lambda \leq 25, \\ 2500 + 80 \cdot \lambda, & 25 < \lambda \leq 100, \\ 12500 - 20 \cdot \lambda, & 100 < \lambda. \end{cases}$$

The function is presented graphically in figure 2.3. Despite the fact that the original problem is non-convex (due to the presence of the binary variables $u_i$), the dual function is a concave function of $\lambda$. It can be seen by inspection that the primal optimal solution keeps the first unit off ($u^\star = (1,1,1)$) and produces from units 2 and 3 ($p^\star = (20,100,80)$), incurring a cost 10900. From figure 2.3, the dual optimal value can be seen to equal 10500. Strong duality does not hold for this problem.

## 2.3    KKT conditions

This section introduces the Karush–Kuhn–Tucker (KKT) conditions. The KKT conditions are conditions for characterizing the optimal solution of an optimization problem. The KKT conditions are the main tool used for analyzing the models presented in the following chapters.

Before proceeding to the discussion of KKT conditions, we first define the **complementarity operator**, indicated by $\perp$. Given two real numbers $a, b \in \mathbb{R}$, $a \perp b$ is equivalent to $a \cdot b = 0$. The definition is overloaded in the case of vectors in order to indicate componentwise complementarity: given two vectors $x, y \in \mathbb{R}^n$, $x \perp y$ is equivalent to $x_i \cdot y_i = 0$ for all $i = 1, \ldots, n$. Finally, the following abbreviated notation is used for the derivation of KKT conditions: given $x, y \in \mathbb{R}^n$, $0 \leq x \perp y \geq 0$ is equivalent to the three conditions $x \geq 0$, $y \geq 0$, and $x \perp y$.

Consider again the following optimization problem:

$$\min_{x} f_0(x)$$

$$\text{s.t. } x \in \text{dom} f_0 \subset \mathbb{R}^n$$

$$(\lambda_i): \quad f_i(x) \leq 0, i = 1, \ldots, m$$

$$(\mu_i): \quad h_i(x) = 0, i = 1, \ldots, p.$$

**Proposition 2.3** *Suppose that strong duality holds, with $x^\star$ optimal for the primal problem, and $\lambda^\star, v^\star$ optimal for the dual problem. Then*

- $x^\star$ *minimizes* $L(x, \lambda^\star, v^\star)$
- $\lambda_i^\star \cdot f_i(x^\star) = 0$ *for* $i = 1, \ldots, m$.

*Proof* Strong duality implies the following:

$$f_0(x^\star) = g(\lambda^\star, v^\star) = \min_{x \in \text{dom} f_0} \left( f_0(x) + \sum_{i=1}^{m} \lambda_i^\star f_i(x) + \sum_{i=1}^{p} v_i^\star h_i(x) \right)$$

$$\leq f_0(x^\star) + \sum_{i=1}^{m} \lambda_i^\star f_i(x^\star) + \sum_{i=1}^{p} v_i^\star h_i(x^\star)$$

$$\leq f_0(x^\star).$$

The first equality holds by strong duality. The second equality holds by definition of the dual function. The first inequality holds because $x^\star$ is not necessarily the minimizer of the Lagrangian function. The second inequality holds because $h_i(x^\star) = 0$ for all $i = 1, \ldots, p$, $f_i(x^\star) \leq 0$ for all $i = 1, \ldots, m$, and $\lambda_i^\star \leq 0$ for all $i = 1, \ldots, m$. Since the left- and right-hand side are identical, the two inequalities above hold with equality. The conclusions follow. □

**Definition 2.4** Consider a problem with a differentiable objective function and constraints. The following four conditions are called **KKT conditions**:

- Primal feasibility: $f_i(x) \leq 0$, $i = 1, \ldots, m$, $h_i(x) = 0, i = 1, \ldots, p$
- Dual feasibility: $\lambda_i \geq 0$, $i = 1, \ldots, m$
- Complementary slackness: $\lambda_i \cdot f_i(x) = 0$, $i = 1, \ldots, m$
- Gradient of the Lagrangian function with respect to $x$ vanishes:

$$\nabla f_0(x) + \sum_{i=1}^{m} \lambda_i \nabla f_i(x) + \sum_{i=1}^{p} v_i \nabla h_i(x) = 0.$$

Proposition 2.3 proves that the KKT conditions are necessary for problems for which strong duality holds. However, in cases where strong duality does not hold, the KKT conditions are not necessary conditions for the optimality of convex optimization problems. The constraint qualifications referred to earlier, whenever they hold, ensure that the KKT conditions are necessary for optimality. For convex optimization problems with differentiable objective functions and constraints and a zero duality gap, the KKT conditions are also sufficient for optimality (see problem 2.1).

The following result describes the KKT conditions of an optimization problem with linear constraints and is used extensively in the textbook.

**Proposition 2.5**  *Consider a maximization problem with differentiable objective function f and linear constraints:*

$$\max_{x,y} f(x,y)$$

$$s.t.$$

$$(\lambda): \quad Ax + By \leq b$$

$$(\mu): \quad Cx + Dy = d$$

$$(\lambda_2): \quad x \geq 0.$$

*Then the KKT conditions have the following form:*

$$Cx + Dy - d = 0,$$
$$0 \leq \lambda \perp Ax + By - b \leq 0,$$
$$0 \leq x \perp A^T \lambda + C^T \mu - \nabla_x f(x,y) \geq 0,$$
$$B^T \lambda + D^T \mu - \nabla_y f(x,y) = 0.$$

*Proof*    The Lagrangian function can be expressed as

$$L(x,y,\lambda,\mu,\lambda_2) = f(x,y) + \lambda^T(b - Ax - By) + \mu^T(d - Cx - Dy) + \lambda_2^T x.$$

In order for the Lagrangian function to be an upper bound to the optimal value, it must be the case that $\lambda, \lambda_2 \geq 0$. These are the dual feasibility conditions. The complementarity condition can be stated as:

$$\lambda^T(b - Ax - By) = 0 \Leftrightarrow \lambda \perp b - Ax - By,$$
$$\lambda_2^T x = 0 \Leftrightarrow \lambda_2 \perp x.$$

The stationarity condition can be stated as

$$\nabla_x L(x,y,\lambda,\mu,\lambda_2) = \nabla_x f(x,y) - C^T \mu - A^T \lambda + \lambda_2 = 0,$$
$$\nabla_y L(x,y,\lambda,\mu,\lambda_2) = \nabla_y f(x,y) - D^T \mu - B^T \lambda = 0.$$

Therefore, the KKT conditions can be expressed as

$$0 \leq \lambda \perp b - Ax - By \geq 0,$$
$$Cx + Dy - d = 0,$$
$$0 \leq \lambda_2 \perp x \geq 0,$$
$$\lambda_2 = A^T \lambda + C^T \mu - \nabla_x f(x,y),$$
$$B^T \lambda + D^T \mu - \nabla_y f(x,y) = 0,$$

or, equivalently,

$$Cx + Dy - d = 0,$$
$$0 \leq \lambda \perp Ax + By - b \leq 0,$$
$$0 \leq x \perp A^T \lambda + C^T \mu - \nabla_x f(x,y) \geq 0,$$
$$B^T \lambda + D^T \mu - \nabla_y f(x,y) = 0. \qquad \square$$

**Example 2.4** *KKT conditions of an economic dispatch problem.* Consider the problem of example 2.3, where activation costs are ignored:

$$\min_{p} 10 \cdot p_2 + 80 \cdot p_3$$

s.t.

$$(\lambda): \quad p_1 + p_2 + p_3 = 200$$
$$(\mu_1): \quad p_1 \le 20$$
$$(\mu_2): \quad p_2 \le 100$$
$$(\mu_3): \quad p_3 \le 100$$
$$p_i \ge 0, i = 1, 2, 3.$$

The KKT conditions of the problem can be expressed by using the mnemonic of proposition 2.5. In order to derive the KKT conditions, it is easiest to start from rewriting the equality constraints:

$$p_1 + p_2 + p_3 = 200. \tag{2.3}$$

Inequality constraints are expressed along with their complementary nonnegative dual multipliers:

$$0 \le \mu_1 \perp 20 - p_1 \ge 0, \tag{2.4}$$
$$0 \le \mu_2 \perp 100 - p_2 \ge 0, \tag{2.5}$$
$$0 \le \mu_3 \perp 100 - p_3 \ge 0. \tag{2.6}$$

The nonnegative primal variables are complementary to a *nonnegative* expression that can be derived by using the mnemonic of proposition 2.5: (i) whenever a primal variable shows up in the objective function, *add* its coefficient to the expression (and *subtract* in a minimization problem), and (ii) whenever the primal variable shows up in a constraint, add its coefficient to the expression, weighted by the dual multiplier of the corresponding constraint:

$$0 \le p_1 \perp \lambda + \mu_1 \ge 0, \tag{2.7}$$
$$0 \le p_2 \perp 10 + \lambda + \mu_2 \ge 0, \tag{2.8}$$
$$0 \le p_3 \perp 80 + \lambda + \mu_3 \ge 0. \tag{2.9}$$

Equations (2.3)–(2.9) constitute the KKT conditions of the optimization problem. Note that the three last conditions can be replaced by the following conditions:

$$0 \le p_1 \perp -\lambda + \mu_1 \ge 0, \tag{2.10}$$
$$0 \le p_2 \perp 10 - \lambda + \mu_2 \ge 0, \tag{2.11}$$
$$0 \le p_3 \perp 80 - \lambda + \mu_3 \ge 0, \tag{2.12}$$

since equation (2.3) can be expressed equivalently as $-p_1 - p_2 - p_3 = -200$. The primal optimal solution of the problem is easily determined to be $(p^\star)^T = (20, 100, 80)$. The dual optimal multipliers of the problem are equal to $\lambda^\star = 80,$

$(\mu^{\star})^{T} = (80, 70, 0)$. The reader should verify that $p^{\star}, \lambda^{\star}$, and $\mu^{\star}$ satisfy equations (2.3)–(2.6) and (2.10)–(2.12).

---

Note that, for more general convex optimization problems where the constraints or objective function of the problem are non-differentiable, the KKT conditions can be generalized by replacing the gradient of the Lagrangian function with its subgradient[1] in the fourth condition. If strong duality holds, then these conditions remain necessary and sufficient for optimality.

## 2.4   Sensitivity

An important property of dual optimal multipliers is that they can be interpreted as the sensitivity of the objective function value with respect to a change in the right-hand-side of the constraint that they correspond to. This result is stated formally in proposition 2.8. Before stating this result, it is necessary to define the "slope" of a function when the function is not differentiable by introducing subgradients.

**Definition 2.6**   Consider a function $g$, $\pi$ is a **subgradient** of $g$ at $u$ if

$$g(w) \geq g(u) + \pi^{T}(w - u) \text{ for all } w. \tag{2.13}$$

If $g(w) \leq g(u) + \pi^{T}(w - u)$ for all $w$, then $\pi$ is a **supergradient** of $g$.

The geometric interpretation of a subgradient is that it is a vector that defines a linear under-estimator of a function $g$, as shown in figure 2.4. Subgradients generalize the notion of gradients for non-differentiable functions. Subgradients are useful for generalizing the KKT conditions to the case of non-differentiable optimization problems, and also for deriving sensitivity results.

**Definition 2.7**   The set of all subgradients of $g$ at $u$ is called the **subdifferential** of $g$ at $u$, and is denoted as $\partial g(u)$.

As one can observe in figure 2.4, $\partial g(u)$ is a closed convex set. It should be intuitively obvious that if a function $g$ is differentiable at a certain point $u$, then $\partial g(u) = \{\nabla g(u)\}$, i.e. the subdifferential consists of a single element which is the gradient of the function at that point. It is also easy to show that, for convex functions, $0 \in \partial g(u)$ implies that $u$ is a global minimum of $g$.

The definition of subgradients is useful for describing the slope of the following function:

$$c(u) = \min_{x} f_0(x)$$
$$\text{s.t.} f_i(x) \leq u_i, i = 1, \ldots, m$$
$$x \in \text{dom} f_0.$$

---

[1] Subgradients are defined in section 2.4.

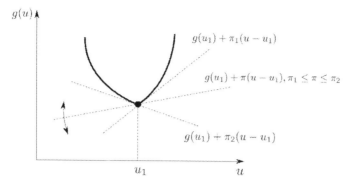

**Figure 2.4** Consider the function $g \colon \mathbb{R} \to \mathbb{R}$ and notice that the function is non-differentiable at $u_1$. $\pi_1$ and $\pi_2$ are both subgradients at $u_1$. Any line $y = g(u_1) + \pi(u - u_1)$ passing through $(u_1, g(u_1))$ with a slope of $\pi \in \partial g(u_1)$ is below $g(u)$ for all $u$, whereas any line with a different slope "cuts into" $g(u)$. $\partial g(u_1) = [\pi_1, \pi_2]$ is the subdifferential of $g$ at $u_1$.

The function $c(u)$, which is often referred to as the *value function* of the optimization problem $\{\min_x f_0(x) : x \in \operatorname{dom} f, f_i(x) \le 0, i = 1, \dots, m\}$, has a natural interpretation. Given a perturbation $u$ to the right-hand side of the constraints $f_i(x) \le 0$, the value function describes how the optimal objective function value is impacted. Therefore, the slope of $c(u)$ at $u = 0$ can be interpreted as the sensitivity of the optimal objective function value of the original problem to an infinitesimal perturbation of the right-hand side of its constraints. It turns out that this sensitivity is equal to the dual optimal multipliers of the original problem, as the following proposition demonstrates.

**Proposition 2.8** *Define $c(u)$ as the optimal value of the following mathematical program:*

$$c(u) = \min_x f_0(x)$$
$$f_i(x) \le u_i, i = 1, \dots, m$$
$$x \in \operatorname{dom} f_0,$$

*and suppose that $\operatorname{dom} f_0$ is a convex set and $f_0, f_i$ are convex functions.*

- *$c(u)$ is a convex function.*
- *Suppose strong duality holds, and denote $\lambda^\star$ as the maximizer of the dual function $\min_{x \in \operatorname{dom} f_0} (f_0(x) - \lambda^T (f(x) - u))$ for $\lambda \le 0$. Then $\lambda^\star \in \partial c(u)$.*

*Proof* To show convexity of $c(u)$, consider any $u_1, u_2$ and denote $x_1, x_2$ as the optimal solution when $u_1, u_2$ is used as input respectively. Similarly, consider any $a \in [0, 1]$ and denote $x_a$ as the optimal solution when $au_1 + (1 - a)u_2$ is used as input. Since $f_i(x_1) \le u_{1,i}$ and $f_i(x_2) \le u_{2,i}$, it follows from the convexity of $f_i$ that $f_i(ax_1 + (1 - a)x_2) \le au_{1,i} + (1 - a)u_{2,i}$. Since $\operatorname{dom} f_0$ is convex, it follows that $ax_1 + (1 - a)x_2$

is an admissible solution when $au_1 + (1 - a)u_2$ is used as input. From the optimality of $x_a$ with respect to $au_1 + (1 - a)u_2$ it follows that $f_0(x_a) \leq f_0(ax_1 + (1 - a)x_2)$. It then follows from the convexity of $f_0$ that $c(au_1 + (1 - a)u_2) \leq ac(u_1) + (1 - a)c(u_2)$.

Consider any $\bar{u}$ and denote $\bar{x}$ as the optimal solution when $\bar{u}$ is used as input. Also, denote $x^\star \in \arg\min_{x \in \mathrm{dom}\, f_0}(f_0(x) - (\lambda^\star)^T (f(x) - u))$. The following relationships hold:

$$\begin{aligned} c(u) &= f_0(x^\star) - (\lambda^\star)^T(f(x^\star) - u) \\ &\leq f_0(\bar{x}) - (\lambda^\star)^T(f(\bar{x}) - u) \\ &= f_0(\bar{x}) - (\lambda^\star)^T(f(\bar{x}) - \bar{u}) - (\lambda^\star)^T(\bar{u} - u) \\ &\leq f_0(\bar{x}) - (\lambda^\star)^T(\bar{u} - u) \\ &= c(\bar{u}) - (\lambda^\star)^T(\bar{u} - u), \end{aligned}$$

where the first relationship follows from strong duality, the second relationship follows from the definition of $x^\star$, the fourth relationship results from the fact that $f(\bar{x}) \leq \bar{u}$ and $\lambda^\star \leq 0$ and the last relationship follows from the definition of $\bar{x}$. $\square$

The following example demonstrates the convexity of $c(u)$ on a simple mathematical program.

---

**Example 2.5** *Convexity of $c(u)$.* Consider the problem of example 2.4. The optimal value of the problem as a function of the capacity of generator 1 can be easily determined by noticing that, in the optimal solution, generator 1 will be used to the greatest possible extent, followed by generator 2, followed by generator 3. As long as the capacity $u$ of generator 1 is below 100 MW, generator 2 is fully utilized and generator 3 is utilized to the least possible extent. Hence, for generator 1 capacity in the range $0 \leq u \leq 100$, the optimal cost $c(u)$ is equal to $c(u) = 10 \cdot 100 + 80 \cdot (100 - u)$. Following the same reasoning, the following expression can be derived for $c(u)$:

$$c(u) = \begin{cases} 9000 - 80 \cdot u, & 0 \leq u < 100, \\ 2000 - 10 \cdot u, & 100 \leq u < 200, \\ 0, & 200 \leq u. \end{cases}$$

The function $c(u)$ is shown in figure 2.5, and is clearly convex.

---

Notice that, if $c(u)$ is differentiable at a certain point $u$, then the dual multiplier $\lambda_i$ of a certain constraint is equal to $\nabla_{u_i} c(u)$, therefore $\lambda_i$ is equal to the *sensitivity* of the objective function $c(u)$ to a marginal change in the right-hand side of the constraint corresponding to $\lambda_i$.

---

**Example 2.6** *The slope of $c(u)$.* Recall the solution of the KKT conditions (equations (2.3)–(2.6) and (2.10)–(2.12)) in example 2.4: $(p^\star)^T = (20, 100, 80)$, $\lambda^\star = 80$, $(\mu^\star)^T = (80, 70, 0)$. The value of $\lambda^\star$ has the following sensitivity interpretation: if the right-hand side of the corresponding constraint, $p_1 + p_2 + p_3 = 200$, is increased by one unit, then this is equivalent to having to satisfy one additional unit of demand. Since the only generator that can increase its output in response to an increase in demand

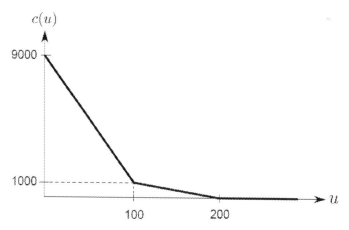

**Figure 2.5** A graphical illustration of the optimal cost $c(u)$ as a function of generator 1 capacity $u$ in example 2.5.

is generator 3, the additional cost imposed to the system is $80, which is exactly the value of $\lambda^\star$. It is important to note that the KKT conditions of the problem can also be expressed using equations (2.3)–(2.9), in which case the solution of the KKT system is $(p^\star)^T = (20, 100, 80)$, $\lambda^\star = -80$, $(\mu^\star)^T = (80, 70, 0)$. Note the change in the sign of $\lambda^\star$.

---

The previous example highlights that the dual optimal multiplier may be equal to the sensitivity, or minus the sensitivity, of the objective function value. As a general rule (as indicated by the proof of proposition 2.8), whether a Lagrange multiplier is equal to the sensitivity, or minus the sensitivity, of the objective function of an optimization problem depends on how the Lagrangian function of the problem is constructed. If the constraint $f_i(x) \leq 0$ (likewise for $h_i(x) = 0$) is relaxed by *subtracting* its weighted violation from the objective function (i.e. $L(x, \lambda) = f_0(x) - \sum_{i=1}^{m} \lambda_i \cdot f_i(x)$), then the dual optimal multiplier is equal to the sensitivity of the objective function. By contrast, if the weighted violation is *added* to the objective function (i.e. $L(x, \lambda) = f_0(x) + \sum_{i=1}^{m} \lambda_i \cdot f_i(x)$), then the dual optimal multiplier is equal to *minus* the sensitivity of the objective function to a change in the right-hand side of the constraint $f_i(x) \leq 0$ (likewise for $h_i(x) = 0$).

## Problems

**2.1** *Sufficiency of KKT conditions in convex problems.* Prove that, for convex optimization problems with differentiable objective functions and constraints, for which there is a zero duality gap, the KKT conditions are sufficient.

**2.2** *Sign of sensitivity.* In the maximization problem of proposition 2.5, the dual multipliers $\mu$ are (prove your claim):

1. equal to the sensitivities of the optimal objective function value to the right-hand side of the $Cx + Dy - d = 0$ constraints
2. equal to the sensitivities of the optimal objective function value to the right-hand side of the $d - Cx - Dy = 0$ constraints.

The dual multipliers $\lambda$ are (prove your claim):

1. equal to the sensitivities of the optimal objective function value to the right-hand side of the $Ax + By - b \leq 0$ constraints
2. equal to the sensitivities of the optimal objective function value to the right-hand side of the $b - Ax - By \geq 0$ constraints.

**2.3** *Eyeballing primal-dual optimal solutions.* Consider the following optimization problem:

$$\max_{x,y} x + 2y$$

$$(\lambda_1): \quad x \geq 0$$

$$(\lambda_2): \quad x \leq 2$$

$$(\mu): \quad -y = -1.$$

What are the KKT conditions of the problem? What is the solution to the KKT system?

**2.4** *Subgradient oracle.* Consider the following optimization problem:

$$\max_x f_0(x)$$

$$f(x) \leq 0$$

$$h(x) = 0,$$

where $x \in \text{dom} f_0$. Consider the following Lagrangian function:

$$L(x, u, v) = f_0(x) - u^T f(x) - v^T h(x)$$

and the associated dual function $g(u, v) = \max_{x \in \text{dom} f_0} L(x, u, v)$. Denote

$$x_0 \in \arg\max_x L(x, u_0, v_0)$$

as the optimal solution of the Lagrangian function given Lagrange multipliers $(u_0, v_0)$. Prove that $(-f(x_0), -h(x_0)) \in \partial g(u_0, v_0)$. This result is algorithmically relevant because it allows us to use first-order optimization methods for maximizing dual functions, using an oracle that furnishes $g(u, v)$ and its subgradient at every $(u, v)$.

**2.5** *KKT conditions for convex optimization problems.* Prove proposition 2.5 for the more general case where $Ax + By \leq b$ is replaced by inequality constraints $g(x, y) \leq 0$, where $g_i, i = 1, \ldots, m$, are convex functions.

**2.6** *Long-run marginal cost.* Consider a power producer with constant investment cost and constant marginal fuel cost. Long-run cost is defined as the minimum cost at which the producer can produce a given output $p$, which can be expressed as the following function $TC(p)$:

$$TC(p) = \min_{q,x} IC \cdot x + MC \cdot q$$

$$(\mu): \quad q \leq x$$

$$(\pi): \quad p - q = 0.$$

The **long-run marginal cost** is a subgradient of the long run cost with respect to output. Prove that, if $q > 0$, then the long-run marginal cost is equal to $MC+IC$.

## Bibliography

**Section 2.1**   The material is based on Boyd and Vandenberghe (2008).

**Section 2.2**   The material is based on Boyd and Vandenberghe (2008). The application of Lagrange relaxation to unit commitment, which is presented in example 2.3, was introduced originally by Muckstadt and Koenig (1977) and further developed by Bertsekas et al. (1983). The mnemonic table 2.1 is sourced from Bertsimas and Tsitsiklis (1997).

**Section 2.3**   The material is based on Boyd and Vandenberghe (2008). The importance of KKT conditions in analyzing electricity market models is discussed by Ehrenmann (2004).

**Section 2.4**   The material is based on Boyd and Vandenberghe (2008). The subgradient oracle result of problem 2.4 is used in dual decomposition algorithms for power systems scheduling (Baldick 1995). This result is also presented in section II.5.4 of Wolsey and Nemhauser (1999).

# 3    Power system operations and power market operations

The development of electricity market models requires an understanding of the engineering principles that govern the operation of electric power systems, as well as the institutional organization of electricity markets. Electricity markets exhibit certain unique characteristics that result in a tight connection between power system operations and power market operations, as opposed to most commonly encountered markets, where the physical action of exchanging goods is often only loosely connected to the financial transactions involved in trading those goods. Unique features of electricity markets that necessitate the tight coordination of physical operations and market transactions include the facts that consumers are largely unresponsive to price and that storage is limited, the fact that supply and demand need to be balanced instantaneously, and the fact that it is difficult to physically exclude consumers from consuming electricity. These factors complicate the design of electricity markets in ways that will be clarified in subsequent chapters of this book. Section 3.1 provides an overview of power system operations. Section 3.2 provides an overview of electricity markets.

## 3.1    Power system operations

Power systems are complex supply chains that are used for transferring electric energy. The typical structure of power systems is presented in figure 3.1. This structure consists of production, high voltage transmission, medium and low voltage distribution, and consumption. Consumers are typically classified into three categories: industrial, commercial, and residential consumers. The following paragraphs provide a brief overview of the key features of each component of the supply chain, including technical and economic characteristics.

### 3.1.1    Production

Electric energy is produced in various ways. The most common means of non-renewable energy production are (i) the combustion of fossil fuels (such as coal, oil, and natural gas) and the conversion of the resulting heat into electric energy through an electromechanical generator, and (ii) the fission of uranium and the conversion

**Table 3.1 Common conversions of energy units.**

| | |
|---|---|
| 1 megajoule | 238.8 kilocalories |
| | 947.8 Btu |
| | 0.278 kilowatt hours |
| 1 kilocalorie | 3.968 Btu |
| 1 kilowatt hour (kWh) | 359.8 kilocalories |
| | 3411 Btu |
| 1 megawatt hour (MWh) | 3.411 million Btu |
| | 3.411 thousand cubic feet (mcf) gas |
| | 0.097 thousand cubic meters gas |
| 1 million Btu (MMBtu) | 1055 megajoules |
| | 2520 megacalories |
| | 293.1 kilowatt hours |
| | 1000 cubic feet gas |
| 1 cubic meter gas | 35.315 cubic feet gas |

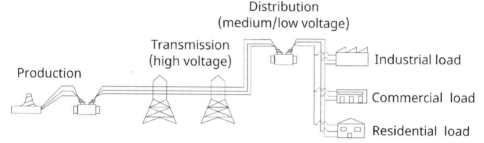

**Figure 3.1** An electric power system consists of production, transmission, distribution, and consumption.

of the resulting heat into electric energy through the heating of water that drives steam turbines. The most common means of renewable energy production include (iii) the conversion of the potential energy of water in rivers and dams into electric energy through a water turbine, (iv) the conversion of geothermal heat into electric energy through steam generators, (v) the conversion of the kinetic energy of wind to electric energy through electromechanical wind turbines, (vi) the conversion of solar electromagnetic energy to electric energy through solar panels, (vii) the conversion of the chemical energy of biomass to electric energy through combustion turbines, and (viii) certain types of renewable energy production (such as wave and tidal energy) with marginal contributions to existing power systems.

Energy is measured using a variety of units, including Joules, calories, watt hours and British thermal units (Btu). Table 3.1 lists some conversions that are commonly used in practice. The most common units of measurement in electric energy are the megawatt hour, which is denoted MWh, and the million Btu, which is denoted MMBtu.

The rate of change in energy is measured by **power**. Power can measure the rate of production, consumption, or flow of energy. The most commonly used unit of measurement for power in electricity systems is the megawatt, denoted MW. One megawatt hour corresponds to the amount of energy that is accumulated by the flow of one megawatt for one hour. It is important to emphasize that the output of generators at any moment in time is measured in MW, not MWh.

Generators are required to obey a variety of operating constraints that are developed in detail throughout the text. Capacity constraints limit the maximum rate at which generators can produce energy. Thermal generators obey minimum run levels, ramp constraints, and minimum up and down times. Hydroelectric generators are subject to capacity constraints that limit the total amount of power that they can produce. Hydroelectric dams abide to storage constraints that limit the total amount of water (and therefore energy) that they can store. A number of constraints are associated with the water levels of hydroelectric dams. The flow of water along a river network also creates linkages among the operation of hydroelectric dams that are operated in cascade.

Production costs are also developed in detail later in this section and also in section 7.1. Here, the focus is on highlighting the difference between variable/operating/fuel cost, marginal cost, average cost, and fixed/investment cost.

**Variable/operating/fuel cost** is the cost that depends on the amount of power that is produced by the plant. Fuel cost is measured in $/h and expresses the hourly cost of producing a certain amount of energy.

**Marginal cost** is the derivative of variable cost with respect to power output, and is measured in $/MWh. If fuel cost is not differentiable but is continuous, one defines lower and upper marginal costs, as well as the marginal cost range. **Left-hand marginal cost** is the savings from producing one less unit of power (equivalently, the left-hand side derivative of fuel cost). **Right-hand marginal cost** is the cost of producing one more unit of power (equivalently, the right-hand side derivative of fuel cost). When the capacity of a plant has been reached, right-hand marginal cost is equal to infinity. The marginal cost range is the set of values between and including the left- and right-hand marginal costs.

When a collection of generators is considered, **aggregate variable cost** is defined as the cheapest way to produce a certain quantity of power among a collection of producers. The above definitions are demonstrated in example 3.1, and formalized mathematically in section 4.2. The **aggregate marginal cost** function is the derivative of aggregate cost, and is obtained by sorting producers in order of increasing marginal cost since this is the most efficient way of producing a certain amount of output.

The system-wide marginal cost curve is obtained by summing horizontally the marginal cost functions of all units in the system. This is demonstrated graphically in figure 3.2. This curve is also called the **merit order curve**, since it determines the order of merit of the units, i.e. the order in which the units should be used in order to satisfy demand at minimum cost.

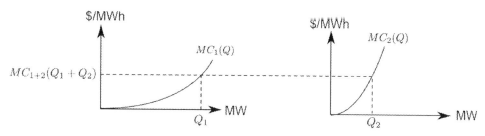

**Figure 3.2** Aggregating individual marginal cost functions: $MC_i(Q), i = 1, 2$, is the marginal cost function of generators 1 and 2, $MC_{1+2}(Q)$ is the aggregate marginal cost function.

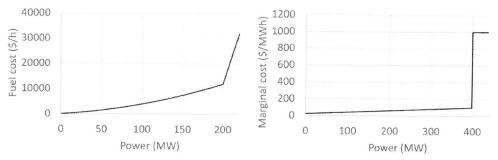

**Figure 3.3** (Left) Variable cost in example 3.1. (Right) The marginal cost of two identical units whose variable cost is provided in the left panel. The aggregate marginal cost curve is obtained as the horizontal summation of the individual marginal cost curves.

**Example 3.1** *Marginal cost of a unit with two ranges of operation.* Suppose that a gas generator has a quadratic fuel cost at a normal operating range of 0–200 MW. The marginal cost is 20 $/MWh at 0 MW and 100 $/MWh at 200 MW. The marginal cost is constant and equal to 1000 $/MWh beyond 200 MW due to stress. The unit cannot produce more than 220 MW. The fuel cost is 0 $/h when the generator is not producing power. The marginal cost of the generator is shown in figure 3.3. The marginal cost range at 200 MW is [100, 1000] $/MWh. The marginal cost range at 220 MW is [1000, +∞] $/MWh.

**Example 3.2** *Merging marginal costs of multiple units.* Consider two identical power plants, the marginal cost curve of which is described in example 3.1. The aggregate marginal cost curve of both units is shown in figure 3.3.

**Fixed cost/investment cost** is the cost that is incurred by putting infrastructure in place that allows a firm to produce, regardless of how much the firm produces. The investment cost of different technologies is often stated as overnight cost in $/kW or some other measure of cost per capacity. **Overnight cost** is the cost that would have to be paid upfront in order to build a plant. For example, the overnight cost of a coal plant might be 1000 $/kW, which implies that a 200 MW plant would require $200 million to be built.

**Table 3.2 Conversion of overnight cost to hourly payment in example 3.3.**

|             | $OC$ ($/kW) | $FC$ ($/kWy) | $FC$ ($/MWh) |
|-------------|-------------|--------------|--------------|
| Gas turbine | 400         | 50.5         | 5.8          |
| Coal        | 1200        | 144.7        | 16.5         |

It is useful to convert overnight cost to a continuous cash flow that is required in order to pay back the investment. This conversion should account for the time value of money through discount rates. For a technology with a lifetime of $T$ years, an overnight cost $OC$ and given an interest rate $r$, the **annualized fixed cost** of investment,[1] given annual compounding, is given by

$$FC = \frac{r \cdot OC}{1 - 1/(1+r)^T}. \tag{3.1}$$

The annualized fixed cost of investment, given continuous discounting, is given by

$$FC = \frac{r \cdot OC}{1 - e^{-rT}}. \tag{3.2}$$

$FC$ represents a cash flow and is measured in $/kWy when the overnight cost is expressed in $/kW. This is a lump sum that needs to be paid *yearly* for having 1 kW of a certain technology available. We can convert this to an *hourly* cash flow for having 1 MW of the technology available by dividing by 8.76. This gives $FC$ expressed in $/MWh, the unit of measurement that is commonly used for variable cost.

When a firm is financed by multiple sources, $r$ is replaced by the **weighted average cost of capital** (WACC). In WACC, the rate of return of each source of capital is weighted by the contribution of that source of capital to the financing of the firm.

---

**Example 3.3** *Converting overnight cost to fixed cost.* Consider a gas turbine with an overnight cost of 400 $/kW and a lifetime of 25 years, and a coal generator with an overnight cost of 1200 $/kW and a lifetime of 45 years. The continuous discounting formula of equation (3.2) with an interest rate of $r = 12\%$ converts the overnight cost to an annual payment per kW of capacity ($/kWy). Dividing by 8.76 results in an hourly payment per MW of capacity ($/MWh). The results are presented in table 3.2.

---

[1] The precise definition of fixed cost in economics is cost that relates to production factors that cannot be affected by the firm at the time scale of the economic model in question. Our reference to investment cost as a fixed cost in this chapter suggests that we are in the time scale where investment has been decided, but production has not, i.e. we find ourselves in the time frame of short-term (e.g. day-ahead, intraday, or real-time) operations. In the long-term models of chapter 11, there are no fixed costs, since even capacity investment is a decision variable for the firm. In the economic dispatch model of chapter 4, minimum load costs can be considered a fixed cost because they have already been decided for a unit that has been brought online, and the only thing left for the firm to decide at such a short time scale is how much to produce.

Generators are typically represented in investment models by marginal variable costs and investment costs. The goal of the above conversion of overnight investment costs to hourly payments is to be able to compare the two on equal terms. A technology that is economically competitive is typically not dominated, meaning that given two technologies, if one has a higher fixed cost it should typically be characterized by a lower marginal cost.

Average cost is the total (variable plus fixed) cost per unit of output. It is defined mathematically as follows:

$$AC(Q) = \frac{TC(Q)}{Q} = \frac{FC + VC(Q)}{Q}.$$

The marginal cost curve of a given plant intersects the average cost curve of the plant at the point where the average cost curve is minimized. To see this, note that the point of minimum average cost is characterized by the following first-order optimality condition:

$$AC'(Q^\star) = 0$$
$$\Rightarrow \frac{MC(Q^\star) \cdot Q^\star - (FC + VC(Q^\star))}{(Q^\star)^2} = 0$$
$$\Rightarrow MC(Q^\star) = \frac{FC + VC(Q^\star)}{Q^\star}$$
$$\Rightarrow MC(Q^\star) = AC(Q^\star).$$

The result is confirmed in the left panel of figure 3.4.

---

**Example 3.4** *Intersection of marginal and average cost curves.* Suppose that gas generators are characterized by a fixed cost of 5.8 \$/MWh, that they are constructed in batches of 200 MW, and that their fuel cost curve is given in figure 3.3. The total cost for each gas generator is as follows:

$$TC(Q) = \begin{cases} 5.8 \cdot 200 + 20 \cdot Q + 0.2 \cdot Q^2 \text{ \$/h,} & 0 \text{ MW} \leq Q < 200 \text{ MW,} \\ 5.8 \cdot 200 + 12000 + 1000 \cdot (Q - 200) \text{ \$/h,} & 200 \text{ MW} \leq Q \leq 220 \text{ MW,} \\ +\infty \text{ \$/h,} & Q > 220 \text{ MW.} \end{cases}$$

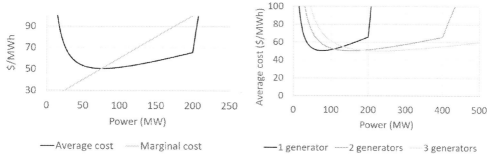

Figure 3.4 (Left) Average cost and marginal cost for example 3.4. (Right) Average cost for multiple units (see example 3.5).

The average and marginal cost curves are given in figure 3.4. Note that, as anticipated by the previous analysis, the marginal cost curve intersects the average cost curve at the point where average cost is minimized.

**Example 3.5** *Average cost curves of multiple identical units.* Consider a collection of identical gas units that correspond to those presented in example 3.4. The average cost curve of a group of 1, 2, and 3 generators is shown in figure 3.4.

It is interesting to interpret the result of the right panel of figure 3.4. It indicates that:

- The owner of a single gas plant would have a cost advantage whenever the total cost in the single gas generator curve is below all competing curves, i.e. in the range 0 MW to 108 MW.
- Two gas plants would have a cost advantage whenever demand is between 108 MW and 186 MW, because in this case two identical gas plants could serve the system demand more efficiently than any alternative configuration.
- Three gas plants would have a cost advantage above 186 MW, and so on.

When we consider identical gas plants with the characteristics of example 3.4, we find that the minimum possible cost that can be attained for any number of units is equal to 50.46 $/MWh (see problem 3.8). If the demand were constant and in the order of a few hundred MWs it could be efficiently supported by *multiple* gas plants, which is conducive of a competitive industry.

Suppose, however, that the industry demand were constant and equal to 1000 MW, and that natural gas plants were competing with nuclear plants that are built in batches of 1000 MW. Suppose, further, that a nuclear plant obeys the cost characteristics of table 1.1, i.e. is characterized by an investment cost of 32 $/MWh and a marginal variable cost of 6.5 $/MWh. Then, at a constant level of output equal to 1000 MW, the single nuclear plant would achieve its minimum average cost, which is equal to the sum of its investment and fuel costs, namely 38.5 $/MWh. Note, however, that this minimum average cost is also attained at a level where production is equal to the total demand in the system, and thus a *single* nuclear plant would be the most cost-efficient way to support the constant demand in the system. Such a situation would undermine competition.

Of course, reality is more complex. For instance, as indicated in the motivating example of section 1.2, demand can vary over time, and this implies that a mix of technologies may be the most efficient way to support system demand. Nevertheless, the point at which average cost is minimized (i.e. the intersection of the marginal cost and average cost curves), which is defined as the **minimum efficient scale** of the industry, carries a special weight in economic analysis. Comparing minimum efficient scale to the level of demand in the industry is an indicator of the prospect for an industry to be competitive. If the minimum efficient scale is comparable to the level of demand in the system, then this is an indication of a **natural monopoly**, where government regulation or intervention is typically warranted in order to ensure that pricing can ensure the economic viability of a sector without leading to abuse of the monopolistic position of a firm. As the previous example suggests, natural monopolies tend to emerge in industries where fixed costs dominate over variable costs. Until recently, electric power markets adhered to the characteristics of a natural monopoly (although this is changing, due to changes in the cost of various technologies as well as a transformation in the demand side of the sector).

### 3.1.2 Transmission and distribution

The transmission network is used for transporting bulk quantities of power from generators to substations that serve large populations of loads. The distribution network is used for transferring power in smaller quantities to commercial and residential consumers. The key differentiating features of transmission and distribution are the amount of power that these networks carry, as well as the voltage level. High voltage reduces losses, which is why transmission networks operate at high voltage (115–765 kilovolts), whereas distribution networks typically operate at voltages ranging from 120 Volts to 20 kilovolts. Transformers are placed on the interface between the transmission and distribution systems in order to reduce the voltage level.

Power transmission is best understood by visualizing the electric system as a graph. The nodes of the graph correspond to locations where producers inject power, consumers withdraw power, where wires meet, or a combination of the above. The arcs of the graph correspond to wires that connect nodes. Figure 3.5 depicts a 3-node network.

An important feature of power transmission can be noted immediately from figure 3.5. Physics and operational security requires power balance: the total power injected into the system must equal the total amount of power consumed in the network (including losses). In fact, power balance must be maintained at each node of the network. In this sense, power grids resemble transportation networks with flow balance constraints that are commonly encountered in operations research. However, the situation is somewhat more complex in power systems.

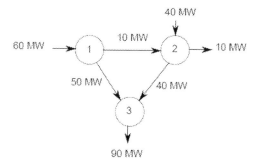

**Figure 3.5** Power refers to the rate of production, consumption, or flow of energy. Power supply and demand must balance at each node of the network.

One feature that differentiates power systems from most other supply chains is the fact that power cannot be routed arbitrarily over transmission lines, but instead obeys a set of physical laws: **Ohm's law** and **Kirchhoff's laws**. These laws determine a mapping of power injections and withdrawals at each location of the electricity network to flows over transmission lines. This mapping is determined by the **power flow equations** and it is relevant from a perspective of resource allocation because transmission lines are constrained in terms of the amount of power that they can carry. A useful analogy for a power system is that it functions like a water tank. Water can be injected into the tank from a hose and can be withdrawn from the tank via a sink, but there is no direct shipment of water from the hose to the sink.

An approximate description of the power flow equations can be obtained by linearizing the mapping, resulting in the **direct current (DC) power flow equations**. The DC power flow model of transmission is operationally adequate for some, but not all, functions of power system operations, including the clearing of certain markets. The DC power flow equations are derived in appendix B. The DC power flow equations establish a *linear* mapping between power injections in each node of the network and the flows over lines of the network. The power flow equations have an intuitive interpretation. Current splits itself along the branches of the network in a way that is inversely proportional to the electrical resistance that is encountered along branches. Since the mapping of power injections to flows is linear, it satisfies the properties of proportionality and additivity. Proportionality is demonstrated in figure 3.6: doubling injections in the nodes of the network results in twice the flow of power over lines. Additivity is demonstrated in figure 3.7: if the injection of power in the nodes of the network is expressed as the sum of two vectors of injections, then the total flow of power over the lines is the sum of the flows caused by these vectors of injections.

Distribution system operations ensure that power is delivered to consumers at an acceptable quality. The key concern is to ensure that the voltage of the distribution system remains within acceptable limits, and that flow limits on lines are not exceeded. Distribution system operations are not developed in detail in the text.

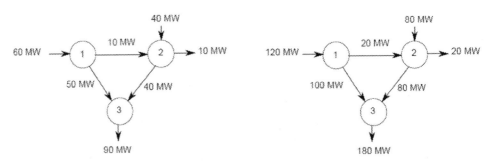

**Figure 3.6** All lines have identical physical characteristics. Flows are proportional to injections.

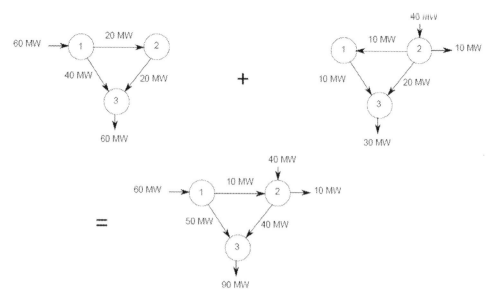

**Figure 3.7** All lines have identical physical characteristics. Flows are additive.

### 3.1.3 Consumption

Consumption is typically categorized between industrial, commercial, and residential. Residential and commercial consumers are treated differently both in operations as well as in electricity markets. Industrial consumers may be connected to either the transmission or the distribution network. Residential and commercial customers receive power from the distribution network. Moreover, industrial consumers have the option of participating directly in the wholesale electricity market, whereas commercial and residential customers buy power in the retail market.

Whereas it is intuitive to first define total fuel cost and then marginal cost when discussing the supply side, it is instead more intuitive to discuss marginal notions first when it comes to the demand side. Although consumers are not accustomed to the notion of procuring power dynamically, **valuation** or **marginal benefit** for power can be defined meaningfully. Valuation is measured in $/MWh. This is the willingness to pay for 1 MW of supply over an hour, regardless of how this amount of power is allocated among consumer usages. As in the case of marginal cost, valuation depends on the current level of consumption of a consumer. **Consumer benefit** is the total benefit of consumers for a given amount of power consumption, and valuation is the derivative of consumer benefit with respect to power consumption.

The **inverse demand function** or **marginal benefit function** $MB(Q)$ maps power consumption $Q$ to the *additional* value that a consumer gains from marginally increasing its consumption above $Q$. The concept is demonstrated in figure 3.8. This is the willingness of the consumer to pay for an additional MW of consumption above $Q$. The marginal benefit function is the demand-side analog of the marginal cost function. It is intuitive that the marginal benefit function should be decreasing, since a consumer will allocate the first MW of power to its most valuable use, and

**Table 3.3 The valuations for example 3.6.**

| Tranche | Demand morning (kW) | Valuation morning ($/MWh) | Demand evening (kW) | Valuation evening ($/MWh) |
|---|---|---|---|---|
| Inflexible | 0.2 | 1500 | 0.3 | 1000 |
| Medium | 0.5 | 120 | 0.3 | 80 |
| Flexible | 0.2 | 30 | 0.1 | 20 |

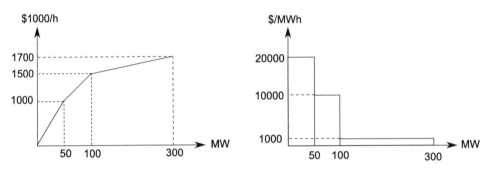

**Figure 3.8** The left graph is the consumer benefit, the right graph is its derivative, the inverse demand function, or willingness-to-pay of consumers. The curves are read as follows: loads are willing to pay 20000 $/MWh for the first 50 MW of power, 10000 $/MWh for the next 50 MW, and 1000 $/MWh for the last 200 MW. The left graph is the integral of the right graph.

allocate additional MW of power to less valuable processes. Consider a hospital as an example. The first tranche of power is expected to be extremely valuable since it supports critical equipment (surgery lighting, respirators, etc.). The second tranche of power may serve important, but less critical supplies (e.g. doctors' office lighting) and so on, until the last tranche, which serves the least critical equipment (e.g. lights at the parking lot). A counterexample to a decreasing marginal benefit function is tasks that require discrete quantities of power.

**Example 3.6** *Inverse demand function of a household.* Consider a household that follows the valuations for power that are presented in table 3.3. The marginal benefit functions are presented in figure 3.9. The inflexible tranche corresponds to the dark gray block, the medium tranche corresponds to the light gray block and the flexible tranche corresponds to the white block. It should be noted that the power slices are not necessarily identified with specific devices. For example, the first 0.1 kW of power may be allocated to lighting in the evening, because this is the consumption of greatest importance, whereas that same load slice may correspond to refrigeration in the morning, at a time of day when lighting is one of the least important uses.

It is worth reflecting on what information is gained from the **demand function**, which is the inverse mapping of the inverse demand function. The demand function maps the price of power $v = MB(Q)$ to the quantity $Q$ that a consumer would procure at that price. To see this, recall that all load slices prior to $Q$ achieve a benefit that is

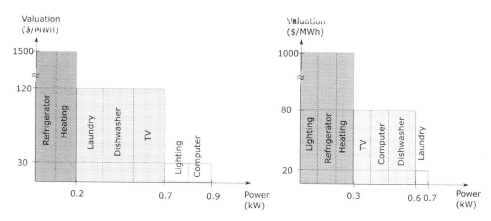

**Figure 3.9** The allocation of load slices to household devices in example 3.6. The left panel presents the valuations of demand in the morning; the right panel corresponds to valuations in the evening for the same household.

at least $MB(Q)$, and therefore all these load slices would be procured at any price less than or equal to $MB(Q)$.

**Demand elasticity** is the sensitivity of demand to changes in price. Denoting the demand function as $Q(v)$, where $v$ is price, demand elasticity is defined as

$$\epsilon = \frac{dQ/dv}{Q/v}.$$

A steep marginal benefit function (or a flat demand function) corresponds to *inelastic demand*, which is largely insensitive to price.

Given a certain state of the system (configuration of generation capacity and transmission network), the **average value of lost load** is the long-run average value of load that is shed in the system due to random disturbances (failures of generators and lines, forecast errors of renewable resources and load, etc.). The average value of lost load obviously depends on the amount of available generation capacity in the system. In capacity expansion planning studies, it is important to quantify the marginal benefit of an additional unit of generation capacity in order to compare it with the investment cost of installing that capacity. A marginal increase in capacity will decrease the average value of lost load. The **value of lost load (VOLL)** is the marginal change in average value of lost load resulting from a marginal increase in system capacity, divided by the marginal decrease in the amount of shed load.

---

**Example 3.7** *Computation of VOLL.* Consider a system with the following demand function:

$$Q(v) = 30000 - 2v.$$

In case there is inadequate capacity, rationing is random. Suppose that each vertical slice of load suffers a 1% decrease. The VOLL is equal to the lost value per MWh of load shed. When 1% of demand is rationed randomly, the lost value is given by

$$\int_{v=0}^{15000} Q(v)dv - \int_{v=0}^{15000} 0.99 Q(v)dv = 0.01 \cdot \frac{15000 \cdot 30000}{2}$$
$$= \$2.25 \cdot 10^6$$

The total rationing, which is equal to 1% of the peak energy consumption, amounts to 300 MWh. This implies that the value of lost load is

$$VOLL = \frac{2250000}{300} = 7500 \text{ \$/MWh},$$

and could have been computed equivalently as the average valuation of the consumers, $\int_0^{15000} Q(v)dv / 30000$. In more sophisticated planning studies where transmission constraints and outages are accounted for, the computation of VOLL is performed for a given level of capacity and is based on simulation.

### 3.1.4   Actors

This section presents the landscape of actors that participate in power system operations and electricity markets. An overview of the actors is presented in figure 3.10.

Generators (also referred to sometimes as **generating companies**) own power generation assets, and typically sell energy and various ancillary services to the system.

The transmission system is managed by the **system operator**. System operators are typically referred to as TSOs (transmission system operators) in Europe and ISOs (independent system operators) in the United States. Among its numerous responsibilities, the system operator is responsible for "keeping the lights on," and is therefore in control of the generators and demand in the system, as well as the transportation system, even if these assets are not owned by the system operator. The system operator is also responsible for operating some of the electricity markets.

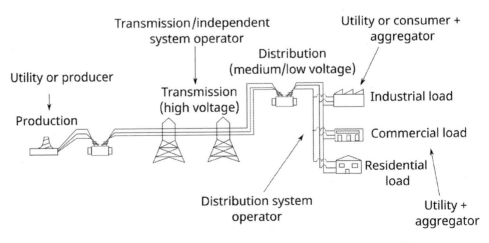

**Figure 3.10** The actors of electric power system operations and electricity markets.

Which markets are managed by the system operator depends on market design and varies from region to region. It should be noted that even though the system operator may operate some electricity markets, this is not a for-profit activity, and any positive or negative surplus that is generated by these markets is allocated back to society. In systems that distinguish system and market operation, some markets are managed by a **power exchange**.

The distribution system is managed by **distribution system operators** (DSOs). The DSO ensures that power is delivered with acceptable quality to consumers that are connected to distribution systems. Keeping voltage in acceptable levels is a particularly relevant concern.

Retail consumers are represented by **utilities**, also referred to as **load-serving entities**. Wholesale consumers may operate independently from utilities or may also be controlled by utilities. Utilities may also own generators, in which case they participate in the system as both producers and consumers. Otherwise, producers may participate independently in the system. **Retailers** are utilities that serve loads that do not participate in the wholesale market. Retailers thus buy power from the wholesale market and resell it to consumers who do not participate in the wholesale market at a retail price.

Aggregators are a new entity in electricity markets. The goal of aggregators is to coordinate large collections of consumers and other assets (e.g. renewable supply assets) such that their coordinated control offers a standard interface or even useful services to the system. The assets controlled by aggregators are often connected to either the medium- or low-voltage system.

**Transmission companies** are entities that own parts of the network and follow the instructions of the system operator in operating them. They are not represented explicitly in the models of this textbook.

### 3.1.5 Uncertainty and reserves

The production side of the supply chain introduces a certain degree of uncertainty at different time scales. At the monthly time scale, the amount of rainfall, which is uncertain, results in an uncertain amount of water in hydroelectric dams, which influences the operations of hydroelectric generators. At the daily time scale, uncertainty can be due to the failure of conventional generators and forecast errors in renewable energy production. The maintenance of conventional generators is different from failure, since it can be scheduled at periods of low stress to the system. The transmission system also introduces uncertainty, since transmission lines may also fail, although this occurs less frequently than generator failure. On the demand side, the major source of uncertainty is load forecast errors. Industrial loads or distribution transformers that serve a large number of customers may also fail. A **contingency** is defined as the failure of any system component (generator, line, transformer, load).

Adjustments due to forecast errors are quite different from adjustments due to contingencies. Forecast errors are part of the normal operation of the system, and

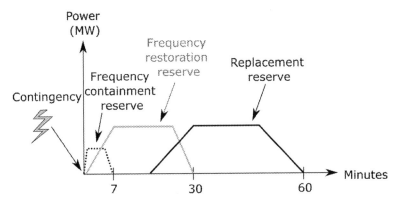

**Figure 3.11** Reaction of different types of reserve.

the system moves smoothly in position to deal with them. Instead, contingencies are shocks to the system that require the intervention of standby resources. Forecast errors are therefore dealt with by adapting the position of each generator in the system to the newly revealed conditions of the system, whereas in contingencies resources are typically pulled in, stand by for a certain amount of time, and are pulled out again.

**Reserves** are used in order to balance the system in case of gradual or steep disturbances. Reserves are a subset of **ancillary services**, which is a set of services that ensure the proper functioning of the electric power system. Reserves are classified according to the speed at which they can react,[2] as shown in figure 3.11. The classification of reserves is discussed in detail in chapter 6. Energy, transmission, and ancillary services, including reserve, comprise the majority of products and services that are traded in electricity markets.

---

**Example 3.8** *Modeling reserves in a linear program.* Consider a system with $n$ generators, with the operating cost of generator $i$ denoted by $f_i$ and its capacity denoted by $C_i$. Formulate a problem where demand is served at minimum cost while ensuring that the system is protected against the worst-case contingency.

**Solution**
In order to ensure that enough capacity is reserved to protect the system from a contingency, an auxiliary variable $r_i$ needs to be introduced in addition to the production decision variable $p_i$. The model can then be formulated as follows:

$$\min_{p,r} \sum_{i=1}^{n} f_i(p_i)$$
$$\text{s.t. } p_i + r_i \leq C_i, i = 1, \ldots, n$$

---

[2] More specifically, a major determinant of reserve type is the **full activation time**. This is the time that is required for the unit to reach its promised output.

$$\sum_{i=1}^{n} r_i \geq \max_{i=1,\ldots n} C_i$$

$$\sum_{i=1}^{n} p_i = D$$

$$p, r \geq 0,$$

where $D$ is the load of the system. The introduction of ramp rate constraints would complicate this model. Ramp and other more sophisticated features are developed in subsequent chapters.

## 3.1.6 Stages of decision making

Under normal operating conditions (i.e. in the absence of contingencies), the system needs to adapt its state over time, since forecasts are never exactly accurate. Forecast deviations are handled by adjusting the position of the system so as to maintain reliable operation at minimum cost. Power system operations are therefore a moving horizon planning problem. The heart of operations is the system operator. The major constraints of operations are (i) the requirement to instantaneously balance supply and demand, (ii) the requirement to operate the system securely, (iii) the technical constraints of generators, and (iv) the technical constraints of the transmission network.

A representative flow chart of power system operations is presented in figure 3.12. The flow chart does not account for capacity expansion planning, which is a long-term planning decision, or for hydro planning, which is a medium-term planning decision. Instead, the flow chart focuses on short-term (day-ahead and real-time) operations. Although this flow chart is not representative of the exact timing of operations in all systems, it does highlight the point in time at which certain operational decisions need to be finalized because updating them later in time is physically impossible. The flow chart corresponds to a centralized organization of electricity markets; nevertheless, similar decisions need to be reached in a decentralized market design (the differences between the two designs are highlighted in subsequent sections).

The first column describes operations that are performed in advance of the day-ahead time frame. The day-ahead market is open a few days prior to day-ahead operations. At this stage, outage schedules are planned. In this respect, planned outages are very different from contingencies because their timing can be optimized during hours of minimum stress to the system. Demand forecasts and ancillary services requirements are also determined in advance of the day-ahead time frame. At this time stage, the decisions that are being fixed are scheduled outages.

The second column describes day-ahead operations. The day-ahead market closes a few hours prior to the beginning of the day in question. Once the market closes, the day-ahead market model is run. This does not produce physically binding com-

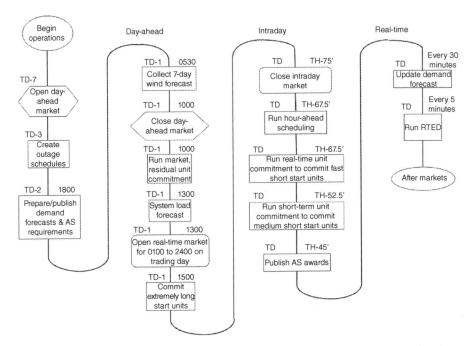

**Figure 3.12** A typical flow chart of system operations. TD and TH stand for trading day and trading hour, respectively.

mitments, but it does produce financially binding commitments. For some systems, residual unit commitment is also performed at this stage in order to ensure that the system operator can guarantee enough reserves in real time (this would be necessary, for example, if the demand forecast of the system operator is substantially higher than the amount of energy that is traded in the day-ahead market). A physically binding decision is made a few hours after the day-ahead market closes, namely the commitment of long-start units. The real-time market opens after the day-ahead market closes, and prior to the actual day of operations.

The third column describes intraday operations. The intraday market closes about one hour in advance of the operating hour in question. At this stage, long-start units have been committed and it is necessary to decide whether fast-start units need to be committed. This is decided by running a real-time unit commitment model. This produces a physically binding decision of which fast-start units should be started. At this point the flow chart moves to the last column, which is real-time operations, and the only flexibility that remains in the system at this stage is the ability to change set-points of units that have already been started (certain system operators also use the transmission network and/or demand response in order to respond to real-time conditions; however, this function is not represented in the flow chart of figure 3.12). Every few minutes, an economic dispatch model is run in order to adjust the set point of generators to the load in the system.

**Table 3.4 List of generators in example 3.9.**

| Generator | Marg. cost ($/MWh) | Max (MW) | Ramp (MW/min) |
|---|---|---|---|
| Cheap | 0 | 20 | $+\infty$ |
| Moderate | 10 | $+\infty$ | 1 |
| Expensive | 80 | $+\infty$ | 5 |

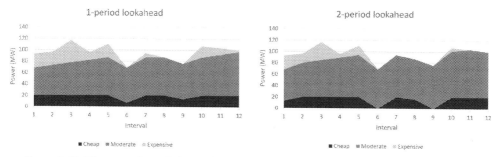

**Figure 3.13** The set-points of the 5-minute and 10-minute look-ahead policies for example 3.9.

All of the aforementioned models (day-ahead unit commitment, hour-ahead unit commitment, real-time unit commitment, economic dispatch) optimize with a look-ahead beyond the time interval in question in order to ensure that the system is prepared to respond to the forecast state in subsequent hours.

---

**Example 3.9** *Looking ahead in forward planning.* Consider a system with three generators, the characteristics of which are presented in table 3.4. Consider the real-time economic dispatch model, which is solved every 5 minutes for the next 5 minutes. Figure 3.13 presents the dispatch of the generators when the real-time economic dispatch model is solved with a look-ahead horizon of 5 minutes and 10 minutes. The initial state of the system is with the moderately expensive generator producing 50 MW and the expensive generator producing 50 MW. The cost in the case of 5-minute look-ahead for the entire hour is $1738.4. The cost for the 10-minute look-ahead is $1406.2. The difference is mainly due to the ability of the second policy to replace the expensive technology with the moderately expensive technology, by properly accounting for the ramp rate limits of the moderately expensive technology. For example, in interval 9 the first policy uses the cheap resource in order to serve demand to the greatest extent possible. Instead, the second policy foresees the increase in demand in interval 10 and pushes down the cheap technology in order to make sure that the moderately expensive technology is used to the greatest possible extent in interval 10. The first policy is forced to use the expensive technology in interval 10 because it is not using enough of the moderately expensive technology in interval 9 and the moderately expensive technology cannot keep up with the demand increase from interval 9 to interval 10.

## 3.2    Power market operations

The motivation of moving from centralized operations to markets stems from a number of advantages that properly functioning markets can deliver relative to centralized operations. These advantages relate to information, scalable operation, and efficient long-term investment signals.

*Information.* The centralized operation of the system requires that the system operator has access to detailed economic and technical information about all producers and consumers in the system. This is practically impossible for consumers due to their enormous number. Such information requirements are manageable on the supply side since even in the largest systems there are at most a few thousand producers. One attraction of a market is that information is decentralized. Producers only require access to the market price and their private technical and economic information in order to operate their generators at an optimal level. Similarly, consumers only require access to the market price and their private information about valuation in order to decide how much power to consume and when to consume it.

*Computation.* Operating a moderately sized power system requires the coordination of tens to hundreds of generators with hundreds of thousands to millions of consumers over multiple time periods. The system has to allocate multiple resources (energy, reserve, transmission) while accounting for complex constraints. It is technically challenging to solve this allocation problem centrally in an operationally acceptable time frame (e.g. a few minutes). A market solves the problem equivalently in a decentralized way. Although the current paradigm of power system operation requires this computation to take place centrally, in principle it could be decentralized, and academic research is underway in this direction. Progress in this direction would enable the scalable operation of power systems in regimes where smaller-scale producers (e.g. rooftop solar resources) and consumers (e.g. electric vehicles) could actively engage in the operation of the system. Adam Smith refers to the ability of markets to lead to efficient outcomes as the "invisible hand" of the market. An optimization interpretation of the "invisible hand" is that markets decompose the global allocation problem to surplus maximization subproblems with market prices functioning as the coordinating signal of self-interested market agents that maximize private surplus.

*Long-term signals.* Regulated monopolies offer limited incentives for innovation, since the profit at which power is sold is regulated by the government and adjusted to provide a reasonable return on capital investment. Technological innovations in a regulated environment have no clear impact on the long-run profits of the innovator; therefore the incentive to innovate may be impacted. If anything, certain incentive regulation schemes can induce some market participants to expand their capital base beyond efficient levels. Instead, a market environment favors innovations that offer market participants a competitive edge. For instance, by squeezing down costs, producers increase profit margins. Short-term incentives for optimally operating existing equipment are accompanied by long-term incentives for optimally expanding the system. The short-run efficiency argument is developed in chapter 4, while the long-run efficiency argument is formalized in chapter 11.

## 3.2.1 Exchanges and pools

The goal of setting up a market is to discover an efficient equilibrium while providing the proper long-run incentives. An important question is how rules can be designed for reaching this equilibrium. The setup of the rules of trade is a question of market design.

The most decentralized form of market organization is based on bilateral trading. More centralized forms of markets include exchanges and pools. Exchanges are markets based on auctions for trading products with simple rules for bidding and clearing. Pools use complex multi-part bids.

---

**Example 3.10** *Internalizing fixed costs.* Consider a generator with a startup cost of $2400, a capacity of 10 MW, and a fuel cost of 20 $/MWh. In a pool, the generator reports its startup cost and fuel cost and receives an energy price, as well as a side-payment if this is needed for the generator not to suffer a loss. Instead, the generator only receives an energy price in an exchange. Suppose that the generator is considering whether or not to produce for the following day. In a pool, the profit of the generator as a function of energy price $\lambda$ is $\max((\lambda - 20) \cdot 10 \cdot 24 - 2400, 0)$ $/day, provided the generator bids truthfully. If the generator is called, the pool has to pay the generator a side payment of $\max(2400 - (20 - \lambda) \cdot 24 \cdot 10, 0)$ $ for the entire day in order to ensure that the generator does not lose money. In an exchange, the generator would have to bid at least 30 $/MWh in order to ensure that it would not lose money if it is called upon.

---

The advantage of bilateral markets is flexibility. Parties can define contracts on their own terms. The advantages of exchanges and pools are that the price is public, there is no risk of default from counterparts since the default counterpart is the exchange or pool, and the markets are liquid since many participants are involved. Note that markets can be organized in order to combine one or more types of trade. For example, certain energy exchanges and pools combine bilateral trading of forward contracts with day-ahead auctions.

Electricity markets tend to be less centralized in advance of operations and more centralized as the system moves closer to real time, as shown in figure 3.14. Power is traded bilaterally, through financial contracts, in the months or years before operations. In the day-ahead time frame, agents adjust their position based on day-ahead forecasts. The intraday market is used for further adjusting positions a few hours in advance of operations. The real-time market is used for balancing supply with demand in real time by activating reserves and other fast-moving resources that are available in real time. The bilateral trading of electricity is impossible in the real-time market due to the fact that power supply and demand need to be balanced at every moment in time, therefore delays related to trading are not possible.

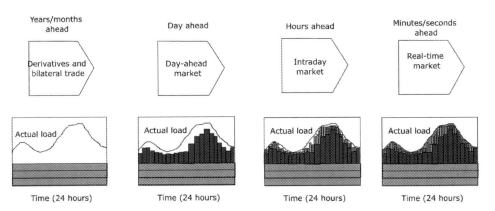

**Figure 3.14** The sequential operation of electricity markets.

## 3.2.2 Uniform and pay-as-bid auctions

Electricity markets rely extensively on auctions. Auctions can be organized in various ways. The two most prevalent forms of auctions used in electricity markets are uniform and pay-as-bid auctions.

In **uniform price auctions**, sellers submit sell bids: price-quantity pairs, representing the price at which they are willing to supply a certain quantity of energy. Buyers submit buy bids: price-quantity pairs representing the price they are willing to pay for the specific quantity of energy. The intersection of the supply and demand curves yields the market clearing price $\lambda$ and the quantities that need to be supplied and consumed. Supply bids that are *in the money* (i.e. bid at a price below $\lambda$) have to be offered and receive a payoff of $\lambda$ $/MWh. Supply bids that are *out of the money* should not be produced and receive no payoff. Demand bids that are *in the money* (i.e. bid at a price above $\lambda$) must consume and pay $\lambda$ $/MWh. Demand bids that are *out of the money* cannot consume and have to pay nothing.

---

**Example 3.11** *Settlement of a uniform price auction.* Consider the bids of figure 3.15 for trading power for 5 minutes. The producer places three sell bids: 30 MW at 12 $/MWh, 35 MW at 28 $/MWh, and 25 MW at 80 $/MWh. The consumer places three buy bids: 10 MW at 90 $/MWh, 40 MW at 40 $/MWh, 25 MW at 20 $/MWh. The market clearing price is $\lambda^\star = 28$ $/MWh. The producer is obliged to offer 50 MW of power for 5 minutes and receives a payment of $(50\text{ MW}) \times (28\text{ \$/MWh}) \times (\frac{1}{12}\text{h}) = \$$ 117. The consumer pays $117 and has to consume 50 MW of power for five minutes. Note that if the same bids had been placed by three separate producers (owning three separate generators) and three separate consumers (serving three different loads), the auction outcome (price and cleared bids) would have been identical.

---

The uniform auction that is used for clearing most day-ahead and real-time electricity markets is a generalization of Vickrey's second-price auction to multiple homogeneous goods. In a second-price auction for supplying a single item, the lowest bidder is paid for supplying the auctioned item, and is paid the price bid by the

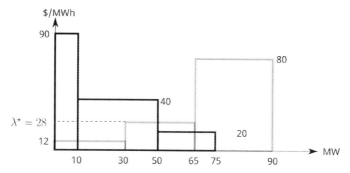

**Figure 3.15** The uniform price auction of example 3.11. Sell bids are represented in gray, buy bids are represented in black. Here, $\lambda^\star$ is the market clearing price.

cheapest losing bidder. This mechanism induces a truthful revelation of supplier costs. The intuition of this result is the following. If a bidder attempts to understate its cost (in the hope of making a more competitive offer), then the bidder should hope that it is not selected, since it runs the risk of supplying at a loss. Moreover, the bidder has no incentive to bid above its true cost (in order to push up the price), since this action has no impact on price (it is the price of the next highest bidder that determines the price) whereas the supplier increases its chances of losing the auction.

---

**Example 3.12** *Truth-telling in a second-price auction.* Consider a potential supplier with a privately known cost of $1000. Prove that an optimal strategy of the supplier is to bid at its true cost in a second-price auction.

**Solution**
The proof strategy is to consider two possible cases: (i) one in which the supplier underbids, for instance at $900, and (ii) one in which the supplier overbids, for instance at $1100. We would like to show that, in either case, the supplier is worse off by deviating from its truthful offer. We can prove this by considering two sub-cases within each case: (i) The state of the supplier does not change (where state means whether the auction is won or not), and (ii) the state of the supplier changes.

The argument is presented graphically in figure 3.16. The color coding is as follows: the winning bidder (i.e. the one with the lowest bid) is indicated in black accompanied by a star. The closest losing bidder (i.e. the one with the second lowest bid) is indicated in gray. In case the agent whose bidding strategy we are analyzing is neither a winning bidder or second highest bidder, then it is indicated in black without an accompanying star.

Case 1.1 corresponds to the agent underbidding without changing its state. This can happen, for instance, if the bidder moves from bidding $1000 to bidding $900, and still loses the auction. This case is indicated in the first row of figure 3.16. In this case, the payoff of the agent does not change because the price is set by the closest losing bidder by the design of the second-price auction.

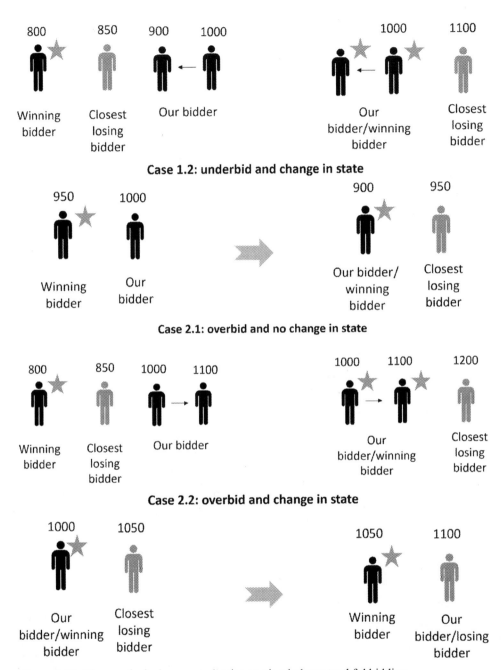

**Figure 3.16** The proof of why a second-price auction induces truthful bidding.

Case 1.2 corresponds to the agent underbidding, and also changing its state. This can happen, for instance, if the bidder moves from bidding $1000 to bidding $900, and moves from losing the auction to winning it. This case is indicated in the second row of figure 3.16. In this case, the payoff of the agent changes; however, it becomes worse. This is because there is a bidder below $1000 who was blocking the bidder in question from winning. Thus, the payoff of the agent changes from zero to a negative value.

Case 2.1 corresponds to the agent overbidding without changing its state. This can happen, for instance, if the bidder moves from bidding $1000 to bidding $1100, and still wins the auction. This case is indicated in the third row of figure 3.16. If the bidder moves from bidding $1000 to bidding $1100 and still wins the auction, then the payoff does not change because the price is set by the closest losing bidder by the design of the second-price auction. Thus, the payoff of the agent in question does not change.

Case 2.2 corresponds to the agent overbidding and also changing its state. This can happen, for instance, if the bidder moves from bidding $1000 to bidding $1100, and moves from winning the auction to losing it. This case is indicated in the fourth row of figure 3.16. If this occurs, it is because there is a bidder above $1000 who is blocking the bidder in question from winning. However, in this case the payoff of the agent changes from a positive quantity to zero.

We thus conclude that, by submitting anything but a truthful offer, the agent either leaves its state unchanged or moves to a new state with a lower payoff.

---

Intersecting supply and demand is a generalization of the Vickrey auction for selling multiple items, since the price that is set by the auction is the one of the lowest losing bid (see figure 3.17). The generalization is not exact, and there in fact exists a far more complicated rule for clearing the auction such that agents have an incentive to bid truthfully[3]; however, the uniform price auction is preferred in practice due to its simplicity.

An alternative auction format to uniform pricing that is often used in electricity markets is the **pay-as-bid auction**. In pay-as-bid auctions, bids are selected in order to maximize the benefits of trade and bidders receive the price that they bid. One concern over uniform pricing stems from the fact that prices are more volatile, since

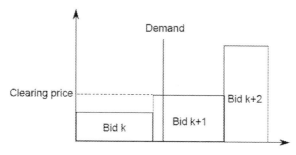

**Figure 3.17** Uniform price auctions are a generalization of second-price auctions when multiple homogeneous items are sold. The "cheapest losing bidder" is the bidder that has its bid only partially accepted.

[3] This mechanism is called the Vickrey–Clarkes–Grove mechanism.

the bid of the cheapest losing bidder sets the reward for all producers. Pay-as-bid pricing counterbalances this effect since even if the cheapest losing bid fluctuates wildly, this will not set the price for all other market participants. An undesirable consequence of uniform pricing has been hockey-stick bidding (which is essentially an exercise of market power), where market participants bid their entire cost curve truthfully, with the exception of a slice in the end of their supply curve which they bid at a high value. The hope of the generators is that once in a while this bid gets accepted, resulting in great profits, even though most of the time this bid will not be accepted. An additional argument against uniform pricing is that low-cost generators are perceived as receiving "unfair" profit margins; however, this argument ignores the need of generators to recover fixed costs. A difficulty associated with uniform pricing is that it is challenging to distinguish high prices due to gaming from high prices due to scarcity, bearing in mind that price spikes are required in order for units to cover their fixed costs. Instead, pay-as-bid pricing is criticized as leading to inefficiencies without keeping prices lower than uniform prices, since bidders try to bid near the market equilibrium price anyway, but sometimes fail to anticipate it accurately, thereby being dispatched inefficiently.

### 3.2.3   Electricity market blueprint

This section describes the structure of a typical electricity market. This includes the most important products that are traded in the market, as well as the actors that participate in the market. A simplified scheme of the actors and products that are traded in electricity markets is presented in figure 3.18.

The electricity market can be classified between the retail and wholesale market. Like other retail markets, the **retail electricity market** is used for selling electric power to small consumers, typically commercial and residential. This function is performed by load-serving entities. Load-serving entities procure the electricity from the wholesale electricity market and sell it to residential and commercial customers at a rate that is typically disconnected from the real-time price of electricity in the

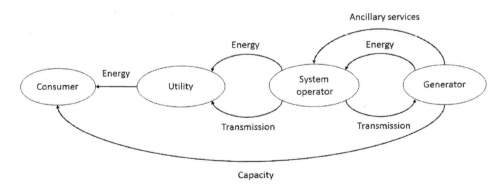

**Figure 3.18** A blueprint of an electricity market. Arrows indicate a sale of the product indicated on the label of the arrow. The agent that sells is at the tail of the arrow, the agent that buys is at the head of the arrow.

wholesale market. Typically, this rate is fixed or varies depending on the time of day, although more sophisticated contracts exist that are discussed in subsequent chapters. Retail tariffs also often include demand charges that bill peak power usage in order to recover the cost of investing in power generation capacity that is used for serving peak demand. Distribution network charges are also an important component of retail bills. Budget imbalances that occur in the wholesale market are commonly passed down to retail consumers through retail bills. Therefore, the retail market is also used for charging operating costs to consumers that are not compensated in the wholesale market (for example, generator side payments).

In the **wholesale electricity market** energy is traded in bulk quantities among producers and load-serving entities. The trading of energy requires transportation, as well as a reliable system. Therefore the wholesale market, in addition to energy, trades transmission and ancillary services. Three actors are mostly involved in the wholesale market: producers, the system operator, and load-serving entities or wholesale consumers that buy power in the wholesale market directly. The system operator is responsible for the reliable and secure operation of the system, and as such it procures reserves on behalf of consumers and also operates the transmission network. In some markets, the system operator also sells transmission capacity directly by running a market for transmission capacity, whereas in other markets the system operator sells transmission capacity implicitly by altering generator schedules in order to ensure transmission constraints are respected. Producers supply energy and ancillary services to the wholesale market. These ancillary services include a range of reserves with varying response times, as discussed in the previous section. In order to sell energy to the market, generators need to buy transmission line capacity in order to ensure that the energy that is injected into the system can be transferred over the transmission network to consumers. Similarly, utilities procure energy and also need to buy access to the transmission lines that are delivering energy to the consumer locations. In the future, it is envisioned that certain ancillary services could be offered by wholesale consumers or collections of residential and commercial loads that are coordinated by an aggregator.

In certain electricity markets, capacity is traded separately and in addition to energy, ancillary services, and transmission. The goal of defining an additional market for generator capacity is to ensure **resource adequacy**. Resource adequacy refers to the existence of sufficient generating capacity in order to meet the long-run needs of the system. Since the wholesale electricity market may not generate sufficient profits for generators in order to ensure adequate investment, a separate market for capacity creates an additional stream of revenue for generators that are rewarded for each MW of reliable capacity that they can offer to the system. The cost of procuring capacity through capacity remuneration mechanisms is often passed on to consumers.

The diagram presented in figure 3.18 is not meant to be all-encompassing; it serves instead as a rough sketch of an archetype market, a starting point relative to which variants or more detailed schemes can be compared. For example, (i) a direct, bilateral sale of energy from generators to utilities could exist, without going through the market; (ii) a trade of ancillary services from the system operator or from generators to utilities could exist, indicating that utilities might have obligations to buy a certain part of the ancillary services that are needed to run the system; (iii) a sale of capacity from generators to utilities might exist, indicating that utilities need to buy a certain part of system capacity directly from generators in order to contribute towards system adequacy; (iv) a sale of ancillary services from utilities to the system operator might be included in the diagram, indicating that utilities are able to control the consumption of consumers in response to emergency needs of the system. It should also be noted that certain ancillary services that are offered by market participants (e.g. reactive compensation) are ignored in this diagram due to their secondary role.

### 3.2.4   Example: California and Central Western Europe

Electricity markets differ not only in terms of how agents interact (who sells what to whom), but also in terms of the exact rules of trade. These variations can have important implications in terms of market performance. In this section two examples of notably different organizations are compared: the California market, and the Central Western European (CWE) market.

The principal differences of these market designs are (i) the degree of centralization in the trading of energy (one is a pool, the other is an exchange), (ii) the degree of coordination in the trading of energy, reserves, and transmission, and (iii) the handling of transmission constraints. The second and third point relate to another important difference between the two markets, namely the separation between the trading of energy and the operation of the transmission network (both of these tasks are performed by the system operator in the case of California, whereas they are performed separately in the case of CWE).

*Pools versus exchanges.* The California day-ahead and intraday market operates as a pool. This means that generators submit detailed information about their technical constraints and costs. In contrast, in the European power exchange, generators submit, in principle, simple supply bids that attempt to internalize their constraints and costs. Important reasons for the existence of these two variants are the complex unit commitment and dynamic constraints of generators, as well as their non-convex costs (for example startup and minimum load cost). Both an exchange and a pool solve a dynamic optimization problem for the following day and result in a schedule that aims at maximizing social surplus, given the declared costs and operating constraints of the bidders. The key in the clearing of an exchange or pool is to determine an energy price, paid per unit of power that is supplied for each hour, such that market agents would voluntarily react to this price in a way that is consistent with the socially optimal solution. This task is simplified to some extent in exchanges with simple bids because the determination of price is simply a matter of intersecting supply and demand curves. However, the task of the bidders becomes more challenging because they are called to internalize, in the simple bids that they submit to the market, the complex reality underlying the operation of their machines.

In power pools, the complexity is transferred to a day-ahead scheduling algorithm that determines power production and consumption, and determines a price that urges bidders to respond optimally. The pool is more complex from the point of view of the market operator because (i) the scheduling problem that the market operator solves represents the complex operating constraints and costs of generators, and (ii) it may be the case that no price exists[4] that would urge all bidders to react optimally. In this case, **uplift payments** are required in order to make up for the negative surplus (or lost surplus) of the bidders. Exchanges are compared to pools in chapter 7.

*Simultaneous versus sequential clearing.* A second major difference between the California and the CWE market relates to the coordination of energy with transmission and reserves. The day-ahead unit commitment model that is solved in the California pool determines generator schedules, reserve schedules, and transmission line usage simultaneously. In contrast, the European power exchange clears energy without accounting for reserves and only roughly accounting for transmission constraints. As shown previously in this chapter, the transmission network cannot be represented adequately as a transportation network, since Kirchhoff's laws impose additional constraints to its operation. The European power exchange accounts in an approximate way for inter-country transmission constraints by using a simplified approximation of the physical laws that govern power flow, and ignores transmission constraints within national borders. For both of these reasons, day-ahead trades can violate physical transmission constraints. These violations are corrected by the system operator on the day of operations through **redispatch** (i.e. changing generator schedules in order to alleviate congestion). Similarly, producers decide how much reserve they should offer *before* their energy schedules have been determined by the exchange. In general, the simultaneous optimization of multiple resources is expected to perform better than the sequential optimization of these resources, which is why pools are expected to perform better in theory in terms of efficiency. However, the introduction of pools results in a multi-unit, multi-period auction where bidder costs and constraints are complex. In such a setting, no simple set of rules exists that offers the appropriate incentives to bidders for truthfully revealing their cost and technical information. It therefore comes as no surprise that pools have been gamed in the past, and that close monitoring and the mitigation of gaming is needed. The proper representation of transmission constraints is discussed in chapter 5, while the trading of reserves is discussed in chapter 6.

*Nodal versus zonal pricing.* The California market design has adopted **nodal pricing**, whereby the constraints imposed by transmission capacity are represented at the resolution of each physical node of the network, resulting in different prices between different nodes. Instead, the European power exchange has adopted **zonal pricing**. A **zone** is a collection of buses, or nodes, at which electric energy is sold at the same price. The motivation for zonal pricing is to simplify the trading of energy by reducing the

---

[4] Modern power exchanges include products that involve non-convex constraints and costs, which implies that they too are susceptible to the fact that no clearing price may exist.

number of markets in which energy is sold. The two designs are compared in detail in chapter 5.

## Problems

**3.1** *Feasibility of power flows.* Which of the power injections in figure 3.19 is feasible for the network of the figure, where all lines have identical characteristics and where the thermal limit of line AB is 50 MW?

1. 100 MW at node B, −100 MW at node C
2. 100 MW at node A, −100 MW at node C
3. 100 MW at node B, −100 MW at node A

**3.2** *Optimal buy bid in a second-price auction.* In keeping with the tradition of his mother, King Charles is a fervent supporter of horse riding. Charles attends an auction where a racing horse is being sold, for which Charles' valuation is $50000 (meaning that he is indifferent between owning the horse and paying $50000, or not owning the horse). The horse is being auctioned off in a second-price auction. What price would you advise the king to submit to the auction, and how would you convince him?

**3.3** *Rolling optimization with look-ahead.* Implement example 3.9 in a mathematical programming language and confirm the results of the example. The demand vector for intervals 1 to 24 is given in table 3.5. What happens when the look-ahead increases even further (e.g. three or more hours)?

**3.4** *Clearing a uniform price auction.* Consider the following offers in a uniform price auction:

- Offer B1: a buy offer for 100 MWh at 100 $/MWh

**Table 3.5 The demand for example 3.9 and problem 3.3.**

| Hour | MW | Hour | MW | Hour | MW | Hour | MW |
|------|--------|------|--------|------|--------|------|-------|
| 1 | 92.97 | 7 | 94.69 | 13 | 79.99 | 19 | 73.75 |
| 2 | 95.91 | 8 | 87.64 | 14 | 116.91 | 20 | 95.72 |
| 3 | 116.48 | 9 | 76.34 | 15 | 105.25 | 21 | 87.53 |
| 4 | 95.83 | 10 | 107.62 | 16 | 95.51 | 22 | 85.31 |
| 5 | 110.52 | 11 | 104.23 | 17 | 100.34 | 23 | 82.65 |
| 6 | 69.22 | 12 | 100.50 | 18 | 96.07 | 24 | 92.00 |

**Figure 3.19** The three-node network of problem 3.1.

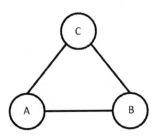

- Offer B2: a buy offer for 100 MWh at 50 $/MWh
- Offer S1: a sell offer for 20 MWh at 20 $/MWh
- Offer S2: a sell offer for 50 MWh at 60 $/MWh
- Offer S3: a sell offer for 40 MWh at 85 $/MWh
- Offer S4: a sell offer for 30 MWh at 125 $/MWh

What is the market clearing price? What are the offers paying/being paid?

**3.5** *Uniform pricing fallacies.* True or false: uniform pricing is unfair because cheap units are paid above their marginal cost.

**3.6** *Budget balance in uniform price auctions.* True or false: in certain cases, uniform price auctions require buyers to pay more than the amount that sellers collect.

**3.7** *Aggregating cost curves.* Provide a closed-form equation for the total cost of the two-generator and three-generator fleet in example 3.5, as well as the average cost, which is depicted in the right panel of figure 3.4.

**3.8** *Constant minimum average cost.* Prove that the minimum average cost that can be achieved by a fleet of $n$ gas plants such as those described in problem 3.7 is equal to 50.46 $/MWh.

**3.9** *Short-run marginal cost equals long-run marginal cost at nominal output.* Consider a power generating unit with a constant marginal fuel cost $MC$ that is producing at its nominal capacity. What is its marginal cost range? How does it compare to the long-run marginal cost that is computed in problem 2.6?

**3.10** *Annualized fixed cost given annual compounding.* Prove equation (3.1).

# Bibliography

**Section 3.1** The material on fixed, variable and average cost, as well as the discussion on natural monopolies, is inspired by Stoft (2002) and Varian (2014). The definition of VOLL in example 3.7 is inspired by Stoft (2002). Figure 3.12 is based on exhibit 2-1 of CAISO (2013). The rolling look-ahead of power system operations is widely applied in practice (Mickey 2015), and has led to interesting challenges in the context of pricing. These challenges are analyzed in a deterministic framework by Hogan (2016a), Schiro (2017), Hua et al. (2019), Zhao, Zheng, and Litvinov (2019), Hogan (2020), Guo, Chen, and Tong (2021), and Biggar and Hesamzadeh (2022), and extended to a stochastic framework by Cho and Papavasiliou (2022).

**Section 3.2** The material in this section is largely inspired by Stoft (2002). The fact that second-price auctions induce truthful revelation of supplier costs is demonstrated formally by Vickrey (1961). Practical difficulties related to the adoption of Vickrey auctions in practice are discussed in Ausubel et al. (2006). Fabra, von der Fehr, and Harbord (2006) compare uniform price auctions to pay-as-bid auctions in a stylized duopoly model with a known inelastic demand, and set a methodological framework for discussions on electricity auction design. This framework is used, for instance, in analyzing uniform price auctions with make-whole payments by Sioshansi and Nicholson (2011). The fact that the minimum average cost remains constant in problem 3.8 is a feature of **constant returns to scale**[5]; the reader is referred to Varian (2014) for a detailed discussion.

---

[5] Constant returns to scale refers to the property that a multiplication of output by a factor of $n$ requires $n$ times more input. **Increasing returns to scale** occur when a multiplication of output by a factor of $n$ requires *less* than $n$ times the input, while **decreasing returns to scale** occur when a multiplication of output by a factor of $n$ requires *more* than $n$ times the input.

# 4   Economic dispatch

This chapter introduces the economic dispatch problem, which is the simplest resource allocation problem around which most other models of this textbook are built. The economic dispatch problem corresponds to the traditional crossing of supply and demand curves in introductory microeconomics textbooks. The detailed exposition of the model in this text serves a number of purposes: (i) The KKT conditions of the model are analyzed in detail. Richer models are based on a generalization of this KKT analysis, so it is important to clarify concepts in a simpler model first. This KKT analysis is developed in section 4.1. (ii) The introduction of the economic dispatch model leads to the definition of certain fundamental economic concepts. Central among these is competitive equilibrium and a number of economically significant measures. These definitions are presented in section 4.2. This section attempts to translate the discussion of economic equilibrium in Stoft (2002) into a language that can be understood by readers with a mathematical programming background and an interest in modeling. (iii) The study of the economic dispatch model enables a transparent development of the equivalence between competitive market equilibrium models and optimization models. This equivalence is established in section 4.3 for a general resource allocation problem and is invoked repeatedly throughout the text.

From a practical standpoint, economic dispatch is the simplest and most central task in electricity markets. Economic dispatch amounts to matching consumers with the highest valuation for power with the suppliers that can offer that power at the lowest possible cost. In practice, the economic dispatch problem is solved by the market operator for operating real-time electricity markets. Real-time markets are typically operated as uniform price auctions, where agents submit sell and buy bids to the market operator. The operator solves the economic dispatch problem and determines the uniform market clearing price according to which sellers and buyers are billed. This process is repeated every few (most commonly five to fifteen) minutes.

## 4.1   The economic dispatch model

Consider a set of generators $G$ and a set of loads $L$. Generators are characterized by an increasing integrable marginal cost function $MC_g : \mathbb{R} \to \mathbb{R}$, which maps power production to marginal cost for each generator $g \in G$. Analogously, loads are characterized by a decreasing integrable demand function $MB_l: \mathbb{R} \to \mathbb{R}$, which maps consumption to marginal benefit for each consumer $l \in L$. The economic dispatch problem optimizes the amount of production $p_g$ among generators and the amount of consumption $d_l$ among consumers with a goal of maximizing *welfare*, which is consumer benefit minus generator cost. The problem can then be formulated as follows:

$$\max_{p,d} \sum_{l \in L} \int_0^{d_l} MB_l(x)dx - \sum_{g \in G} \int_0^{p_g} MC_g(x)dx$$

$$(\lambda): \quad \sum_{l \in L} d_l - \sum_{g \in G} p_g \leq 0$$

$$(v_l): \quad d_l \leq D_l, l \in L$$

$$(\mu_g): \quad p_g \leq P_g, g \in G$$

$$p_g \geq 0, g \in G, d_l \geq 0, l \in L.$$

Here, $P_g$ is the maximum power that a generator can produce and $D_l$ is the maximum power that a load can consume.

Dual variables are indicated to the left of the constraints. Note that welfare maximization is equivalent to cost minimization when demand is inelastic, which is to say when the demand function is infinite up to some fixed value and drops to zero thereafter. The KKT conditions of the economic dispatch problem can be stated as follows:

$$0 \leq p_g \perp -\lambda + MC_g(p_g) + \mu_g \geq 0, g \in G, \tag{4.1}$$

$$0 \leq d_l \perp -MB_l(d_l) + \lambda + v_l \geq 0, l \in L, \tag{4.2}$$

$$0 \leq \mu_g \perp P_g - p_g \geq 0, g \in G, \tag{4.3}$$

$$0 \leq v_l \perp D_l - d_l \geq 0, l \in L, \tag{4.4}$$

$$0 \leq \lambda \perp \sum_{g \in G} p_g - \sum_{l \in L} d_l \geq 0. \tag{4.5}$$

**Proposition 4.1** *Given an optimal solution of the economic dispatch problem, there exists a threshold $\lambda$ such that:*

1. *If a generator is operating strictly within its operating limits $(0 < p_g < P_g)$, then $MC_g(p_g) = \lambda$. If a load is consuming strictly within its dispatch interval $(0 < d_l < D_l)$, then $MB_l(d_l) = \lambda$.*
2. *If a generator is producing zero $(p_g = 0)$, then $MC_g(p_g) \geq \lambda$. If a load is consuming zero $(d_l = 0)$, then $MB_l(d_l) \leq \lambda$.*
3. *If a generator is producing at peak capacity $(p_g = P_g)$, then $MC_g(p_g) \leq \lambda$. If a load is consuming at peak capacity $(d_l = D_l)$, then $MB_l(d_l) \geq \lambda$.*

*Proof*   The result follows from the KKT conditions of the problem. For example, consider a generator for which $0 < p_g < P_g$. From the condition of equation (4.3) it follows that $\mu_g = 0$. From the condition of equation (4.1) it follows that $\lambda = MC_g(p_g) + \mu_g = MC_g(p_g)$. The proof for the other cases follows the same logic. $\square$

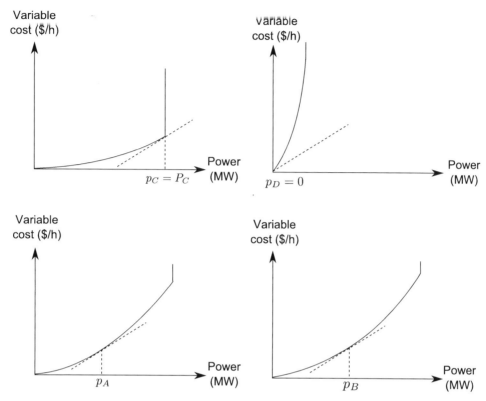

**Figure 4.1** An illustration of the KKT conditions for the economic dispatch problem. The slope that appears in all graphs corresponds to the system lambda at the optimal solution. Generators $A$ and $B$ are operating strictly within their operating range ($0 < p_g < P_g$), and the point at which they operate is such that the slope of the variable cost curve is equal to the system lambda. Generator $C$ is operating at its technical capacity, and the slope of its marginal cost curve at the peak capacity $P_C$ does not exceed the system lambda. Generator $D$ has a slope at zero that is greater than the system lambda, so the generator is producing zero.

The threshold that characterizes the optimal solution, which is the marginal cost of the marginal generating unit (i.e. the generating unit that will supply the next unit of power at lowest cost), is referred to as the **system lambda**. A graphical illustration of the optimal solution is presented in figure 4.1.

The KKT conditions provide an intuitive characterization for the optimal solution: by always picking the cheapest increment of production and the most valuable increment of consumption, a solution is obtained where (i) generators that are operating strictly within their technical limits are operating at the same level of marginal cost, which is equal to the marginal benefit of loads that are consuming strictly within their technical limits, (ii) units operating at their maximum capacity have a marginal cost less than or equal to the system lambda and loads that are consuming at their maximum demand have a valuation greater than or equal to the system lambda, and (iii) units operating at zero have a marginal cost greater than or equal to the system lambda and loads consuming nothing have a valuation less than or equal to the system lambda. Effectively, the optimal solution is matching the cheapest generators with the loads that value most the power that is being produced.

Using the sensitivity result of proposition 2.8, it is known that $\lambda$ is equal to the sensitivity of the objective function to the constraint $\sum_{l \in L} d_l - \sum_{g \in G} p_g \leq 0$. This makes sense intuitively. A unit increase on the right-hand side of this constraint enables a unit increase in power consumption from a load, or a unit decrease in power production from a generator. This increases the objective function value by the system lambda.

## 4.2    Competitive market equilibrium

The economic dispatch problem is used for clearing the real-time electricity market as a uniform price auction. One by-product of the solution is the system lambda, which is used as the clearing price of the real-time market. In this section it is argued that this is the price that would anyways be negotiated between agents who maximize surplus under conditions of perfect competition in bilateral trade. One reason why the real-time market is operated centrally by the market operator, and not bilaterally, is that supply needs to be constantly balanced with demand in order to prevent system blackouts. Bilateral trade is not fast enough to keep up with this requirement.[1] Therefore, the system operator operates assets physically (or instructs producers and consumers on how much power to produce and consume respectively), and the real-time market is used for financial transactions, i.e. for billing producers and consumers. In this way, the physical balance of power supply and demand is preserved, while also ensuring that the "right" price is charged for the traded power through the real-time electricity market auction.

It is worth reflecting about why the system lambda is a reasonable price to charge for power. In order to formalize arguments, the first step is to characterize how producers and consumers would behave in a competitive environment. A competitive environment is defined as follows:

**Definition 4.2**    A **competitive market** is a market in which the following conditions hold:

- Agents are price-taking. Mathematically, this means that when agents maximize surplus they consider price as a fixed parameter in their optimization, rather than a decision variable that can be influenced by their actions.
- The variable costs of producers are convex; equivalently, their marginal costs are increasing. Analogously, the total benefit of consumers is concave; equivalently, the marginal benefit of consumers is decreasing.
- Agents have access to market prices.

---

[1]  It took 2.5 minutes from the moment of frequency deviation until system collapse in the 2003 blackout in Italy, which is possibly faster than most agents can negotiate a good price in the market.

Newton's laws of motion describe the dynamic behavior of a mechanical system. Markets also abide to "laws of motion," driven by the behavior of agents that maximize surplus. These laws of motion dictate the equilibrium point at which an economic system settles, which is of paramount interest to analysts. The laws of motion of markets are referred to as *quantity* and *price adjustment*. These concepts are first introduced intuitively, and then defined mathematically.

In order to develop intuition about the notion of price adjustment, it is useful to reason about the marginal cost function of the system in terms of vertical slices of tiny suppliers. The supplier willing to provide the $Q$th increment of power is willing to do so at a price of $MC_G(Q)$ or more, where $MC_G(\cdot)$ is the aggregate marginal cost. Aggregate marginal cost is defined mathematically as follows:

**Definition 4.3    Aggregate variable cost** $TC(Q)$ is the cheapest way for a set $G$ of generators to produce $Q$ units of power:

$$TC_G(Q) = \min_p \sum_{g \in G} \int_0^{p_g} MC_g(x)dx$$

$$\text{s.t. } \sum_{g \in G} p_g = Q$$

$$p_g \in \text{dom } MC_g, g \in G,$$

where $MC_g$ is the marginal cost of generator $g$, $p_g$ is the power production, and dom $MC_g$ is the feasible set of each producer. **Aggregate marginal cost** is the derivative of aggregate variable cost, provided it exists: $MC_G(Q) = TC'_G(Q)$.

Reasoning similarly about consumers, the marginal benefit function of the system should be thought of as a sequence of tiny consumers, stacked as vertical slices in order of decreasing valuation, with the consumer willing to consume the $Q$th increment of power being prepared to pay a price of $MB_L(Q)$ or less. Here, $MB_L(\cdot)$ is the aggregate marginal benefit. The formal definition of aggregate marginal benefit is analogous to that of aggregate marginal cost.

**Definition 4.4    Aggregate benefit** is the most beneficial way to consume a certain quantity of power among a collection of consumers. Mathematically, it is given by:

$$TB_L(Q) = \max_d \sum_{l \in L} \int_0^{d_l} MB_l(x)dx$$

$$\text{s.t. } \sum_{l \in L} d_l = Q$$

$$d_l \in \text{dom } MB_l, l \in L,$$

where $L$ is a set of consumers, $MB_l$ is the marginal benefit of consumer $l$, $d_l$ is the power consumption, and dom $MB_l$ is the feasible set of each consumer. **Aggregate marginal benefit** is the derivative of aggregate benefit, provided it exists: $MB_L(Q) = TB'_L(Q)$.

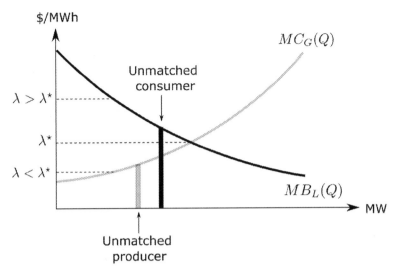

**Figure 4.2** A market price above or below $\lambda^*$ leaves unmatched producers and consumers who can trade profitably with each other.

The system lambda that was introduced in the previous section is the intersection of the aggregate marginal cost curve with the aggregate marginal benefit curve. It is denoted as $\lambda^*$ in figure 4.2. To understand the process of price adjustment, consider the case where the market price is lower than $\lambda^*$. Then, as indicated in the figure, there are suppliers who refuse to produce at such a low price, but who are able to produce at a price between their marginal cost and $\lambda^*$. Similarly, there are consumers willing to pay a price of $\lambda^*$ or more who have not bought power. Unmatched suppliers are then able to profit by quoting a higher price, creating an upward pressure on prices. Similarly, if the price is above $\lambda^*$ there are buyers with a valuation below the current price but above $\lambda^*$ who are not consuming. These buyers should be able to discover suppliers whose cost is below $\lambda^*$ and who have not found buyers to sell their power to. This implies that quoting a price below the existing price can lead to profitable trade, thereby creating downward pressure on prices. The only price at which the system balances out is $\lambda^*$, the intersection of the supply and demand curves.

For any given price, agents that maximize surplus adjust their quantity such that they maximize surplus. This process is referred to as quantity adjustment. The forces of price and quantity adjustment are those that drive a competitive market into equilibrium: surplus maximization on behalf of market agents, and the pressure on prices when the system is out of equilibrium, i.e. when there are opportunities for profitable trade. These notions are now defined formally.

**Quantity adjustment** is the process by which a price-taking producer increases its output if its marginal cost is less than the market price and decreases output if it is greater. In the case of economic dispatch, the producer solves the following optimization problem:

$$\max_{p} \lambda \cdot p_g - \int_0^{p_g} MC_g(x)dx \qquad (4.6)$$

$$(\mu_g): \quad p_g \leq \Gamma_g \tag{4.7}$$

$$p_g \geq 0. \tag{4.8}$$

Similarly, a price-taking consumer decreases consumption if its marginal benefit is less than the market price and increases consumption if it is greater, namely the consumer solves the following optimization problem:

$$\max_{d} \int_0^{d_l} MB_l(x)dx - \lambda \cdot d_l \tag{4.9}$$

$$(v_l): \quad d_l \leq D_l \tag{4.10}$$

$$d_l \geq 0. \tag{4.11}$$

**Price adjustment** refers to the process by which prices rise whenever demand exceeds supply, and prices drop whenever supply exceeds demand. This implies that, for each product that is traded in a market, either supply for the product equals demand for the product, or the price of the product is zero whenever supply for the product exceeds demand for the product. Mathematically, this is stated as a complementarity condition called the **market clearing condition**:

$$0 \leq \sum_{g \in G} p_g - \sum_{l \in L} d_l \perp \lambda \geq 0. \tag{4.12}$$

A market is in equilibrium when no profitable opportunities for trade exist. This is accomplished when prices and quantities are such that agents maximize their surplus and the market clearing conditions hold. The **market clearing price** is the price that clears a market. This definition applies to any market, even if the market is not competitive. An equilibrium in a competitive market is called a **competitive equilibrium**, and the price at which it is achieved is called the **competitive price**. The formal definition follows:

**Definition 4.5**    A competitive equilibrium is a combination of prices $\lambda^*$ and allocations $p^*$ and $d^*$ such that

- given the price $\lambda^*$, producers and consumers maximize their surplus, and
- the market clears, i.e. $0 \leq \lambda^* \perp \sum_{g \in G} p_g^* - \sum_{l \in L} d_l^* \geq 0.$

The workings of the invisible hand of the market are revealed when the KKT conditions of surplus-maximizing producers and consumers (quantity adjustment) and the market clearing condition (price adjustment) are gathered into a single system of complementarity constraints:

$$\text{Producers: } 0 \leq p_g \perp -\lambda + \mu_g + MC_g(p_g) \geq 0, g \in G, \tag{4.13}$$

$$0 \leq \mu_g \perp P_g - p_g \geq 0, g \in G, \tag{4.14}$$

$$\text{Consumers: } 0 \leq d_l \perp \lambda + v_l - MB_l(d_l) \geq 0, l \in L, \tag{4.15}$$

$$0 \leq v_l \perp D_l - d_l \geq 0, l \in L, \tag{4.16}$$

$$\text{Market Clearing: } 0 \leq \lambda \perp \sum_{g \in G} p_g - \sum_{l \in L} d_l \geq 0. \tag{4.17}$$

It can be observed that these conditions are identical to the KKT conditions of the economic dispatch problem, equations (4.1)–(4.5). This proves the following proposition:

**Proposition 4.6**    *The competitive equilibrium results in an allocation that is optimal for the economic dispatch problem.*

*Proof*    The KKT conditions of the economic dispatch problem are identical to the conditions that define a competitive market equilibrium. Since the KKT conditions are necessary and sufficient for optimality, the proposition follows.                □

The implications of this observation are profound, and perhaps surprising at first glance. The reason that the result may evade intuition when first encountered is because a competitive market is driven by self-interested market agents who are not striving to maximize the welfare of society, but instead their individual surplus. It is also interesting to note that agents reproduce the solution to a centralized optimization problem, although the decision-making process of a market is decentralized. More specifically, agents do not communicate directly with each other, rather they merely coordinate through the market price. The next section generalizes proposition 4.6. This generalized result is used repeatedly in the text.

Having defined the equilibrium price, it is now possible to define a number of other measures of economic interest Performance in markets is measured in terms of welfare. Welfare is the total benefit of trade, which can be broken down into producer and consumer surplus. Producer surplus, consumer surplus, and welfare are demonstrated graphically in figure 4.3.

**Definition 4.7**    For a given price, **producer surplus** is the profit of producers who are willing to sell at that price, given by

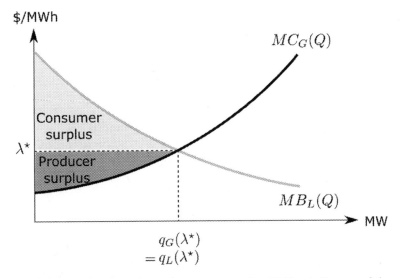

**Figure 4.3** Producer surplus and consumer surplus. Welfare is the sum of the light and dark gray surface.

$$\lambda \cdot q_G(\lambda) - \int_0^{q_G(\lambda)} MC_G(x)dx,$$

where $q_G(\lambda)$ is the amount of power sold at price $\lambda$.

**Definition 4.8**   For a given price, **consumer surplus** is the surplus of consumers who are willing to buy at that price, given by

$$\int_0^{q_L(\lambda)} MB_L(x)dx - \lambda \cdot q_L(\lambda),$$

where $q_L(\lambda)$ is the amount of power bought at price $\lambda$.

**Definition 4.9**   **Total surplus**, or **welfare**, is the sum of producer and consumer surplus.

A market is said to be efficient when total surplus is maximized. This includes minimizing the cost of what is produced and maximizing the value of what is consumed, as well as producing and consuming the right amount.

## 4.3   Modeling market equilibrium as an optimization problem

The central result of the previous section, proposition 4.6, implies that a competitive market can be analyzed by solving a single optimization problem, and that the dual optimal solution of this optimization problem has an economic interpretation and can be used for computing market performance metrics such as agent profits and welfare. This section generalizes proposition 4.6 to multiple products and discusses the use of optimization models for simulating markets.

The economic dispatch problem presented in the previous section, as well as a number of other models encountered in this text, obey the following format:

$$(Sep): \quad \max_x \sum_{i=1}^n f_i(x_i)$$
$$(\rho_i): \quad g_i(x_i) \le 0, i = 1, \dots, n$$
$$(\lambda): \quad \sum_{i=1}^n h_i(x_i) \le 0.$$

According to this format, there are $n$ agents who decide on private actions $x_i \in \mathbb{R}^{n_i}$. The system goal is to maximize a total objective that is summable over individual agent benefits $f_i$, which are assumed to be *concave* differentiable functions. The benefit of each agent depends *only* on the decision of that agent, i.e. $f_i$ is a function of $x_i$ only, with $f_i : \mathbb{R}^{n_i} \to \mathbb{R}$. Each agent has to obey a set of private constraints $g_i(x_i) \le 0$. However, as a collective all agents also have to respect a set of constraints $\sum_{i=1}^n h_i(x_i) \le 0$. The functions $g_i$ and $h_i$ are assumed convex differentiable, with $h_i : \mathbb{R}^{n_i} \to \mathbb{R}^m$ and $g_i : \mathbb{R}^{n_i} \to \mathbb{R}^{a_i}$. The coordination constraints $\sum_{i=1}^n h_i(x_i) \le 0$ are additively separable over agents, and can therefore be interpreted in a specific way: (i) there is a set of $m$ resources that are limited; (ii) each agent, by deciding $x_i$, uses up (or produces, if $h_i(x_i)$ is negative) an amount $h_i(x_i)$ of each of the $m$ limited resources; (iii) the total amount of the limited resources that is consumed cannot exceed the amount of the resources that is produced.

Consider the KKT conditions of problem (*Sep*), and suppose that these KKT conditions are necessary and sufficient for optimality. These conditions can be expressed as:

$$-\nabla_{x_i} f_i(x_i) + (\nabla_{x_i} g_i(x_i))^T \rho_i - (\nabla_{x_i} h_i(x_i))^T \lambda = 0, i = 1, \ldots, n, \quad (4.18)$$

$$0 \leq \rho_i \perp -g_i(x_i) \geq 0, i = 1, \ldots, n, \quad (4.19)$$

$$0 \leq \lambda \perp -\sum_{i=1}^{n} h_i(x_i) \geq 0, \quad (4.20)$$

where $\nabla_{x_i} f_i(x_i) \in \mathbb{R}^{n_i}$ is the gradient of $f_i$ and $\nabla_{x_i} g_i(x_i) \in \mathbb{R}^{a_i} \times \mathbb{R}^{n_i}$ is the Jacobian matrix of $g_i(x_i)$ (likewise for $\nabla_{x_i} h_i(x_i)$).

Instead of solving the above problem centrally, consider how a competitive market would perform if each of the $m$ resources were traded among the $n$ agents. Each product that is traded has a price $\lambda_i$ associated to it, and producers are paid $\lambda_i$ for selling the commodity whereas consumers pay $\lambda_i$ for buying the commodity. Therefore, in a market each agent solves the following surplus maximization problem for a given price vector $\lambda^\star$:

$$(Surplus - i): \quad \max_{x_i, q_i} (f_i(x_i) - (\lambda^\star)^T q_i)$$

$$(\rho_i): \qquad g_i(x_i) \leq 0$$

$$(\lambda_i): \qquad h_i(x_i) = q_i,$$

where $q_i$ represents the vector of resources procured (or sold, if negative) by agent $i$.

Extending the definition of a competitive market equilibrium introduced in the previous section, a competitive market equilibrium for multiple products is defined:

**Definition 4.10** A **competitive equilibrium** over multiple products is defined as a set of prices $\lambda^\star$, agent decisions $x_i^\star$, and product procurement $q_i^\star$ such that:

- $(x_i^\star, q_i^\star)$ maximize surplus given $\lambda^\star$, i.e. they solve the problem (*Surplus* − $i$) defined above, and
- market clearing holds:

$$0 \leq \lambda^\star \perp \sum_{i=1}^{n} q_i^\star \leq 0.$$

A competitive equilibrium therefore describes a set of prices and agent actions such that given the prices no agent has an interest in altering its actions, and such that the supply of products is equal to the demand for products. The qualifier *competitive* refers to the fact that agents accept product prices as given, meaning that these prices cannot be influenced by the actions of the agents.

One way to model a competitive market equilibrium is by writing out the KKT conditions of the agent surplus maximization problems and adding the market equilibrium conditions. It is then possible to solve the resulting KKT system in order to obtain prices $\lambda$, agent decisions $x_i$, and product allocations $q_i$. KKT systems are generally difficult to solve, and typically amount to solving nonlinear problems (the nonlinearity arising from the fact that the complementarity condition $a \perp b$ can be expressed equivalently as $a \cdot b = 0$). Instead, a competitive market equilibrium can be described equivalently as a *single* optimization problem, as shown in the following proposition.

**Proposition 4.11** *Suppose that KKT conditions are necessary and sufficient for the optimality of (Sep) and (Surplus − i). A competitive market equilibrium satisfies the KKT conditions of problem (Sep), equations (4.18)–(4.20). Therefore, a competitive market equilibrium results in an optimal solution of (Sep), i.e. a globally optimal allocation of resources. The converse also holds, namely a primal-dual solution to the KKT conditions of (Sep) is a competitive equilibrium.*

*Proof* The optimal solution of the agent surplus maximization problem (Surplus-i) satisfies the following KKT conditions:

$$-\nabla_{x_i} f_i(x_i) + (\nabla_{x_i} g_i(x_i))^T \rho_i - (\nabla_{x_i} h_i(x_i))^T \lambda = 0,$$
$$\lambda^\star - \lambda = 0,$$
$$0 \le \rho_i \perp -g_i(x_i) \ge 0,$$
$$h_i(x_i) = q_i.$$

Collecting the KKT conditions of all agent surplus maximization problems and the market clearing condition, it is straightforward to verify that the market equilibrium satisfies equations (4.18)–(4.20). Since, by assumption, the KKT conditions of (*Sep*) are necessary and sufficient for optimality, it follows that a competitive market equilibrium results in a globally optimal allocation of resources. The converse result is also obtained by comparing KKT conditions. □

In pool-based markets, the system operator solves an optimization problem that determines the allocation of transmission, the production and consumption of power, and the provision of ancillary services, and uses the dual multipliers of the optimal solution to set market prices. This process determines thousands of prices every 5–15 minutes in dozens of markets that trade billions of dollars in products per year. The pricing of transmission, energy, and ancillary services through simultaneous auctions in pools is justified theoretically by proposition 4.11.

Another important consequence of this result relates to modeling and policy analysis. A one-shot optimization problem, like (*Sep*), can be handled computationally by more efficient methods than the large system of KKT conditions that would be required in order to represent an agent-based market equilibrium model explicitly. The insight of proposition 4.11 was used in some of the earliest sector-wide models of the US and world economy by the United States Department of Energy, including the National Energy Modeling System (NEMS) and the Project Independence Evaluation System (PIES).

Proposition 4.11 justifies the use of optimization models for analyzing competitive electricity markets; however, there are circumstances where the assumptions of competitive markets are violated and for which coordinated optimization is not the appropriate modeling choice. Consider, for example, the case of **Cournot oligopoly**,[2] where a producer realizes that its choice of quantity can influence market price.

---

**Example 4.1** *Cournot duopoly.* Consider a market with a linear marginal benefit function, $MB(Q) = a - b \cdot Q$, and two agents with identical variable cost functions $TC_1$ and $TC_2$, respectively. Welfare maximization corresponds to solving the following optimization problem:

$$\max_{p_1, p_2, d} a \cdot d - 0.5 \cdot b \cdot d^2 - TC_1(p_1) - TC_2(p_2)$$

$$p_1 + p_2 = d$$

$$p_1 \in \text{dom } TC_1, p_2 \in \text{dom } TC_2$$

$$p_1, p_2, d \geq 0,$$

where the technical limits of the generators are implicitly represented in the domain of the total cost functions. Assuming an interior solution (i.e. $p_1$ and $p_2$ are positive), then by using KKT conditions it can be shown that the solution is optimal when marginal costs are equal to marginal benefit:

$$MC_1(p_1) = MC_2(p_2) = a - b \cdot (p_1 + p_2) \Rightarrow p_i = \frac{1}{2b}(a - MC_i(p_i)).$$

The implication follows from the additional fact that $p_i = p_{-i}$ due to symmetry.

Consider, instead, the case where each agent realizes that its choice of quantity influences price according to the marginal benefit function. Then, agent $i$ solves the following profit maximization problem:

$$\max_{p_i}(a - b \cdot (p_i + p_{-i})) \cdot p_i - TC_i(p_i).$$

$$p_i \geq 0.$$

Assuming an interior solution $p_i > 0$, results in a first-order condition

$$p_i = \frac{1}{2b}(a - MC_i(p_i)) - \frac{1}{2}p_{-i}.$$

Here, $p_{-i}$ denotes the decision of the competing agent. Note that each player decreases its output in order to lift the price, up to the point where marginal revenue equals marginal cost. This solution is different from the solution of the centralized optimization problem and therefore suboptimal from the point of view of welfare maximization. The solution can be expressed in closed form by observing that, due to symmetry, $p_i = p_{-i}$, which implies that

$$p_i = \frac{1}{3b}(a - MC_i(p_i)).$$

---

[2]  The situation where there are only few sellers in the market that can influence prices through their actions is referred to as an **oligopoly**. The mirror case where a small number of consumers represent a substantial portion of the market and are likely capable of decreasing prices is referred to as an **oligopsony**.

**Table 4.1 The demand bids of problem 4.2.**

| Utility 1 | Utility 2 | Utility 3 |
| --- | --- | --- |
| 200 MW at 800 $/MWh | 300 MW at 600 $/MWh | 300 MW at 500 $/MWh |
| 100 MW at 300 $/MWh | 100 MW at 300 $/MWh | 300 MW at 250 $/MWh |
| 50 MW at 80 $/MWh | | 10 MW at 40 $/MWh |

Example 4.1 demonstrates that deviations from efficiency can be expected in markets with few producers who are aware that their decisions can have a direct impact on price. The strategic withholding of production from electricity markets by producers with the intention of *profitably* increasing prices is defined as **market power**. Market power is a real, as opposed to merely theoretical, concern in electricity markets. For this reason, regulators vigilantly monitor market participant behavior and mitigate market power through various regulatory interventions such as bid mitigation or price caps. Unfortunately, such interventions may interfere with the long-run efficiency of the market because the price signals generated by the market are influenced by regulatory intervention, rather than being determined by competitive forces.

Market power is an important phenomenon in electricity markets, and is best analyzed through game theory. A vast literature exists on the subject of game-theoretical models for analyzing electricity markets. Optimization models are generally not adequate for analyzing strategic interactions between agents. Since the text focuses on optimization models, the assumption throughout most of the text is that agents behave competitively (i.e. as price takers), and game-theoretic models are not touched upon except in few occasions, and often for the purpose of comparison with competitive market models.

## Problems

**4.1** *Technical minimum.* Generalize proposition 4.1 for the case where a minimum run constraint is added to the model: $p_g \geq P_g^-$.

**4.2** *Economic dispatch as a linear program.* Consider a system with three generators. One is a base-load nuclear unit with marginal cost of 10 $/MWh and a maximum run capacity of 800 MW. The other is a coal unit with marginal cost of 25 $/MWh and a maximum capacity of 400 MW. The third is a peaking gas unit with a marginal cost of 35 $/MWh and a capacity of 200 MW. The demand consists of three utilities.

1. Suppose the total system demand is 1360 MW. Solve the economic dispatch model in a mathematical programming language. What is the cost? What is the market clearing price assuming demand is inelastic?

2. Suppose that the system demand in the previous problem is in fact elastic, and described by the demand bids in table 4.1.

   1. What is the system cost? What is the consumer benefit? What is the generator profit? What is the consumer surplus? What is the welfare?

2. Suppose that the gas unit fails. What is the new market clearing price? How many MWs of demand are "in the money"? How many are "out of the money"? Could the market clearing price be 249 $/MWh? Could it be 251 $/MWh?

**4.3** *Economic dispatch as a complementarity problem.* Write down the KKT conditions of the economic dispatch problem for the system given in the previous problem (with all units operational). Solve the complementarity equations using a nonlinear programming solver. Split the set of total KKT conditions into seven groups, depending on whether they correspond to the profit maximization problem of one of the three generators, the surplus maximization problem of one of the three utilities, or the market clearing constraint.

**4.4** *Asymptotic behavior of Cournot equilibrium.* Consider an extension of example 4.1 with $n$ identical suppliers. Derive the Cournot equilibrium conditions and show that as $n \rightarrow +\infty$ the equilibrium converges to the competitive market equilibrium.

**4.5** *The law of one price.* In response to the 2022 natural gas crisis in Europe, various Member States of the European Union proposed splitting the European electricity market into two parts in order to "contain" the high prices to a certain segment of the market. According to such proposals (one of which is explored in problem 11.4), one part of the market would include renewable resources, nuclear plants, and certain hydro resources, whereas the other part would include thermal resources such as coal and natural gas. We represent a simple market splitting model as follows:

$$\max_{p, d_1, d_2} V \cdot (d_1 + d_2) - \sum_{g \in G_2} MC_g \cdot p_g$$

$$(\lambda_1): \quad d_1 - \sum_{g \in G_1} p_g = 0$$

$$(\lambda_2): \quad d_2 - \sum_{g \in G_2} p_g = 0$$

$$(\mu_g): \quad p_g \leq P_g, g \in G = G_1 \cup G_2$$

$$(\nu): \quad d_1 + d_2 \leq D$$

$$p \geq 0, d_1 \geq 0, d_2 \geq 0.$$

Consumers have a total demand of $D$ MWh valued at $V$ $/MWh, and decide the amount of energy that they procure from market 1 ($d_1$) and market 2 ($d_2$), respectively. The set of units that sell in market 1 is indicated by $G_1$, these units are assumed to have a zero marginal cost. The set of units in market 2 is indicated by $G_2$, and these units have a marginal cost of $MC_g$. The production decision of all units $g \in G$ is denoted as $p_g$.

1. Write the KKT conditions of the problem.
2. If demand is sufficiently high so that consumers need to buy electricity in both markets ($d_1 > 0$ and $d_2 > 0$), then prove that the price is equal in both

markets ($\lambda_1 = \lambda_2$). Can you provide an interpretation of your proof in terms of economic intuition?

## Bibliography

**Section 4.1**    The maximization of welfare may appear to be an obvious goal from the perspective of using resources as efficiently as possible. Nevertheless, it is not the only objective that has been considered for electricity auctions. This debate relates to broader welfare considerations, and alternative goals have been analyzed in the literature, such as payment cost minimization (Hao et al. 1998, Zhao et al. 2008, Litvinov, Zhao, and Zheng 2009, Zhao et al. 2010). For instance, the minimization of procurement cost is sometimes supported by system operators in the context of the procurement of reserves, since system operators are the ones procuring the reserves. As pointed out in the literature, such a design can distort truthful bidding incentives (Litvinov et al. 2009, Zhao et al. 2010), can exacerbate efficiency losses in the case of strategic bidding behavior (Zhao et al. 2010), and typically leads to computationally hard market clearing models (Luh et al. 2006, Fernandez-Blanco, Arroyo, and Alguacil 2011).

The term *system lambda* occurs in Schweppe et al. (1988) and Ring (1995).

**Section 4.2**    The discussion about the motivation for the existence of markets for electricity is based on Green (2000), Stoft (2002), and Oren (2004).

Problem 4.5 highlights one of the challenges that emerge in the context of market splitting proposals (Keay and Robinson 2017a, Keay and Robinson 2017b) that have recently re-emerged in response to the 2022 European natural gas crisis (Greek delegation, EU Council 2022, Hellenic Republic of Environment and Energy 2022). Problem 4.5 is a demonstration of the **law of one price** (Jevons 1879, Cramton & Stoft 2006): "*In the same open market, at any moment, there cannot be two prices for the same kind of article.*" The law of one price is pertinent in the case of **commodities**: scarce resources that are traded at a single market price without differentiation.

**Section 4.3**    The connection between optimization and market equilibrium is discussed by Harker (1993). The lecture series includes a discussion of the cases where market models cannot be represented as equivalent optimization problems. The effect of regulatory intervention in electricity markets, which implies that the resulting market equilibrium cannot be represented by an equivalent optimization model, is discussed by Ehrenmann and Smeers (2011b). The application of proposition 4.11 in the context of long-term investment and transmission is discussed in Özdemir (2013). The reader is referred to Gabriel et al. (2012) for game-theoretical models for analyzing electricity markets.

# 5     The transmission network

The economic dispatch model presented in chapter 4 is the simplest model of operations that may be encountered in electricity markets. This model forms the basis for a family of increasingly complex variants. Two key features that are ignored in the economic dispatch model are transmission constraints, which are discussed in the present chapter, and ancillary services, which are discussed in chapter 6.

The allocation of transmission capacity to market agents is complicated by the fact that the transmission of power is governed by physical laws that complicate the trading of power. These laws are described by the power flow equations, and the task of operating the system optimally while accounting for these physical laws is referred to as the optimal power flow problem. The optimal power flow problem is formulated in section 5.1.

Energy, transmission, and ancillary services represent the most important products traded in electricity markets. Transmission and ancillary services are traded in day-ahead power pools and in some real-time electricity markets. The detailed rules for trading these products differ among different markets and correspond to different electricity market design philosophies. The two most widespread approaches towards pricing transmission are discussed in this chapter. Concretely, section 5.2 discusses nodal pricing, while zonal pricing is presented in section 5.3.

## 5.1    Direct current optimal power flow

The **optimal power flow (OPF)** problem is the problem of dispatching generators and loads in an electric power system optimally, while respecting the physical constraints that are imposed by Kirchhoff's voltage and current laws, as well as the physical limits of the transmission network. Kirchhoff's laws create a nonlinear mapping between injections at each node of the network and the flow of power over the lines of the network. This nonlinear mapping is described by the **power flow equations** and is developed in detail in appendix B.

The **alternating current optimal power flow problem (ACOPF)** is the problem of optimizing dispatch while accounting for the power flow equations. It is often the case in market operations that the power flow equations are approximated by a linear model in order to simplify computation and pricing. The linearized ACOPF is referred to as the **direct current optimal power flow problem (DCOPF)**.

Transmission lines impose constraints that limit the amount of power that can be carried from node to node due to (i) thermal limits (limits that, if exceeded, result in the physical damage of transmission lines), (ii) stability limits (limits that, if exceeded, render the system vulnerable to unexpected disturbances), and (iii) voltage limits (limits that, if exceeded, result in voltage levels that can affect the proper functioning of electrical equipment and voltage collapse).

DCOPF can be described with various equivalent models, depending on how the power flow equations are approximated. Two different linear approximations are presented in appendix B. The resulting optimal power flow models come with one of two types of information about the physical characteristics of transmission lines: (i) susceptance, or (ii) power transfer distribution factors. It is possible to obtain one set of parameters from the other, as described in appendix B. The two equivalent models of optimal power flow that can be formulated based on this input data are analyzed in this section.

In this section, the power transmission network is described as a *directed* graph. The set of nodes is indicated by $N$, the set of lines is indicated by $K$, and lines are denoted either by an index $k$ or by the adjacent nodes of an arc, $k = (m,n)$, where $m$ is the origin of $k$ and $n$ is the destination of $k$. The set of generators that are located in node $n$ is denoted as $G_n$, the set of loads located at node $n$ as $L_n$, and the set of generators and loads are denoted, respectively, by $G$ and $L$, with $G = \cup_{n \in N} G_n$ and $L = \cup_{n \in N} L_n$.

In the optimal power flow problem, the decision variables to be optimized are the amount of power produced by each generator, $p_g$, and the amount of power consumed by each load, $d_l$. The state of the system is fully determined by these decisions. This dependency is depicted in figures 5.1 and 5.5, by using black font to depict the essential decision variables, and gray font to depict the variables of the system that result from fixing the decisions of production and consumption.

## 5.1.1  DCOPF based on power transfer distribution factors

Figure 5.1 presents the model of power flow based on *power transfer distribution factors*, which are defined later in this section. The definition of this model requires the determination of a reference node, called the **hub node**. Once the hub node is determined, it is possible to define an injection decision $r_n$ for each node of the network (including the hub node), which indicates the amount of power shipped from node $n$ to the hub node. This is not to be confused with the amount of power that flows over the line connecting a node to the hub node, but instead as the *net* amount of power that a node injects into the system, and which is absorbed by the hub node. As shown in figure 5.1, the hub node may or may not be connected to a node via a physical transmission line (for example, in the figure the hub node is physically connected to node $n$, but not node $m$), but the injection decision $r$ is defined

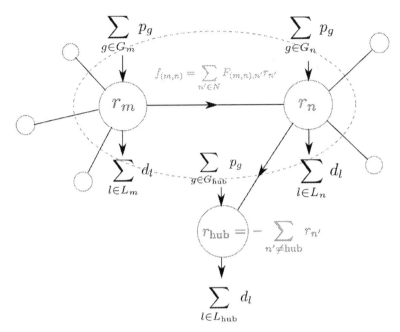

**Figure 5.1** A model of power flow based on power transfer distribution factors. The decision variables (black) are $p_g$ and $d_l$. The variables $r_n$ and $f_{(m,n)}$ (gray) are fully determined once $p_g$ and $d_l$ are fixed.

for every node. By definition, the injection is described as

$$r_n = \sum_{g \in G_n} p_g - \sum_{l \in L_n} d_l, n \in N.$$

Therefore, once production and consumption have been decided, the amount of power shipped to the hub node is also determined. Conservation of energy implies that the total amount of power injected from all nodes to the hub node sums to zero:

$$\sum_{n \in N} r_n = 0.$$

Once injection variables have been determined, it is possible to determine the flow of power on each transmission line from the *power transfer distribution factors*.

**Definition 5.1  Power transfer distribution factors (PTDFs)**, defined for each pair of lines and nodes and denoted as $F_{kn}$, represent the amount of power that flows over line $k$ as a result of shipping 1 MW from node $n$ to the hub node. By definition, $F_{k,\text{hub}} = 0$.

PTDFs are provided as input data to the DCOPF model and depend on the physical characteristics of the transmission lines. Moreover, PTDFs depend on the choice of the hub node; therefore, a hub node needs to be specified in order to have a well-defined DCOPF model based on PTDFs. By the definition of PTDFs, the flow over line $k$ is provided by the following equality:

$$f_k = \sum_{n \in N} F_{kn} r_n, k \in K.$$

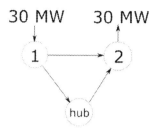

**Figure 5.2** The network of example 5.1.

Note that $f_k$ may be either positive or negative. Positive flow corresponds to power that is flowing in the reference direction of the line, whereas negative flow corresponds to power that is flowing opposite to the reference direction. Once flows over lines have been determined, it is possible to impose constraints on the amount of power flowing over each line. Denoting $T_k$ as the limit on the amount of power that each line can carry, the following constraint is obtained:

$$-T_k \leq f_k \leq T_k, k \in K.$$

---

**Example 5.1** *Computing flows by decomposing to bilateral transactions.* Consider the network of figure 5.2. Suppose that all lines have identical electrical characteristics. From Kirchhoff's laws (see appendix B) it follows that whenever 1 MW is shipped from node 1 to node 2, two thirds flow through the path $1 - 2$, and the remaining one third flows through the path $1 - \text{hub} - 2$. Stated mathematically, $F_{1-2,1} = 1/3$ and $F_{1-2,2} = -1/3$. The shipment of 30 MW from node 1 to node 2 can then be decomposed to a shipment of 30 MW from node 1 to the hub, and the shipment of $-30$ MW from node 2 to the hub. This translates to a flow of

$$f_{1-2} = \frac{1}{3} \cdot 30 - \frac{1}{3} \cdot (-30) = 20 \text{ MW}.$$

Note that the injection of node 1 ($r_1 = 30$ MW) is different from the amount of power over the line connecting node 1 and the hub ($f_{1-\text{hub}} = 10$ MW).

---

The full DCOPF model can be stated as follows, along with its associated dual multipliers:

$$(DCOPF): \quad \max_{p,g,r,f} \sum_{l \in L} \int_0^{d_l} MB_l(x)dx - \sum_{g \in G} \int_0^{p_g} MC_g(x)dx$$

$$(\lambda_k^+): \quad f_k \leq T_k, k \in K$$

$$(\lambda_k^-): \quad -f_k \leq T_k, k \in K$$

$$(\psi_k): \quad f_k - \sum_{n \in N} F_{kn}r_n = 0, k \in K$$

$$(\rho_n): \quad r_n - \sum_{g \in G_n} p_g + \sum_{l \in L_n} d_l = 0, n \in N$$

$$(-\phi): \quad \sum_{n \in N} r_n = 0$$

$$p_g \geq 0, g \in G$$

$$d_l \geq 0, l \in L.$$

Note the minus sign in the multiplier of the power balance constraint, $-\phi$. This pre-scaling means that, whatever the dual multiplier of this constraint is, $\phi$ is equal to its negative. This allows us to interpret $\phi$ as the price of energy in the hub node, as we explain later.

Recalling the notation of the previous section, $MB_l \colon \mathbb{R} \to \mathbb{R}$ and $MC_g \colon \mathbb{R} \to \mathbb{R}$ correspond to the marginal benefit of consumers and the marginal cost of generators, respectively. The objective function of the DCOPF, as in the case of the economic dispatch problem, is welfare. Technical limits on production and consumption can be imposed through the domain of the marginal cost and marginal benefit functions. For example, a technical maximum $p_g \leq P_g$ can be imposed equivalently by $MC_g(x) = +\infty$ for $x > P_g$. Alternatively, technical limits can be added explicitly to the DCOPF:

$$(\mu_g): \quad p_g \leq P_g, g \in G,$$

$$(\nu_l): \quad d_l \leq D_l, l \in L.$$

**Proposition 5.2** *Suppose that $MC_g(\cdot)$ and $MB_l(\cdot)$ are differentiable. For the optimal solution of the DCOPF, there exists a threshold $\rho_n$ for each node $n$ such that:*

- *If a generator is producing strictly within its operating region ($0 < p_g < P_g$), then $\rho_n = MC_g(p_g)$. Similarly for loads, if $0 < d_l < D_l$, then $\rho_n = MB_l(d_l)$.*
- *If a generator is producing at its technical maximum ($p_g = P_g$), then $\rho_n \geq MC_g(P_g)$. Similarly for loads, if $d_l = D_l$, then $\rho_n \leq MB_l(D_l)$.*
- *If a generator is producing zero ($p_g = 0$), then $\rho_n \leq MC_g(0)$. Similarly for loads, if $d_l = 0$, then $\rho_n \geq MB_l(0)$.*

*Proof*    Denote $n(g)$ and $n(l)$ as the nodes where generator $g$ and load $l$ are located, respectively. Applying proposition 2.5 to the DCOPF, the KKT conditions of the problem can be partitioned according to conditions that relate to generators, loads, and the transmission network, as shown in figure 5.3. The proof of the proposition relies on the following subset of KKT conditions that relate to a given generator $g$:

$$0 \leq p_g \perp MC_g(p_g) - \rho_{n(g)} + \mu_g \geq 0,$$

$$0 \leq \mu_g \perp P_g - p_g \geq 0,$$

and the following subset of KKT conditions that relate to a given load $l$:

$$0 \leq d_l \perp -MB_l(d_l) + \rho_{n(l)} + \nu_l \geq 0,$$

$$0 \leq \nu_l \perp D_l - d_l \geq 0.$$

**Producers**

$$0 \leq p_g \perp MC_{(p_g)} - \rho_{n(g)} + \mu_g \geq 0$$
$$0 \leq \mu_g \perp P_g - p_g \geq 0$$
$$\Leftrightarrow$$
$$\max \rho_{n(g)} p_g - \int_0^{p_g} MC_g(x)dx$$
$$(\mu_g): \quad p_g \leq P_g$$
$$p_g \geq 0$$

**Consumers**

$$0 \leq d_l \perp -MB_l(d_l) + \rho_{n(l)} + \nu_l \geq 0$$
$$0 \leq \nu_l \perp D_l - d_l \geq 0$$
$$\Leftrightarrow$$
$$\max \int_0^{d_l} MB_l(x)dx - \rho_{n(l)} d_l$$
$$(\nu_l): \quad d_l \leq D_l$$
$$d_l \geq 0$$

**Transmission**

$$\max_{f,r} \sum_{n \in N} \rho_n \cdot (-r_n)$$
$$(\psi_k): \quad f_k - \sum_{n \in N} F_{kn} \cdot r_n = 0, k \in K$$
$$(-\phi): \quad \sum_{n \in N} r_n = 0$$
$$(\lambda_k^+): \quad f_k \leq T_k, k \in K$$
$$(\lambda_k^-): \quad -f_k \leq T_k, k \in K$$
$$r_n - \sum_{g \in G_n} p_g + \sum_{l \in L_n} d_l = 0, n \in N$$

$$\Leftrightarrow$$

$$f_k - \sum_{n \in N} F_{kn} \cdot r_n = 0, k \in K$$
$$\sum_{n \in N} r_n = 0$$
$$\lambda_k^+ - \lambda_k^- + \psi_k = 0, k \in K$$
$$-\sum_{k \in K} F_{kn} \cdot \psi_k + \rho_n - \phi = 0, n \in N$$
$$0 \leq \lambda_k^+ \perp T_k - f_k \geq 0, k \in K$$
$$0 \leq \lambda_k^- \perp T_k + f_k \geq 0, k \in K$$

**Market clearing**

$$r_n - \sum_{g \in G_n} p_g + \sum_{l \in L_n} d_l = 0, n \in N$$

**Figure 5.3** The decomposition of the KKT conditions of DCOPF.

The proof of the proposition follows from these KKT conditions, as in the case of proposition 4.1. For example, consider a generator for which $0 < p_g < P_g$. Since $p_g < P_g$, it follows from $\mu_g \perp P_g - p_g$ that $\mu_g = 0$. From the complementarity condition $p_g \perp MC_g(p_g) - \rho_{n(g)}$ and the fact that $p_g > 0$, it follows that $\rho_{n(g)} = MC_g$. The other cases are proven similarly. □

Proposition 5.2 generalizes the result that is derived for the economic dispatch problem, with the role of system lambda replaced by a marginal cost threshold $\rho_n$, which is different for each location. Generators with marginal costs below this threshold should produce, and loads with marginal benefit above this threshold should consume. Similarly, generators with marginal cost above this threshold should not produce, and loads with marginal benefit below this threshold should not consume.

The sensitivity interpretation of the dual variables of the DCOPF is useful in understanding the pricing of energy in networks with transmission constraints. The dual variable $\phi$ is equal to the marginal change in welfare from a marginal requirement to produce, or a marginal requirement to decrease consumption, in the hub node. The dual variables $\lambda_k^+$ and $\lambda_k^-$ are equal to the marginal benefit of increasing the capacity of lines. The dual variable $\rho_n$ can be interpreted as the marginal increase in welfare from a requirement to marginally increase consumption, or a requirement to marginally decrease generation, in node $n$. In the special case of a system with inelastic demand the consumer benefit term is removed from the objective function. In this case, $\rho_n$ is equal to the marginal cost increase that is incurred by marginally increasing the amount of power required in a certain node, or the marginal cost savings achieved by marginally decreasing the amount of power consumed in that location.

---

**Example 5.2** *Computing $\rho$ in a network without congestion.* Consider the network of figure 5.4. Suppose that $D_2 = 50$ MW and that there is no capacity limit on the line. The optimal solution is to utilize unit 1 to the greatest extent possible. From

**Figure 5.4** The network of examples 5.2–5.4.

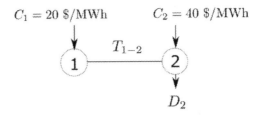

$C_1 = 20$ \$/MWh    $C_2 = 40$ \$/MWh

proposition 5.2 it follows that $\rho_1 = 20$ \$/MWh since unit 1 is operating strictly within its operating region. Since there is no line congestion, nodes 1 and 2 are indistinguishable, and therefore $\rho_2 = 20$ \$/MWh. Notice the sensitivity interpretation of $\rho_1$: a marginal increase of demand in location 1 by 1 MW results in an increase of cost by 20 \$/h.

---

**Proposition 5.3**   *The following identity holds for the dual optimal multipliers of the DCOPF:*

$$\rho_n = \phi + \sum_{k \in K} F_{kn} \lambda_k^- - \sum_{k \in K} F_{kn} \lambda_k^+, n \in N.$$

*Proof*   The proof follows from the KKT conditions of the DCOPF that relate to the transmission network, and are presented in figure 5.3. For a given node $n$, use the following KKT conditions:

$$\lambda_k^+ - \lambda_k^- + \psi_k = 0, k \in K,$$
$$-\sum_{k \in K} F_{kn} \psi_k + \rho_n - \phi = 0.$$

Then the conclusion of the proposition follows.    □

---

**Example 5.3** *Computing $\rho$ in a network on the verge of congestion.* Consider the network of figure 5.4. Suppose that $D_2 = 50$ MW and $T_{1-2} = 50$ MW. The optimal dispatch of generation does not change. Since unit 1 is strictly within its operating range, $\rho_1 = 20$ \$/MWh. If node 1 is chosen as the hub node and the reference direction of the line is from node 1 to node 2, then $F_{1-2,2} = -1$. From sensitivity, it follows that $\lambda_{1-2}^+$ is the marginal benefit of a marginal change in the capacity of line 1–2. By marginally increasing the capacity of the line, there is no additional benefit. By marginally decreasing its capacity, unit 2 needs to produce and unit 1 needs to decrease output, resulting in a marginal cost increase of 20 \$/h. Therefore, 0 \$/MWh $\leq \lambda_{1-2}^+ \leq 20$ \$/MWh. Using proposition 5.3, it follows that $\rho_2 = \rho_1 + \lambda_{1-2}^+$; therefore 20 \$/MWh $\leq \rho_2 \leq 40$ \$/MWh. The same conclusion could have been reached by analyzing the sensitivity of cost to a marginal change in the demand of node 2.

**Example 5.4** *Computing $\rho$ in a network with congestion.* Consider the network of figure 5.4. Suppose that the line has a capacity limit of $T_{1-2} = 50$ MW and $D_2 = 60$

MW. The optimal dispatch corresponds to $p_1 = 50$ MW and $p_2 = 10$ MW. It follows that $\rho_1 = 20$ \$/MWh and $\rho_2 = 40$ \$/MWh.

## 5.1.2 DCOPF based on susceptance

An equivalent formulation of the DCOPF problem that is often used relies on susceptance. As in the case of the DCOPF problem based on PTDFs, the decision variables are the production, $p_g$, and consumption, $d_l$, of each agent. The difference between the two models is how flow on transmission lines is computed as a function of implied variables. In the DCOPF model based on PTDFs the implied variables are the net injections, $r_n$. As indicated in figure 5.5, the implied variables in the susceptance model are the phase angles[1] at each node, $\theta_n$. In the DCOPF model based on PTDFs, the input data that depend on the physical characteristics of lines are the PTDFs. The input data that depend on the physical characteristics of the lines in the DCOPF model based on susceptance are the susceptance values $B_{(m,n)}$ of each line $(m,n)$. The flow of power over a line is proportional to the nodal phase angle difference, with susceptance being the proportionality factor:

$$f_{(m,n)} = B_{(m,n)}(\theta_m - \theta_n), (m,n) \in K.$$

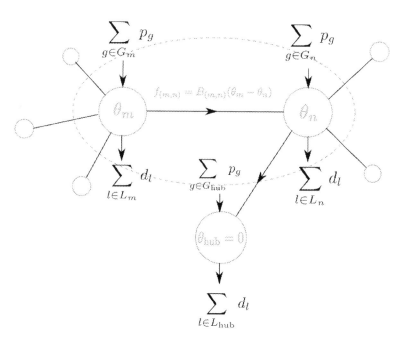

**Figure 5.5** A model of power flow based on susceptance. The decision variables (black) are $p_g$ and $d_l$. The variables $\theta_n$ and $f_{(m,n)}$ (gray) are fully determined once $p_g$ and $d_l$ are fixed.

[1] The physical meaning of phase angles is explained in appendix B.

The right-hand side of this equation also grants this optimal power flow formulation the name **B-theta formulation**.

Translation of phase angles by a constant factor results in identical flows; therefore the phase angle of the hub node can be fixed (e.g. to $\theta_{\text{hub}} = 0$). An energy conservation constraint is imposed at each node of the network:

$$\sum_{g \in G_n} p_g + \sum_{k=(\cdot,n)} f_k = \sum_{k=(n,\cdot)} f_k + \sum_{l \in L_n} d_l, n \in N.$$

Note that, in contrast to the DCOPF model based on PTDFs, the input data for this model is independent of network topology. Consequently, this representation of DCOPF is useful for models in which the topology of the network changes (e.g. transmission line investment, transmission line outages, or transmission line switching). The full problem can be described as follows:

$$(DCOPF2): \quad \max_{p,d,f,\theta} \sum_{l \in L} \int_0^{d_l} MB_l(x)dx - \sum_{g \in G} \int_0^{p_g} MC_g(x)dx$$

$$(\rho_n): \quad -\sum_{g \in G_n} p_g - \sum_{k=(\cdot,n)} f_k + \sum_{l \in L} d_l + \sum_{k=(n,\cdot)} f_k = 0, n \in N$$

$$(\gamma_k): \quad f_k - B_k(\theta_m - \theta_n) = 0, k = (m,n) \in K$$

$$(\lambda_k^+): \quad f_k \leq T_k, k \in K$$

$$(\lambda_k^-): \quad -f_k \leq T_k, k \in K$$

$$p_g \geq 0, g \in G$$

$$d_l \geq 0, l \in L.$$

The dual multiplier $\rho_n$ of the power balance constraint retains the same sensitivity interpretation as in the previous DCOPF model. Proposition 5.2, which determines how the dispatch of units and loads at each bus depends on $\rho_n$, remains valid for the DCOPF based on susceptance.

In order to see how $(DCOPF2)$ can be used for adapting the topology of the network, note that we can introduce a binary variable $z_k$ that indicates whether or not a line is kept in service. If the line is operational ($z_k = 1$), then the constraint $f_k = B_k(\theta_m - \theta_n)$ is active. In the opposite case ($z_k = 0$), the constraint is deactivated by using a so-called big-M parameter. Concretely, the topology reconfiguration model can be formulated as follows:

$$(TR): \quad \max_{p,d,f,z} \sum_{l \in L} \int_0^{d_l} MB_l(x)dx - \sum_{g \in G} \int_0^{p_g} MC_g(x)dx$$

$$-\sum_{g \in G_n} p_g - \sum_{k=(\cdot,n)} f_k + \sum_{l \in L} d_l + \sum_{k=(n,\cdot)} f_k = 0, n \in N$$

$$f_k - B_k(\theta_m - \theta_n) \leq M \cdot (1 - z_k), k = (m,n) \in K \tag{5.1}$$

$$f_k - B_k(\theta_m - \theta_n) \geq -M \cdot (1 - z_k), k = (m,n) \in K \tag{5.2}$$

$$f_k \leq T_k \cdot z_k, k \in K \tag{5.3}$$

$$-f_k \leq T_k \cdot z_k, k \in K \tag{5.4}$$

$$p_g \geq 0, g \in G$$

$$d_l \geq 0, l \in L$$

$$z_k \in \{0, 1\}, k \in K.$$

The model $(TR)$ assumes that *all* lines can be kept in or out of service, although this can be easily adapted to a subset of candidate lines. Note the effect of the big-M formulation:

- If $z_k = 0$, then constraints (5.1) and (5.2) are nonbinding for sufficiently large values of $M$. This is due to the fact that the right-hand side of constraint (5.1) is so large that the inequality becomes irrelevant. Similarly, for sufficiently large values of $M$, the right-hand side of constraint (5.2) becomes so small that the constraint becomes irrelevant. This means that the linear approximation of the power flow equations is no longer relevant for this line. Constraints (5.3) and (5.4) force the flow on the line to be zero; thus effectively, the line does not exist on the network.
- If $z_k = 1$, then constraints (5.1) and (5.2) jointly imply that $f_k = B_k(\theta_m - \theta_n)$ for that line. Moreover, constraints (5.3) and (5.4) imply that the flows must remain within the physical limits of the line. Thus, the line is effectively introduced in the network.

Problem $(TR)$ can be used both for line switching in short-term system operations, as well as for transmission expansion planning in long-term models. Although the model is correct, it is also very challenging to solve. This is due to the use of the big M in constraints (5.1) and (5.2), which poses challenges for branch-and-bound solvers. Heuristics (such as considering one or a limited number of line switches at a time, or using information from dual multipliers of $(DCOPF2)$ for a given network topology) are therefore often employed in practice for recovering feasible solutions with acceptable performance. An interesting empirical observation is that a few line switches can often yield much of the full benefit that can be accomplished from switching all lines. This has been observed empirically on certain test systems in the literature.

## 5.2 Locational marginal pricing

Transmission constraints complicate market design because the transmission capacity of lines is a scarce resource. According to proposition 4.11, in order for a market to maximize efficiency, scarce resources that are shared between agents should be priced. Therefore, the existence of transmission constraints implies that there should exist a market for transmission capacity. In simple transportation networks the usage of transmission capacity is a variable that can be decided directly by agents, and basic economic arguments demonstrate that the socially optimal allocation of resources can be achieved by introducing a market where agents trade the available transportation capacity. This still holds true in electricity markets; however, the complication of electricity markets is that the use of the transmission capacity is dictated by physical laws, namely Kirchhoff's laws. This physical reality complicates the trading of transmission capacity.

Two prevalent market designs have emerged for the trading of transmission capacity in electricity markets: zonal pricing and nodal pricing. The latter is also referred to as locational marginal pricing. Nodal pricing is presented and analyzed in this section; zonal pricing is developed in section 5.3.

**Definition 5.4    Locational marginal pricing**, also referred to as **nodal pricing**, is a uniform price auction that is conducted as follows:

- Sellers submit increasing bids: price-quantity pairs, representing the price at which they are willing to supply a certain quantity. Buyers submit decreasing bids: price-quantity pairs indicating the price they are willing to pay for the specific quantity.
- The market operator solves $(DCOPF)$ and announces $\rho_n$ as the market clearing price for bus $n$. Supply bids that are "in the money" (i.e. bid at a price below $\rho_n$) have to be offered and receive a payoff of $\rho_n$ per MWh from the market operator, whereas supply bids that are "out of the money" should not produce and receive no payoff. Demand bids that are "in the money" (i.e. bid at a price above $\rho_n$) need to consume and pay $\rho_n$ per MWh to the market operator, whereas demand bids that are "out of the money" cannot consume and do not have to pay anything. Supply and demand bids that are "on the money" (i.e. bid price is equal to $\rho_n$) are instructed by the market operator to produce/consume a quantity between 0 and their bid quantity.

The uniform prices that result from locational marginal pricing are called locational marginal prices (LMPs).

---

**Example 5.5** *LMPs can be different between adjacent nodes, even if the line connecting them is not congested.* Consider the network of figure 5.6. Suppose that all lines have the same transmission limit, $T_{1-2} = T_{2-3} = T_{1-3} = 50$ MW. The optimal dispatch is $p_1 = 50$ MW, $p_2 = 150$ MW, and $p_3 = 100$ MW, and the flows are $f_{1-2} = 0$ MW, $f_{2-3} = f_{1-3} = 50$ MW. To confirm that this is the optimal solution of the DCOPF, note that if line 1–2 is not congested then any dispatch that does not fully use the capacity of lines 1–3 and 2–3 is suboptimal because generators 1 and 2 are both cheaper than generator 3. From proposition 5.2 it follows that $\rho_1 = 40$ $/MWh, $\rho_2 = 80$ $/MWh, and $\rho_3 = 140$ $/MWh, since all units are strictly within their operating range. This example demonstrates the fact that it is possible for two buses that are connected by a line that is not congested to have different LMPs. Because of Kirchhoff's laws, the fact that line 1–2 is not congested does not imply that it is possible to transport power over that line from a cheaper location (bus 1) to a more expensive location (bus 2). It is possible that such a transfer results in

Table 5.1 The settlement of the LMP auction in example 5.5.

|  | Bid | Cleared | Payment ($/h) |
|---|---|---|---|
| G1 | $+\infty$ MW at 40 $/MWh | 50 MW at 40 $/MWh | 2000 |
| G2 | $+\infty$ MW at 80 $/MWh | 150 MW at 80 $/MWh | 12000 |
| G3 | $+\infty$ MW at 140 $/MWh | 100 MW at 140 $/MWh | 14000 |
| L2 | 100 MW at $+\infty$ $/MWh | 100 MW at 80 $/MWh | −8000 |
| L3 | 200 MW at $+\infty$ $/MWh | 200 MW at 140 $/MWh | −28000 |

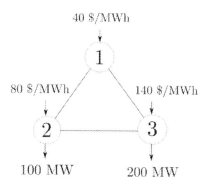

Figure 5.6 The network of examples 5.5–5.7. All lines have identical electrical characteristics (susceptance)

transmission constraint violations in another line of the network. The settlement of the LMP auction is shown in table 5.1. Note that the total payments from buyers to the auctioneer amount to $36000 for a single hour. The payments from the auctioneer to the sellers amount to $28000 for a single hour. The remaining amount, 12000 $/h, is kept by the auctioneer and can be used for other purposes (e.g. paying back the investment costs of lines).

**Example 5.6** *LMPs can be different from the marginal cost of any given generator.* Consider again the network of figure 5.6. Suppose that $T_{1-2} = 50$ MW, $T_{2-3} = 100$ MW, and $T_{1-3} = 120$ MW. The optimal solution now becomes $p_1 = 160$ MW, $p_2 = 140$ MW, $p_3 = 0$ MW, and the flows become $f_{1-2} = 40$ MW, $f_{2-3} = 80$ MW, and $f_{1-3} = 120$ MW. The price in node 3 becomes $\rho_3 = 120$ $/MWh. To understand the price at node 3, consider an increase in demand at node 3 by 1 MW. The new optimal dispatch decreases output from generator 1 by 1 MW (savings of 40 $/MWh), and increases output from generator 2 by 2 MW (cost increase of 160 $/MWh). The net change, 120 $/MWh, is the LMP at node 3. The LMP may therefore be different from the marginal cost of any generator in the market.

**Example 5.7** *Multiplicity of LMPs.* Consider the network of figure 5.6. Suppose that $T_{1-2} = 50$ MW, $T_{2-3} = 100$ MW, and $T_{1-3} = 100$ MW. The optimal dispatch is $p_1 = 100$ MW, $p_2 = 200$ MW, and $p_3 = 0$ MW. The flows are $f_{1-2} = 0$ MW, $f_{2-3} = f_{1-3} = 100$ MW. To compute the price at node 3, note that if the load at node 3 were to increase, the only way to serve it would be by an increase in the output of generator 3, since lines 2–3 and 1–3 are already fully congested. Therefore, $\rho_3 = 140$

$/MWh is a valid LMP. Note that if consumption in node 3 is reduced by 1 MW, then the system can be redispatched such that generator 1 increases its output by 1 MW (cost increase of 40 $/MWh) and generator 2 decreases output by 2 MW (cost savings of 160 $/MWh), resulting in a net saving of 120 $/MWh. Therefore, $\rho_3 = 120$ $/MWh is also an acceptable price.

It is worth posing the question *why is nodal pricing a good idea?* One motivation[2] behind nodal pricing is surprisingly simple:

**Proposition 5.5** *If agents bid truthfully, locational marginal pricing maximizes welfare, and the resulting allocation maximizes the surplus of agents* given *the market clearing price.*

*Proof*    The fact that nodal pricing maximizes welfare is a consequence of the fact that, in the LMP auction, the auctioneer is using truthful bids in order to maximize welfare while accounting for network constraints.

The fact that the dispatch resulting from the auction maximizes agent surplus follows from the fact that the KKT conditions for any producer $g \in G$,

$$0 \leq p_g \perp MC_g(p_g) - \rho_{n(g)} \geq 0,$$
$$0 \leq \mu_g \perp P_g - p_g \geq 0,$$

are equivalent to the following profit maximization of producer $g$:

$$\max_p \ \rho_{n(g)} \cdot p_g - \int_0^{p_g} MC_g(x)dx$$
$$(\mu_g): \ p_g \leq P_g$$
$$p_g \geq 0.$$

Similarly, for any consumer $l \in L$, the KKT conditions

$$0 \leq d_l \perp \rho_{n(l)} - MB_l(d_l) \geq 0,$$
$$0 \leq v_l \perp D_l - d_l \geq 0,$$

are equivalent to

$$\max_d \ \int_0^{d_l} MB_l(x)dx - \rho_{n(l)} \cdot d_l$$
$$(v_l): \ d_l \leq D_l$$
$$d_l \geq 0. \qquad \qquad \square$$

It is worth emphasizing that, under nodal pricing, each bus of the network is trading power at a different price. Locational marginal pricing settles how power should be produced, consumed, and priced in real-time and other short-term markets. However, market participants also need to trade financial instruments that grant

---

[2] The reader is asked to verify that the above proposition also applies for the network operator in problem 5.6. The problem shows that the prices generated by an LMP auction are also compatible with the maximization of the value of using the network.

them the right to use transmission capacity in order to manage their financial risk. The organization of financial markets where transmission capacity can be bought or sold *in advance* of short-term operations is the subject of section 9.2.

### 5.2.1 Congestion rent and congestion cost

The real-time auction without transmission constraints, presented in chapter 4, transfers the payments of buyers to the sellers, with no surplus left to the auctioneer. By contrast, in example 5.5 it is shown that in an LMP auction the amount of funds collected from the buyers can be more than the amount that the auctioneer pays to the sellers. The difference is referred to as congestion rent.

**Definition 5.6**   **Congestion rent** is the surplus that is created in an LMP auction as a result of locational price differences:

$$CR = \sum_{n\in N} \rho_n \cdot \left( \sum_{l\in L_n} d_l - \sum_{g\in G_n} p_g \right).$$

The following proposition shows that congestion rent is nonnegative.

**Proposition 5.7**   *Congestion rent is nonnegative and is given by the following expression:*

$$CR = \sum_{n\in N} \rho_n \cdot \left( \sum_{l\in L_n} d_l - \sum_{g\in G_n} p_g \right) = \sum_{k\in K}(\lambda_k^+ + \lambda_k^-) \cdot T_k.$$

*Proof*   By the definition of $r_n$ for any given location $n \in N$,

$$\sum_{n\in N} \rho_n \cdot \left( \sum_{l\in L_n} d_l - \sum_{g\in G_n} p_g \right) = -\sum_{n\in N} \rho_n \cdot r_n.$$

From the KKT conditions $\rho_n = \sum_{k\in K} F_{kn} \cdot \psi_k + \phi, n \in N$, $\psi_k = \lambda_k^- - \lambda_k^+, k \in K$, and the conservation of energy constraint ($\sum_{n\in N} r_n = 0$), it follows that

$$-\sum_{n\in N} \rho_n \cdot r_n = \sum_{k\in K}(\lambda_k^+ - \lambda_k^-) \cdot \sum_{n\in N} F_{kn} r_n.$$

From the definition of $f_k$,

$$\sum_{k\in K}(\lambda_k^+ - \lambda_k^-) \cdot \sum_{n\in N} F_{kn} \cdot r_n = \sum_{k\in K}(\lambda_k^+ - \lambda_k^-) \cdot f_k.$$

Recalling the KKT conditions of $(DCOPF)$,

$$0 \le \lambda_k^+ \perp T_k - f_k \ge 0, k \in K,$$
$$0 \le \lambda_k^- \perp T_k + f_k \ge 0, k \in K,$$

the following identity is then obtained:

$$\sum_{n\in N} \rho_n \cdot \left(\sum_{l\in L_n} d_l - \sum_{g\in G_n} p_g\right) = \sum_{k\in K}(\lambda_k^+ + \lambda_k^-) \cdot T_k.$$

Congestion rent is equal to the sum over all lines of the capacity of lines multiplied by the dual multipliers of the transmission constraints. Since these are nonnegative, the conclusion that congestion rent is nonnegative follows. □

A related notion to congestion rent is that of congestion cost.

**Definition 5.8    Congestion cost** is the excess cost resulting from the finite capacity of transmission lines.

Intuition suggests that congestion rent and congestion cost are related. Although the two may be equal in certain cases, they should not be confused, since in general they are not equal and no general relationship can be established between them.

---

**Example 5.8** *Congestion cost can be equal to congestion rent.* Consider the network of figure 5.4. Suppose that $D_2 = 50$ MW and $T_{1-2} = 50$ MW. It is shown in example 5.3 that $\rho_1 = 20$ \$/MWh and $20$ \$/MWh $\leq \rho_2 \leq 40$ \$/MWh are nodal prices. Therefore, congestion rent is equal to any value between 0 \$/h and 1000 \$/h. Congestion cost is obviously equal to 0 \$/h.

**Example 5.9** *Congestion cost can be less than congestion rent.* Consider the same network with $D_2 = 60$ MW and $T_{1-2} = 50$ MW. It is shown in example 5.4 that $\rho_1 = 20$ \$/MWh and $\rho_2 = 40$ \$/MWh. Therefore, congestion rent becomes equal to 1000 \$/h. Congestion cost is equal to 200 \$/h.

---

In the case of locational marginal pricing the market operator collects congestion rent by effectively charging a higher price for power than it is buying it for. The existence of congestion rent reflects a need for investment in transmission between locations. The use of congestion rent for settling financial contracts for transmission rights is discussed in chapter 9. The following result is used in section 9.2 in order to prove the revenue adequacy of financial transmission rights.

**Proposition 5.9**    *Consider an injection $(\tilde{r}_n, n \in N)$ that is feasible but not necessarily optimal, and denote $(r_n, n \in N)$ as the optimal injection. Then*

$$-\sum_{n\in N} \rho_n \cdot r_n \geq -\sum_{n\in N} \rho_n \cdot \tilde{r}_n,$$

*where $\rho_n, n \in N$ are the dual multipliers of the optimal injection.*

*Proof*    Following the arguments of the previous proof, it can be seen that

$$-\sum_{n\in N} \rho_n \cdot (r_n - \tilde{r}_n) = \sum_{k\in K}(\lambda_k^+ - \lambda_k^-) \cdot (f_k - \tilde{f}_k),$$

where $\lambda_k^+, \lambda_k^-$ are dual optimal multipliers, $f_k$ are the flows corresponding to $r_n$, and $\tilde{f}_k$ are the flows corresponding to $\tilde{r}_n$. Suppose a line is congested in the positive direction in the optimal dispatch. Then $\lambda_k^+ - \lambda_k^- = \lambda_k^+ > 0$ and $f_k - \tilde{f}_k = T_k - \tilde{f}_k \geq 0$. Similarly, if a line is congested in the negative direction in the optimal dispatch, then $\lambda_k^+ - \lambda_k^- = -\lambda_k^- < 0$ and $f_k - \tilde{f}_k = -T_k - \tilde{f}_k \leq 0$. Hence, $\sum_{k \in K} (\lambda_k^+ - \lambda_k^-) \cdot (f_k - \tilde{f}_k) \geq 0$ and the result follows.                                                                                    □

## 5.2.2 Competitive market model for transmission capacity

A further motivation for nodal pricing is that, assuming truthful bidding, the outcome of the nodal pricing auction coincides with a competitive market equilibrium. Before proceeding with the rigorous proof of this statement, it is necessary to contemplate how one might model competitive markets with transmission constraints.

Recall from section 4.2 that the definition of competitive equilibrium requires (i) defining agents and scarce resources (commodities) that are traded, and (ii) describing the surplus maximization (quantity adjustment) of agents and the market clearing condition (price adjustment) for each commodity.

Any market agent that injects or withdraws power from the network is effectively using the capacity of *all* lines in the network. Therefore, it is apparent that when defining a competitive market for power and transmission there should be at least two types of agents, producers and consumers, and at least two types of products, energy and transmission capacity.

When defining a competitive market equilibrium model, it is necessary that the usage of scarce resources by an agent should depend on the decisions of that agent *alone* (recall the format of problem (*Sep*) in chapter 4). The complication in markets with transmission is that the usage of transmission capacity by an agent does not depend solely on the decision of the seller to produce, but also on where the buyer of the produced power is located. How, then, can a competitive equilibrium be defined?

This dependence can be broken by recalling that a bilateral exchange of power can be decomposed to a shipment of power from the producer location to a reference node, and the shipment of that same amount of power from the reference node to the buyer node. Choosing the hub node as a reference trading node, a convention can be adopted whereby sellers are responsible for buying the transmission capacity that is needed to transport the power they produce to the hub, and buyers are responsible for procuring the transmission capacity that they consume from the hub to their location.

Denote $\phi$ as the price of power, $\lambda_k^+$ as the price of transmission rights in the reference direction of a line, and $\lambda_k^-$ as the price of transmission rights opposite to the reference direction of a line. The following profit maximization model can be defined for each producer $g \in G$:

$$\max_p \phi \cdot p_g - \sum_{k \in K} \lambda_k^+ F_{k,n(g)} p_g + \sum_{k \in K} \lambda_k^- F_{k,n(g)} p_g - \int_0^{P_g} MC_g(x) dx$$

$$p_g \leq P_g$$

$$p_g \geq 0,$$

and for each consumer $l \in L$:

$$\max_d \int_0^{d_l} MB_l(x)dx - \phi \cdot d_l + \sum_{k \in K} \lambda_k^+ F_{k,n(l)}d_l - \sum_{k \in K} \lambda_k^- F_{k,n(l)}d_l$$

$$d_l \leq D_l$$
$$d_l \geq 0.$$

The market clearing condition for power can be expressed as a strict equality:

$$\sum_{g \in G} p_g = \sum_{l \in L} d_l,$$

while the market clearing condition for transmission capacity can be expressed through the following complementarity conditions:

$$0 \leq \lambda_k^+ \perp T_k - f_k \geq 0, k \in K,$$
$$0 \leq \lambda_k^- \perp T_k + f_k \geq 0, k \in K.$$

Having defined a competitive market with transmission capacity, the following result can be established:

**Proposition 5.10**  *Nodal pricing produces an allocation of power and market clearing prices that correspond to a competitive market equilibrium.*

*Proof*   The proposition is a special case of proposition 4.11.     □

### 5.2.3   Losses

Losses are integrated in the optimal power flow model using the approximation that is derived in section B.6. Concretely, we show in section B.6 that losses can be approximated as follows:

$$lo = \sum_{k \in K} \left( L_{k0} + L_{k1} \cdot \sum_{n \in N} PTDF_{kn} \cdot r_n \right),$$

where

$$L_{k0} = -R_k \cdot \bar{P}_k^2, k \in K,$$
$$L_{k1} = 2 \cdot R_k \cdot \bar{P}_k, k \in K,$$

and where $(\bar{P}_k, k \in K)$ is the vector of baseline flows in the system and $R_k$ is the resistance of line $k$.

We now require an assumption about how losses are distributed in different nodes of the network. One possible approach is for contributions to sum up to 1, with each node contributing proportionally to the total losses over all branches that connect to that node. We denote the contribution of node $n$ to losses as $D_n$, with $\sum_{n \in N} D_n = 1$.

The optimal power flow model can then be expressed as follows:

$$(DCOPF - L): \quad \max_{p,d,r,r',f,lo} \sum_{l \in L} \int_0^{d_l} MB_l(x)dx - \sum_{g \in G} \int_0^{p_g} MC_g(x)dx$$

$$lo = \sum_{k \in K}(L_{k0} + L_{k1} \cdot \sum_{n \in N} PTDF_{kn} \cdot r_n)$$

$$-T_k \le f_k \le T_k, k \in K$$

$$f_k - \sum_{n \in N} PTDF_{kn}r'_n = 0, k \in K$$

$$(\rho_n): \quad r_n - \sum_{g \in G_n} p_g + \sum_{l \in L_n} d_l = 0, n \in N$$

$$r'_n = r_n - D_n \cdot lo, n \in N$$

$$\sum_{n \in N} r'_n = 0$$

$$p \ge 0, d \ge 0, lo \ge 0.$$

The interpretation of $\rho_n$ as a market clearing price remains valid. However, the value of the LMP now accounts for the effect of losses. The decision vector $r$ corresponds to net injections before accounting for losses, while $r'$ corresponds to net injections after losses have been accounted for.

---

**Example 5.10** *LMPs with losses on a two-node network.* We return to the two-node network of example B.5, where all quantities are expressed per unit. We assume that there is a generator in node 1 with a marginal cost of 20 $/MWh. The factors $D_n$ are assumed equal, $D_1 = D_2 = 0.5$, since there is only one line adjacent to both nodes; thus the total losses associated with each of the nodes are equal. Let us assume a base dispatch for which $\bar{P}_{12} = 1$. The parameters $L_0$ and $L_1$ are then equal to:

$$L_0 = -R_{12} \cdot \bar{P}_{12}^2 = -0.01,$$
$$L_1 = 2 \cdot R_{12} \cdot \bar{P}_{12} = 0.02.$$

The $(DCOPF)$ model without losses produces an LMP of 20 $/MWh for both nodes, whereas the model that includes losses results in an LMP of 20 $/MWh for node 1 (since the generator is at the money at this node) and 20.41 $/MWh, where the price difference is due to losses. The reader is asked to verify this result in problem 5.12. Note that this produces a price differential between the two nodes, even if the line is not congested. The losses of the model amount to 0.010. This is consistent with the result of example B.5. The economic interpretation of the higher price in node 2 is that, to get the power there, one needs to pay for the marginal cost of both the power itself and also the power that is lost along the way. Since the increment in losses is approximately equal to $2 \cdot R_{12} \cdot P_{12}$, the increment in marginal cost amounts to approximately $MC_{G_1} \cdot 2 \cdot P_{12} = 0.40$ $/MWh for a base dispatch of $P_{12} = 1$.

## 5.3 Zonal pricing

### 5.3.1 Motivations for zonal pricing

Although locational marginal pricing is currently the norm in US markets, this has not always been the case. US markets evolved into their current state after failed, and at times quite painful, attempts to implement zonal pricing followed by redispatch. Zonal pricing remains the prevailing paradigm in Europe, despite the fact that congestion management costs have recently rocketed in a number of European markets, including Germany and the UK. Zonal pricing is also a contender in a number of emerging markets, including India and China. The direct implementation of nodal pricing often stumbles upon similar arguments in different systems, some of which are summarized below.

*Institutional acceptability*: A common impediment towards the implementation of nodal pricing is the fact that it can produce higher prices in locations with greater load. This can result in political frictions. For instance, certain European Member States stand to lose from the implementation of a nodal pricing mechanism because it would imply a higher price for industrial load centers, thereby rendering the industry of said Member States less competitive. Another institutional barrier is the centralization of information and control. Whereas the sharing of information and control of sensitive network infrastructure may be acceptable between States of a single country, surrendering this information and control between countries can be more delicate.

*Implementation complexity*: Nodal pricing is sometimes argued to be "too complicated," or that there are "too many prices." It is fair to say that the sophistication of the nodal pricing infrastructure is impressive, with systems such as PJM consisting of thousands of buses with a potentially different price in each different location. Nevertheless, the fact that nodal pricing has been implemented successfully in systems of such scale proves that the endeavor is technologically and institutionally feasible. Detailed network models can prove to be taxing for market designs that implement specific types of pricing rules for dealing with non-convex constraints or costs (see, for instance, the "no-PAB" pricing rule in chapter 7), nevertheless there are various rules for pricing with non-convexities (see, for instance, IP pricing or linear relaxation pricing in chapter 7) that can function in tandem with nodal pricing models. Interestingly, despite its apparent simplicity, zonal pricing turns out to require quite complex and nontransparent procedures (such as the definition of feasible injection sets $\mathcal{R}$, which is detailed later in the text) that put into question its claimed simplicity.

Nodal pricing further requires[3] **unit-based bidding**: under such a design, each physical generator bids separately in the market. This is contrasted to **portfolio bidding**, whereby numerous assets are offered in the market as a portfolio. In those European markets where portfolio bidding applies,[4] portfolio bidding in the day-ahead market

---

[3] More precisely, units in different nodes of the grid cannot be bid as part of a portfolio.

[4] Not all European markets apply portfolio bidding. Ireland, Greece, Portugal, and Spain are notable exceptions where unit-based bidding applies.

is followed by **nominations**, whereby portfolio owners disaggregate their day-ahead market position to individual units within their portfolios. These individual unit positions are communicated to the system operator a few hours before real time. Unit-based bidding enables nodal pricing, whereas a portfolio that includes assets that are located in different nodes is not consistent with nodal pricing. Whereas unit-based bidding is attractive in terms of market power monitoring and can be handled by modern market clearing algorithms, it is perceived as being disruptive in certain portfolio-based systems.

*Market power*: Locational marginal pricing has been criticized for creating opportunities for the exercise of market power due to the fact that it splits the overall market to smaller local markets, the traded volume of which is substantially lower. One might then argue that, although a firm may represent a relatively tiny fraction of overall volume traded in the market, it may represent a significant fraction of the volume traded in its location under nodal pricing, thereby enhancing its opportunities to profitably manipulate prices. Pretending that a network bottleneck does not exist when it actually exists unfortunately does not take care of the market power problem, and any market design is unfortunately susceptible to market power if conditions for market power exist. It has been shown in the literature that zonal pricing is also exposed to market power, and that it can also introduce opportunities for gaming the market even for fringe firms that do not have market power.

*Budget imbalance*: Early advocates of zonal pricing argued that the nodal pricing alternative leaves too much "money on the table." "Money on the table" refers to a positive budget imbalance, i.e. funds that remain available to the auctioneer after the conclusion of a nodal pricing auction. Instead, zonal pricing can achieve the same dispatch outcome with a much lower budget imbalance (which can typically be negative, i.e. which corresponds to funds that the auctioneer needs to gather from other sources, such as network tariffs). This point is illustrated in example 5.15. However, this argument assumes that market participants do not adapt their offers strategically to the zonal design, which is not true in practice.

*Counterintuitive prices*: When one reasons about pricing by analogy to transportation networks, certain pricing rules might appear intuitive. For instance, one might expect that prices between nodes connected by a line that is not congested should be equal (we show that this may not be the case in nodal pricing in example 5.5, and we show that it must be the case in transportation networks in problem 5.10). Or one might expect that power should always flow from locations with lower to higher prices.[5] These properties are not guaranteed to hold in optimal power flow, and the seemingly counterintuitive price behavior may induce resistance towards institutional acceptance for nodal pricing. The logical error here amounts to reasoning by analogy to transportation networks when Kirchhoff's laws are at work.

---

[5]  In European market design, the phenomenon where power in an optimal power flow can move from high-price to low-price areas has been menacingly baptized as **nonintuitive flows**. Considerable algorithmic effort has been invested in eliminating this behavior in the European day-ahead market model, even though this is a natural behavior of optimal power flow models.

*Risk management*: Hedging the risk of transportation cost (or locational price differences) is an important prerequisite of well-functioning electricity markets. As we show in section 9.2, this can be accomplished through financial contracts that pay back the difference between the energy prices of the two locations between which power is being traded. Under nodal pricing, the pairs of locations increase significantly, which can undermine liquidity in the financial market. One proposal for overcoming this issue has been a **contract network**, which is fairly analogous to airport networks: a small number of (large) trading hubs trade financial transmission rights between each other, and (smaller) spokes trade financial transmission rights with their nearest hub. This decreases the pairs of possible locations that trade financial transmission rights significantly, thereby aiming to enhance liquidity in the financial transmission rights market.

### 5.3.2   Models of zonal pricing

The idea of zonal pricing is to aggregate nodes into larger markets, referred to as zones, and have the system operator redispatch units after the market has cleared in order to ensure that transmission constraints are not violated. The hope is that the same goal as nodal pricing can be achieved with far fewer prices and far less complexity in terms of market operations.

Two prevailing models of zonal pricing are discussed in this textbook. One is based on a transportation network. The other is flow-based zonal pricing. Both can be described with the following generic notation:

$$(ZP): \max_{p,d,r} \sum_{l \in L} \int_0^{d_l} MB_l(x)dx - \sum_{g \in G} \int_0^{p_g} MC_g(x)dx$$

$$(\rho_z): \quad r_z = \sum_{g \in G_z} p_g - \sum_{l \in L_z} d_l, z \in Z$$

$$r \in \mathcal{R}$$

$$p_g \geq 0, g \in G$$

$$d_l \geq 0, l \in L.$$

Here, $\mathcal{R}$ is a set of feasible zonal positions. How this set is precisely defined is what distinguishes flow-based zonal pricing from zonal pricing based on a transportation network. Both the transportation model and the flow-based model involve arbitrary parameter choices, which strongly influence the zonal pricing outcome. This parameter dependence is a major drawback of zonal pricing, since it introduces considerable complexity as well as room for arbitrary and nontransparent choices into the process. The point of parameter choices is elaborated further below.

**Definition 5.11   Zonal pricing** is a uniform price auction that is conducted as follows:

• Sellers submit increasing bids: price-quantity pairs, representing the price at which they are willing to supply a certain quantity. Buyers submit decreasing bids: price-quantity pairs indicating the price they are willing to pay for the specific quantity.

- The market operator solves ($ZP$) and announces $\rho_z$ as the market clearing price for zone $z$.

An important property of this model, compared to nodal pricing, is that there are as many zonal prices, $\rho_z$, as there are zones. As long as the number of zones remains low, the number of prices produced by this model remains low.

**Zonal pricing based on a transportation network**    Zonal pricing based on a transportation network (or **transportation-based zonal pricing**) aggregates the physical network and entirely ignores Kirchhoff's laws. This earns the method its name: in operations research, a transportation network is a network on which we can directly control not only the injection of a commodity at each location, but also the flows of the commodity on the links of the network.

The first step in this process is to define an aggregate representation of the true network in terms of zones and links that connect the zones. The zonal network can be represented as a directed graph, where $Z$ represents the set of zones, and $A$ represents the set of links between the zones. Generators and loads located in zone $z$ are denoted as $G_z$ and $L_z$, respectively.

The process of choosing how to aggregate the system is already nontrivial, and the choice of aggregation influences the market clearing outcome. In European market design, this challenge goes under the name of **bidding zone configuration**, and can be a controversial and highly complex exercise. Although it is often the case that many European Member States correspond to zones (thus the choice of aggregation is based on political rather than technical criteria), it is also the case that certain Member States (such as Sweden and Norway) consist of multiple zones.

Once zones are defined, it is necessary to define the **available transfer capacity (ATC)** over the links that connect the zones. A transportation model is used to represent the capacity constraints over the links of the network. For each zone $z$, an energy conservation constraint is enforced, along with flow constraints on the links. Thus, the set of feasible zonal injections can be described as follows.

$$\mathcal{R} = \left\{ r : r_z = \sum_{a=(z,\cdot)} f_a - \sum_{a=(\cdot,z)} f_a, z \in Z \right.$$
$$\left. -ATC_a^+ - f_a \le ATC_a^+, a \in A \right\}$$

The transportation-based zonal pricing model is then obtained as follows:

$$(ZPT): \max_{p,d,r,f} \sum_{l \in L} \int_0^{d_l} MB_l(x)dx - \sum_{g \in G} \int_0^{p_g} MC_g(x)dx$$

$$(\rho_z): \quad r_z = \sum_{g \in G_z} p_g - \sum_{l \in L_z} d_l, z \in Z$$

$$r_z = \sum_{a=(z,\cdot)} f_a - \sum_{a=(\cdot,z)} f_a, z \in Z$$

$$ATC_a^- \leq f_a \leq ATC_a^+, a \in A$$
$$p_g \geq 0, g \in G$$
$$d_l \geq 0, l \in L.$$

As indicated earlier, the transportation-based model ($ZPT$) does not account for Kirchhoff's laws. Moreover, the model ignores congestion within zones, and injections within a zone are assumed to have an identical influence on the flows of a given interconnector among zones.

---

**Example 5.11** *A six-node network with six different LMPs.* Consider the six-node network of figure 5.7. The PTDFs of the system are presented in table 5.2. Only the PTDFs of capacity constrained lines are reported. Line 1–6 has a capacity of $T_{1-6} = 200$ MW, and line 2–5 has a capacity of $T_{2-5} = 250$ MW. The marginal cost and marginal benefit of suppliers and consumers are linear functions, and are indicated in the figure. The network has cheap generation located in nodes 1 and 2, whereas the demand with highest valuation is located in nodes 5 and 6. Nodal pricing produces a welfare of 23000 \$/h and a different price at each node: $\rho_1 = 25$ \$/MWh, $\rho_2 = 30$ \$/MWh, $\rho_3 = 27.5$ \$/MWh, $\rho_4 = 47.5$ \$/MWh, $\rho_5 = 45$ \$/MWh, and $\rho_6 = 50$ \$/MWh. The line flows are $f_{1-6} = f_{2-5} = 200$ MW. The figure presents a partition of the network into two zones. The northern zone consists of nodes $\{1, 2, 3\}$, the southern zone consists of nodes $\{4, 5, 6\}$. The prices of the zonal model depend on the ATC of the North–South link. Suppose that, in order to prevent overload, the lesser of the line capacities is chosen, i.e. assume an ATC value of 200 MW. Then the price in the North becomes 24.2 \$/MWh and the price in the South becomes 50.8 \$/MWh. The flows are $f_{1-6} = 109.4$ MW, $f_{2-5} = 90.6$ MW. Although this solution respects transmission constraints, the attained welfare amounts to 18520 \$/h because the ATC value is overly conservative. Instead, suppose that the capacity of the link

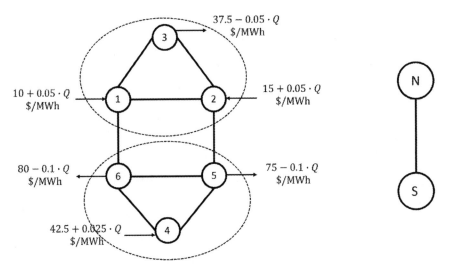

**Figure 5.7** The six-node network of example 5.11 and its aggregation into a network with two zones.

**Table 5.2** PTDFs for the network of figure 5.7.

|          | Node 1 | Node 2 | Node 3 | Node 4  | Node 5 |
|----------|--------|--------|--------|---------|--------|
| Link 1–6 | 0.625  | 0.5    | 0.5625 | 0.0625  | 0.125  |
| Link 2–5 | 0.375  | 0.5    | 0.4375 | −0.0625 | −0.125 |

is assumed equal to the sum of the line capacities, i.e. assume an ATC that is equal to 450 MW. Then the price becomes 28.3 \$/MWh in the North and 46.7 \$/MWh in the South. The welfare becomes 24146 \$/h. This value exceeds the welfare attained by nodal pricing. However, these gains are artificial because the resulting dispatch violates transmission constraints: the flows are $f_{1-6} = 234.4$ MW, $f_{2-5} = 215.6$ MW.

---

Example 5.11 highlights that the choice of ATC capacity is nonobvious, and influences the market clearing outcome. This is a highly problematic aspect of zonal pricing, which introduces scope for arbitrary and nontransparent choices. European legislation requires that ATCs are such that they do not exclude zonal commercial transactions that could be physically supported by the network, but also that ATCs are such that they do not allow commercial transactions that cannot be physically supported by the network.

There is sometimes a distinction in zonal systems between **loop flows** and **transit flows**. Loop flows correspond to flows within a zone that are induced by transactions that take place *within* a neighboring zone. Transit flows correspond to flows within a zone that are induced by transactions that take place *between* neighboring zones. The distinction is illustrated in example 5.12.

---

**Example 5.12** *Loop and transit flows.* Consider the network of figure 5.8. Suppose that all lines have identical physical characteristics. The left part of the figure presents three zones, and corresponds to a transit flow of 1/3 MW in zone C (assuming that either or both of the lines AC and CB belong to its jurisdiction), since the flow that the zone is experiencing is due to transactions between two distinct zones A and B. The right part of the figure corresponds to a loop flow of 1/3 MW since the flow that

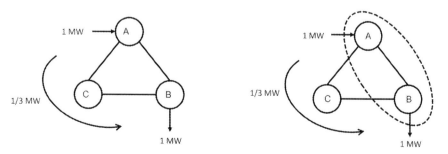

**Figure 5.8** The figure of example 5.12. The left part corresponds to a transit flow in zone C, the right part corresponds to a loop flow in zone C.

zone C is experiencing is due to a trade between different locations in the zone $\{A, B\}$ that is indicated in the dashed figure.

---

**Flow-based zonal pricing**    In an attempt to approximately account for Kirchhoff's laws in zonal pricing, European system operators have implemented so-called **flow-based zonal pricing**. The idea of flow-based models is to express the set $\mathcal{R}$ of feasible zonal net injections as follows:

$$\mathcal{R} = \left\{ r : \sum_{z \in Z} PTDF_{zl} \cdot r_z \leq RAM_l, l \in CB, \right.$$

$$\left. \sum_{z \in Z} r_z = 0 \right\}$$

There are three new objects introduced in this model, relative to the models that we have seen so far:

- The set of **critical branches** $CB$ is the set of zonal network elements for which such constraints are enforced.
- The **zone-to-line PTDFs** $PTDF_{zl}$ are approximations of how a net injection in a zone maps to a flow on a critical branch.
- The **remaining available margin (RAM)** is the amount of capacity that is available in the critical branch, after a *reference flow* and additional headroom is subtracted from the physical capacity of a critical branch. The computation of the reference flow requires assuming a **base case**, i.e. a dispatch of the system when the market clears. This creates circularity problems because the choice of this reference flow, and thus the RAM, influences the dispatch of the system, which influences the reference flow that is being estimated in the first place.

Choosing values for zone-to-line PTDFs and remaining available margins is nonobvious. As in the case of ATCs in transportation-based zonal pricing, this adds to the fact that the mechanism depends on nontransparent and discretionary choices, which is a disadvantage of the design.

Zone-to-line PTDFs are fictional objects (in contrast to node-to-line PTDFs, which depend purely on the physical characteristics of the network), because the flow induced on a critical branch by an injection of a certain zone is a function of which specific node is injecting. One might argue that the node that is injecting is well defined for a given market clearing outcome of the zonal model, but again there is a circularity problem here: the market clearing outcome itself depends on the choice of zone-to-line PTDF. Notwithstanding, certain system operators employ **generation shift keys (GSKs)** for estimating zone-to-line PTDFs. $GSK_{gz}$ maps the contribution of a generator $g$ that belongs to zone $G_z$ towards the net export of that zone. This implies that $PTDF_{lz}$ can be computed using $GSK_{gz}$ as follows:

$$PTDF_{lz} = \sum_{g \in G_z} GSK_{gz} \cdot PTDF_{l,n(g)},$$

where $G_z$ is the set of generators in zone $z$ and $n(g)$ is the node at which generator $g$ is located. Again, notice the circularity: the parameter $GSK_{gz}$ depends on the market dispatch, but the market dispatch depends on the value of $GSK_{gz}$.

The overall flow-based zonal pricing model is expressed as follows:

$$(ZPFB): \quad \max_{p,d,r,f} \sum_{l \in L} \int_0^{d_l} MB_l(x)dx - \sum_{g \in G} \int_0^{p_g} MC_g(x)dx$$

$$(\rho_z): \quad r_z = \sum_{g \in G_z} p_g - \sum_{l \in L_z} d_l, z \in Z$$

$$f_k = \sum_{z \in Z} PTDF_{kz} \cdot r_z, k \in CB$$

$$f_k \leq RAM_k, k \in CB$$

$$\sum_{z \in Z} r_z = 0$$

$$p_g \geq 0, g \in G$$

$$d_l \geq 0, l \in L.$$

The set $\mathcal{R}$ has a characteristic diamond shape. We illustrate this in the following example.

---

**Example 5.13** *Visualizing flow-based domains.* Consider the network of the upper part of figure 5.9. The zone-to-line PTDFs are given in table 5.3. The flow-based domain $\mathcal{R}$ is characterized by the following inequalities:

$$\text{Critical branch AB: } \frac{1}{3}r_A - \frac{1}{3}r_B \leq 1000$$

$$\text{Critical branch BA: } -\frac{1}{3}r_A + \frac{1}{3}r_B \leq 1000$$

$$\text{Critical branch BC: } \frac{1}{3}r_A + \frac{2}{3}r_B \leq 1000$$

$$\text{Critical branch CB: } -\frac{1}{3}r_A - \frac{2}{3}r_B \leq 1000$$

$$\text{Critical branch AC: } \frac{2}{3}r_A + \frac{1}{3}r_B \leq 1000$$

$$\text{Critical branch CA: } -\frac{2}{3}r_A - \frac{1}{3}r_B \leq 1000$$

$$r_A + r_B + r_C = 0$$

The last constraint allows us to view $\mathcal{R}$ in the two-dimensional space $(r_A, r_B)$ without concerning ourselves with the value of $r_C$, since $r_C$ is implied by this last energy balance constraint. The set $\mathcal{R}$ is represented graphically in the lower part of figure 5.9. Each line at the border of the diamond-shaped feasible set corresponds to one critical branch, and the intersection of these lines corresponds to extreme points of the set $\mathcal{R}$. There are six such points, and their coordinates are indicated in the figure.

---

Table 5.3 The zone-to-line PTDFs of example 5.13.

|     | A    | B    |
| --- | ---- | ---- |
| AB  | 1/3  | −1/3 |
| BA  | −1/3 | 1/3  |
| BC  | 1/3  | 2/3  |
| CB  | −1/3 | −2/3 |
| AC  | 2/3  | 1/3  |
| CA  | −2/3 | −1/3 |

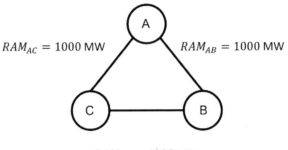

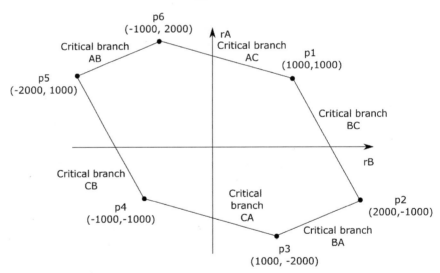

Figure 5.9 The network of example 5.9 (upper panel), and the resulting set of feasible injections $\mathcal{R}$ (lower panel).

A way to overcome the parameter dependence and circularity of the flow-based zonal pricing model is to directly encode Kirchhoff's power flow constraints and the requirement of zonal pricing for a unique price within a given zone. Interestingly, such a model of zonal pricing (which is described below as model (*FBP*)) was conceived in academic work years before the industry implementation of flow-based market coupling in Europe (which is based on model (*ZPFB*)).

In order to develop such a model of zonal pricing, we can introduce a direct flow constraint on every network element:

$$-T_k \leq \sum_{n \in N} F_{kn} \cdot \left( \sum_{g \in G_n} p_g - \sum_{l \in L_n} d_l \right) \leq T_k, k \in K.$$

Given that we require a uniform price at each zone while respecting the flow limits on all lines that belong to $K$, we arrive at the following model:

$$(FBP): \max_{p,d,\rho} \sum_{l \in L} \int_0^{d_l} MB_l(x)dx - \sum_{g \in G} \int_0^{p_g} MC_g(x)dx$$

$$0 \leq p_g \perp MC_g(p_g) - \rho_{z(g)} \geq 0, g \in G$$

$$0 \leq d_l \perp -MB_l(d_l) + \rho_{z(l)} \geq 0, l \in L$$

$$\sum_{g \in G} p_g - \sum_{l \in L} d_l = 0$$

$$-T_k \leq \sum_{n \in N} F_{kn} \cdot \left( \sum_{g \in G_n} p_g - \sum_{l \in L_n} d_l \right) \leq T_k, k \in K.$$

Here, $z(\cdot)$ is the zone at which a market agent is located. The price $\rho_z$ corresponds to the uniform price of zone $z$. Model $(FBP)$ thus strives to reduce the efficiency losses implied by zonal pricing as much as possible, while respecting the true physical constraints of the network.

Model $(FBP)$ is referred to as a **mathematical program with equilibrium constraints (MPEC)**. The equilibrium constraints in the case of $(FBP)$ originate from the KKT conditions that express the rules of a uniform price auction at each zone. MPECs can be cast as nonlinear optimization problems, since a complementarity condition can be expressed equivalently as an equality constraint involving the product of two functions.

---

**Example 5.14** *Model $(FBP)$ applied to the six-node network.* Consider again the six-node network shown in figure 5.7. The price that is obtained from a flow-based model is 27.2 \$/MWh for the North zone and 47.8 \$/MWh for the South zone. The welfare is 22806.6 \$/h. The line flows are $f_{1-6} = 200$ MW and $f_{2-5} = 181.3$ MW. The flows respect the transmission constraints.

---

Note that imposing a constraint on the flow of all network elements in the last set of constraints of $(FBP)$ may very well lead to an infeasible model. Thus, it may be desirable to limit these constraints to a subset of network elements/critical branches, $CB$, as in $(ZPFB)$.

## 5.3.3 Redispatch

Once the zonal pricing model is solved, there is no guarantee that the transmission constraints of the system are respected. This is demonstrated in example 5.11. It is therefore necessary to adjust the dispatch schedule of producers and consumers in order to restore feasibility. This is accomplished through a process named redispatch.

**Definition 5.12  Redispatch** is a py-as-bid auction that is conducted after zonal pricing, as follows:

- Sellers submit increment (INC) and decrement (DEC) bids. INC bids are price–quantity pairs that represent the price at which producers are willing to provide additional power relative to the quantity cleared in the zonal pricing auction. DEC bids represent the price that producers are willing to pay to the market operator for decreasing production relative to the quantity cleared in the zonal pricing auction. Buyers can also submit INC and DEC bids.
- The system operator solves $(DCOPF)$. Sellers and buyers follow the results of the redispatch, and are paid as bid for the *changes* in their production or consumption, compared to the zonal pricing auction.

Redispatch is designed so that the system operator minimizes its payments to INC bids and maximizes its payment from DEC bids. The INC bids are typically paid a higher price than the price charged to DEC bids. It is important to emphasize that the outcomes of the redispatch auction do not have any impact on the revenues from the zonal pricing auction. For example, if DEC bids are accepted, producers still earn the zonal price for the *full* amount of power that is cleared in the zonal pricing auction.

One motivation for zonal pricing followed by redispatch is that, under truthful bidding, it achieves the same result as nodal pricing (relieving congestion) with fewer prices and a lower budget imbalance.[6] This is clarified in the following example.

---

**Example 5.15** *Redispatch on a two-node network.* Consider the network of figure 5.10. Nodal pricing results in the production of 800 MW from node 1 and the production of 400 MW from node 2. The nodal prices are $\rho_1 = 56$ \$/MWh and $\rho_2 = 68$ \$/MWh. The LMP auction results in a net surplus of 9600 \$/h paid to the market operator. Consider the case where both nodes are aggregated into a single zone. In a zonal market, the market clearing price is 62 \$/MWh, with 1100 MW sourced from node 1 and 100 MW sourced from node 2. However, this violates the limit of line 1–2.

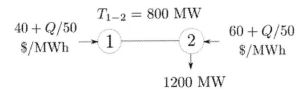

**Figure 5.10** The two-node network of example 5.15.

---

[6] The budget of an auction is balanced when the sum of payments transferred from buyers to the auctioneer is equal to the sum of payments transferred from the auctioneer to sellers. Otherwise, there is a budget imbalance that may be positive or negative.

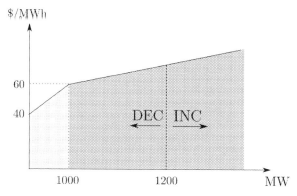

**Figure 5.11** The aggregate inverse supply function of example 5.15. The area in light gray indicates generators that are located in node 1, whereas the area in dark gray indicates generators that are located in both areas. The zonal auction clears at 1200 MW. INC bids are cleared from node 2, and are more expensive than DEC bids that are cleared from node 1. The redispatch auction creates a positive net transfer to generators.

If generators bid truthfully in the redispatch auction, 300 MW of incremental bids are cleared from node 2 and 300 MW of decremental bids are cleared from node 1. The cleared DEC bids in node 1 correspond to the 300 most expensive MW, as shown in figure 5.11. The total payment *to* the market operator for cleared DEC bids is 17700 $/h. The cleared INC bids in node 2 correspond to the 300 cheapest MW. The total payment *from* the market operator for cleared INC bids is 19500 $/h. There is no net surplus left to the market operator after the zonal pricing auction. The net surplus of the redispatch auction is −1800 $/h, and is typically collected from consumers. This budget imbalance is substantially lower than that of nodal pricing, which amounts to +9600 $/h and is directly paid by consumers through the LMP auction.

### 5.3.4  INC-DEC gaming

Despite the apparent attraction of the zonal pricing mechanism in terms of achieving congestion management with fewer transfers relative to nodal pricing, the zonal pricing mechanism followed by redispatch is susceptible to gaming. **INC-DEC gaming**, or **DEC game**, is a gaming strategy that has been devised in order to manipulate the zonal pricing system in a quite straightforward way by exploiting the disconnect between zonal pricing and the physical reality of operation. INC-DEC gaming was applied extensively during the deregulation of the California electricity market, which commenced as a zonal market, and contributed to the meltdown of the original California design in 2001. It is also an endemic problem in various European markets.

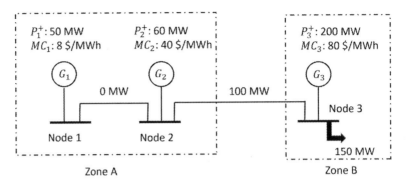

**Figure 5.12** INC-DEC gaming in a three-node, two-zone system.

In order to illustrate the problem, we consider the system of figure 5.12. Zone A consists of nodes 1 and 2, while zone B consists of node 3. There is cheap generation in node 1; however, it is isolated from the rest of the network since there is no more capacity available on the line 1–2 given the existing dispatch of the system. The next cheapest generator in the system is G2, which is still in zone A, while the most expensive unit, G3, is colocated with the system load in zone B at node 3. The physical capacity on line 2–3 is assumed to be equal to 100 MW, and this is also assumed to be the ATC of the link connecting zone A to zone B.

In order to illustrate concepts, suppose that the units in nodes 2 and 3 are bidding truthfully. Then consider a possible strategy whereby the unit in node 1 bids 8 $/MWh in the zonal auction, followed by 0 $/MWh in the redispatch step.

From an operational standpoint, it would be difficult for such a bid to make sense. The redispatch bid is supposed to reflect the marginal cost of the fuel that is saved by the unit if it ends up not producing in real time. There is no reason for this marginal cost to differ from the marginal cost that is bid in the original zonal auction, unless the unit is operating at a different segment of its marginal cost curve. And since in this example we assume that the unit in node 1 has a constant marginal cost, such a bid would clearly misrepresent the true marginal cost of the unit. Allowing the freedom for units to update their bids in the redispatch phase is a practice that is referred to as **market-based redispatch**.[7]

What, then, could G1 possibly gain by submitting such an offer? The key is that it can extract surplus by having the TSO procure its power in the zonal auction at its true marginal cost (or more), and then forcing the TSO to buy the power back at whatever price it declares in the redispatch phase. If this price is lower than the zonal price, the generator profits for every MWh that is redispatched down for doing literally nothing: the MWh is first sold to the system in the zonal auction and then taken back by the generator in the redispatch phase, thus it is never produced. This earns the gaming strategy its name, since the DEC bid is key towards extracting this surplus.

---

[7] The choice of naming is actually quite misleading, since in a true market setting electricity produced in a different location is a different commodity, thus belonging to a different market, and should therefore be traded in a different market price.

Concretely, in the case of figure 5.12, the zonal dispatch is sourcing 50 MWh from G1 and 50 MWh from G2, up to the point where the link from zone A to zone B is saturated. The remainder of the energy is sourced from G3. The zonal auction clears zone A at 40 $/MWh, since G2 is setting the price of the zone. Note the absurdity of the dispatch, since G1 is essentially disconnected from the system, nevertheless it is dispatched in the zonal auction step. In the redispatch phase, G1 is dispatched back to 0 MWh since there is no other way for the TSO to relieve line 1–2. Nevertheless, G1 knows this, and thus claims that its marginal cost is now 0 $/MWh, so that it is required to pay $0 back to the TSO for avoided fuel costs in the redispatch phase. G1 thus walks out of the zonal auction pocketing $2000 (40 $/MWh × 50 MWh), without having to pay anything back to the TSO in the redispatch phase (0 $/MWh × 50 MWh), thus it earns 40 $/MWh for each of the 50 MWh that it never produces in the first place.

This argument can be pushed to the extreme: G1 could submit an *arbitrarily negative* DEC bid in the redispatch phase, and not only pay nothing for being redispatched down but in fact be *paid* by the TSO for doing so! This is not a theoretical argument: it has been observed in practice, with market agents submitting bids that reach the floor of the redispatch auction. A possible remedy that has been considered by certain markets which suffer from this phenomenon is to implement so-called **cost-based redispatch**, whereby there is no bidding taking place at the redispatch phase, but instead INC and DEC price bids are assumed to be equal to marginal costs that are estimated by a regulatory agency. With an increasing penetration of demand response and storage in future electricity systems, such estimates are expected to become increasingly difficult to obtain, since the valuation information of demand response is largely private, and the bids of storage are based on opportunity costs, which can be very difficult to estimate accurately.

INC-DEC gaming was employed by ENRON and municipally owned utilities during the California electricity market crisis in 2001. The strategy relied on scheduling the delivery of power above levels that the system could handle, and being paid at the redispatch auction without ever intending to produce power.

Locational marginal pricing was originally adopted by the Pennsylvania–Jersey–Maryland (PJM) market, and subsequently by the New York ISO, the New England ISO, and the Midwest ISO. Beyond the United States, locational marginal pricing was adopted in New Zealand in 1997. Experience suggests that this has been the most successful framework for organizing electricity markets that can cope with transmission constraints. Locational marginal pricing has not been adopted widely in Europe at the time of writing this text, with the exception of certain countries that attempt to implement it unilaterally.

## Problems

**5.1** *No-arbitrage interpretation of LMP.* How would you interpret proposition 5.3 as a no-arbitrage condition?

**5.2** *LMP differences in the B-theta formulation.* Show that the following relationship holds among dual optimal multipliers for the OPF problem of section 5.1.2:

$$\rho_n - \rho_m = \lambda_k^+ - \lambda_k^- + \gamma_k, k = (m, n).$$

Can a conclusion be drawn about the relative value of $\rho_n$ and $\rho_m$ when power flows from $n$ to $m$?

**5.3** *A six-node network with six different prices.* Implement the nodal model that is described in example 5.11 and confirm the results (welfare, LMPs, flows on the lines connecting zones).

**5.4** *Insecure operation under zonal pricing.* Implement the zonal model of example 5.11 and reproduce the results (welfare, prices, and physical flows implied by the dispatch).

**5.5** *Multiple equilibrium prices.* In example 5.7, show that any value between 120 $/MWh and 140 $/MWh is also a valid LMP.

**5.6** *LMP maximizes network operation surplus.* Prove that LMP prices are consistent with the goal of the network operator to maximize the economic value of the network.

**5.7** *Direction of flow on a two-node network.* Consider a two-node network with a flow of 10 MW from node 1 to node 2, and an LMP of 20 $/MWh in node 1 and 10 $/MWh in node 2. Is this possible? Provide an example of an OPF model if it is, or prove why it is not.

**5.8** *Graphical solution of three-node examples.* Compute the optimal power flow solution of example 5.5 by plotting the feasible set on the two-dimensional space of production in node 1 and node 2. Compute the LMP of node 3, $\rho_3$, using the two-dimensional feasible set and the sensitivity interpretation of dual variables. Keep in mind, when solving the problem, that only lines 1–3 and 2–3 should be considered, in their reference directions, because the other network constraints are not close to being binding in the optimal solution.

**5.9** *Power always flows from cheap to expensive locations in transportation networks.* Derive the KKT conditions of model (*ZPT*), and show that power always flows from cheap to expensive locations in this model. Is this also true for (*OPF*)?

**5.10** *Prices are equal in adjacent zones without congestion in transportation networks.* Show that, in a general transportation network, if there is no congestion on a link then the zones that are adjacent to the link must have equal prices.

**5.11** *Power can flow from more expensive to cheaper areas in locational marginal pricing.* Provide an example in which power flows from a more expensive to a cheaper area in locational marginal pricing, and verify the result with mathematical programming code.

**5.12** *DCOPF with losses.* Verify the results of example 5.10 with mathematical programming code.

**5.13** *Zonal pricing as a cross-subsidy.* Consider a market with one node in the north and one node in the south. The node in the north consists of perfectly competitive producers with a capacity of 500 MW and a marginal cost of 10 $/MWh, while the node in the south consists of consumers with a demand of 400 MW and a valuation of 200 $/MWh. The line that connects the north to the south has a capacity of 300 MW.

1. In a market with locational marginal pricing (LMP), what is the price in the north and in the south? What is the total amount of consumer payments? What is the surplus of consumers in the south?
2. In a market with zonal pricing where the system is considered as one single zone and where the transmission constraint from north to south is ignored, what is the zonal price? Assuming that producers in the north and loads in the south submit truthful marginal costs/valuations at the redispatch stage, what is the redispatch outcome, i.e. how many MW are redispatched upwards/downwards in the north/south? What are the payments from the system operator to consumers in the south, and from the producers in the north to the system operator?
3. Does the final consumption of loads differ in the zonal and nodal designs?
4. What is the consumer surplus in zonal pricing (consumer benefit from final consumption minus payments in the zonal auction plus redispatch payments)? If you represent consumer interests and are therefore aiming at maximizing consumer surplus, which of the two designs would you lobby for, zonal or nodal pricing?

5.14 *Zonal pricing as a cross-subsidy on a triangle network.* Suppose that country B, and the north and south of country G can be represented as a simplified triangular network, with equal reactance $X$ on the three lines of the network. The lines that connect the north of country G to B and the south of country G to country B have no thermal limit, while the line that connects the north of country G to the south of country G has a thermal limit of 8000 MW. We have 35000 MW of cheap producers in the north of country G, with a marginal cost of 20 \$/MWh. The south of country G has an inelastic load of 30000 MW and generators with a capacity of 20000 MW and a marginal cost of 200 \$/MWh. In our simplified model, country B has no production or demand and all producers offer their entire capacity at its actual marginal cost to the market.

1. Compute the outcome of the zonal market (generator dispatch and the zonal price of country G) if each country is a separate zone and if the ATC between country B and country G is equal to 1000 MW in both directions.
2. Is the zonal solution compatible with the thermal limits of the network?
3. Compute the outcome of locational marginal pricing (generator dispatch and prices in the north and south of country G).
4. If your goal is to keep electricity prices low in the south of country G (ignoring redispatch costs), which of the two designs do you prefer, zonal or nodal pricing?

# Bibliography

**Section 5.1**   Some of the examples in this section are inspired by the course notes of Professor Kory Hedman at Arizona State University.

Transmission line switching was originally proposed by Fisher, O'Neill, and Ferris (2008), and subsequently analyzed in depth by Hedman. Hedman's research focuses on quantifying the gains of transmission switching on systems of realistic scale by introducing models of increasing complexity, starting with a static model subject to contingency constraints (Hedman et al. 2009), and culminating in a co-optimization of topological control and unit commitment subject to $N-1$ reliability criteria (Hedman et al. 2010). A discussion on the importance of active network management, and topological control in particular, is provided by Bill Gates (Gates 2023). The potential benefits of topological control for the European power grid are analyzed by Han and Papavasiliou (2015, 2016). Zonal pricing especially stands to benefit from topological optimization, due to the tendency of zonal systems to violate network constraints, and this contributes to the fact that topological control was pioneered by European TSOs well before being studied in detail by the US research community and industry. The interplay between topological control and zonal pricing is studied in a two-part paper (Lété & Papavasiliou, 2020a,b).

The use of dual multipliers for the development of line switching heuristics has been proposed by Fuller, Ramasra, and Cha (2012). Limiting the number of switching actions is also documented in the literature (Hedman et al. 2010), where it has been empirically observed that this can lead to capturing a majority of the potential gains of network reconfiguration. Finding the smallest possible value of big M in model ($TR$) can also prove to be very useful in producing tighter linear programming relaxations of the model, which can accelerate branch-and-bound solvers for tackling the problem noticeably. A method for finding the smallest value of big M that keeps the model valid is presented in Binato, Pereira, and Granville (2001).

**Section 5.2**   Some of the material in this section is inspired by §5-3, 5-4, and 5-7 of Stoft (2002) as well as Boucher and Smeers (2001). The correspondence of competitive market equilibrium and linear programming in transportation networks was recognized as early as 1952 by Samuelson (1952). Locational marginal pricing was proposed originally by Schweppe et al. (1988), with a predecessor of the text published in an MIT report by Caramanis, Tabors & Stevenson (1982). The equivalence between nodal pricing and competitive equilibrium in electricity markets is inspired by the exposition of §5-5 of Stoft (2002). The fact that power can flow from more expensive to cheaper locations in locational marginal pricing is discussed in Wu et al. (1996). The discussion on losses is based on Hogan (1992) and Eldridge, O'Neill, and Castillo (2016).

**Section 5.3**   A review of the evolution of US markets to nodal pricing is provided by Hobbs and Oren (2019). The generic exposition of zonal pricing in ($ZP$) is based on Aravena et al. (2021).

A critical appraisal of nodal pricing is provided by Oren et al. (1995). Harvey and Hogan counter concerns about market power (Harvey & Hogan 2000) and point out that any design is susceptible to market power, thereby underlining that a fair comparison between designs is required. An important enabler of nodal pricing is the implementation of financial transmission rights, which allow hedging against nodal prices. The concept was proposed and pioneered by Hogan (1992), and is discussed in further detail in section 9.2.

The network used in example 5.11 is drawn from Chao et al. (2000). The comparison between nodal pricing and zonal pricing in example 5.11 is based on Ehrenmann and Smeers (2005). The evolution of nodal and zonal pricing is detailed in Ehrenmann and Smeers (2005), Dijk and Willems (2011), and Lété (2022). The (*FBP*) model was originally proposed by Bjorndal and Jornsten (2001).

INC-DEC gaming was a concern early on in the California market (Alaywan, Wu, and Papalexopoulos 2004); the example therein inspired the example of section 5.3.4. A postmortem summary of the California crisis and the gaming strategies of ENRON is presented in Oren (2005a). The discussion about INC-DEC gaming has re-emerged recently in Europe, and Hirth and Schlecht (2018) make the point that market power is not a prerequisite for INC-DEC gaming to take place. Hirth and Schlecht (2018) further point out that INC-DEC gaming due to zonal pricing can exacerbate congestion problems. Germany is a particularly interesting case in point, where redispatch costs have risen to the order of billions of euros annually in recent years (Kunz 2013). Lété (2022) provides quantitative models of market-based redispatch and INC-DEC gaming that corroborate the analysis of Hirth and Schlecht (2018). INC-DEC gaming was observed during the rocky launch of the Greek balancing market, where it is interesting to note that INC and DEC bids that were intended for gaming the market were also used for setting the balancing price of the *entire* market, instead of being settled out of the market (Papavasiliou 2021a).

The short-term operational inefficiencies of zonal pricing due to misguided day-ahead commitment of thermal resources are analyzed in Aravena and Papavasiliou (2017). Although line switching can mitigate such inefficiencies (Han and Papavasiliou 2015), significant inefficiencies relative to nodal pricing remain (Lété & Papavasiliou 2020a). The long-term inefficiencies of zonal pricing in terms of misguided investment signals are analyzed by Lété, Smeers, and Papavasiliou (2022).

# 6      Ancillary services

The reliable production and transmission of power requires a set of supporting services that are traded in electricity markets, in addition to transmission and energy. These services are referred to collectively as ancillary services. This textbook focuses on reserves, which constitute a subset of ancillary services. Reserves are defined as the excess capacity that is withheld by generators or loads in order to ensure that the system can withstand unanticipated disturbances without interrupting service to customers.

Section 6.1 provides an overview of ancillary services and a classification of reserves. The extension of the economic dispatch model to a model that includes reserves is presented in section 6.2. Section 6.3 addresses the design of markets for reserves. Section 6.4 focuses on operating reserve demand curves (ORDCs). Section 6.5 discusses balancing, which is the task of activating reserve capacity in order to balance supply with demand in real time, when uncertainty has been fully revealed. The balancing model is essentially equivalent to the real-time economic dispatch model presented in chapter 4, but carries its own terminology, which is discussed in section 6.5.

## 6.1    Classification of ancillary services and reserves

**Ancillary services** are the services needed to support the transmission of electric power from producers to consumers given the obligations of system operators to maintain reliable operations. The specification of ancillary services is not uniform across electricity markets. The United States Federal Energy Regulatory Commission (FERC) classifies ancillary services as follows: (i) scheduling and dispatch, (ii) reactive power and voltage control, (iii) loss compensation, (iv) load following, (v) system protection, and (vi) energy imbalance.

Scheduling and dispatch refers to functions performed by the system operator. Scheduling refers to the determination of generator production schedules and the allocation of transmission capacity in advance of system operations, with the goal of maximizing the economic benefit of the available infrastructure. Dispatch is the function of real-time generator control. Reactive power and voltage control is the provision of reactive power (see appendix B for a discussion of reactive power) from generators to the system operator such that voltage levels remain at acceptable limits. Loss compensation is the provision of real power by generators in order to ensure that the thermal losses due to the transmission of power are replenished. Load following is the task of adjusting the production level of generators as real time approaches in order to ensure that short-term (five-minute) variations of actual load are satisfied. This is accomplished through the operation of a real-time market, as well as other markets near real time. Energy imbalance refers to the actions required in order to maintain the instantaneous balance of supply and demand in the system.

Having described the classification of ancillary services according to FERC, we proceed with a focus on reserves. Reserves can be defined as follows.

**Definition 6.1**    **Reserve** is excess capacity that is set aside in order to ensure that the system can withstand unpredictable disturbances while maintaining system frequency without involuntarily interrupting service to consumers.

The first clarification regarding the definition of reserve is the classification of short-term operational uncertainty that the system operator faces between continuous and discrete disturbances. Continuous disturbances result from renewable energy production and load forecast errors. Discrete disturbances, or **contingencies**, refer to outages of system elements, namely transformers, transmission lines, generators, and large loads. Different types of reserves are used for responding to different sources of uncertainty.

The second clarification regarding the definition of reserves is the reference to involuntary interruption of loads. Reserves can be offered by generators as well as loads. Price elastic consumers can interrupt their consumption because the real-time price exceeds their valuation: this is voluntary load curtailment. Instead, consumers whose power supply is interrupted despite the fact that their average valuation exceeds the market price experience *involuntary* load shedding. This is exactly the type of incident that reserves are designed to prevent.

As a subset of ancillary services, reserves fit under items (iv)–(vi) of the FERC classification: load following, system protection, and energy imbalance. As in the case of ancillary services, there is no uniform classification of reserves. This text defines the reserves most commonly encountered in practice in US and European markets. The most common reserve products that are traded in US electricity markets include regulation, spinning reserve, non-spinning reserve, and replacement reserve. In Europe, a similar classification exists between frequency containment reserve, automatic and manual frequency restoration reserve, and replacement reserve. There is no exact correspondence between the definition of reserves in US and European markets.

An intuitive way to visualize the classification and function of reserves is presented in figure 6.1: (i) In normal conditions (left-hand side of the figure), day-ahead and hour-ahead scheduling typically determine the set-point of generators with an hourly or fifteen minute time resolution. In various markets, units start moving towards their set-points ten minutes before the end of the hour, and are expected to reach their set-points ten minutes within the hour. Real-time dispatch follows after hour-ahead

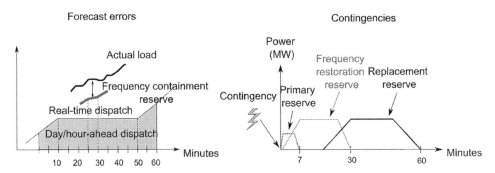

**Figure 6.1** The role of frequency containment, frequency restoration, and replacement reserves under normal (left-hand side) and emergency (right-hand side) conditions.

scheduling and further corrects the output of generators with a resolution of a few (e.g. five) minutes. Frequency containment reserve is used for correcting any remaining deviation that is left over after the real-time dispatch, in case there is a deviation of frequency from its target level. (ii) In the case of a component failure (right-hand side), frequency containment reserve reacts immediately to frequency deviations and arrests the drop in frequency. Frequency restoration reserve is activated within a few seconds or minutes (e.g. seven minutes) in order to restore system frequency to its target value, and replacement reserve is activated after frequency restoration reserve in order to free up frequency restoration reserve, until normal operations are restored.

**Primary reserve**, also known as **primary control** or **frequency containment reserve**, is the fastest responding reserve. In steady-state operations, the current and voltage of the system fluctuate as sinusoids. The system frequency is the frequency of fluctuation of these sinusoidal signals. In case of an instantaneous oversupply in the system, the rotors of large generators accelerate since the excess energy is stored as kinetic energy in these rotors. This causes an increase in the system frequency. In contrast, in the case of an undersupply the generator rotors decelerate and this causes a drop in system frequency. System frequency is therefore an indicator of supply–demand balance in the system and deviations from its target value (50 Hertz in Europe, 60 Hertz in the US) are immediately corrected. The first means of reaction of the system is through the change in the rotational inertia of rotors. The second line of defense is through certain generators that are capable of reacting through frequency-responsive governors. Governors are automatic controllers that react to frequency deviations. The reaction is immediate, though it may take a few seconds for generators to settle to their target set-point. The third line of defense is **automatic generation control** (AGC). This control signal is generated so as to minimize the area control error (ACE), which is a weighted sum of frequency deviations and deviations of inter-area power exchanges. The AGC signal may be updated once every few seconds. AGC is also sometimes referred to as **load frequency control** or **regulation**.

**Secondary reserve**, also known as **frequency restoration reserve, secondary control**, or **operating reserve**, is the reserve that responds immediately after primary reserve. It can be classified between automatic and manual. This service is expected to start a few seconds after requested and should provide the full response within 5 to 10 minutes. This type of reserve is sometimes used in case of component failures. When this occurs, the system operator strives to restore ACE error within 5 to 10 minutes. Operating reserve is sometimes classified between spinning and non-spinning reserve. **Spinning reserve** is offered by generators that are online. **Non-spinning reserve** is offered by generators that are offline but can start rapidly. For interconnected systems this can include reserves that can be imported from neighboring systems. Operating reserve requirements are often dictated by the capacity of the greatest generator in the system and the peak load of the system.

**Tertiary control, tertiary reserve**, or **replacement reserve**, is the reserve that responds after frequency restoration reserve. This reserve is expected to be available within a few minutes. The term operating reserve sometimes includes replacement reserve.

Other terms cross between the above categories. **Frequency responsive reserve** is reserve that responds to frequency. **Contingency reserve** is also a term that is used in order to describe reserve that is used in case of contingencies. An ancillary service product that has been introduced relatively recently in certain markets (e.g. in CAISO and MISO) due to the increasing requirements for ramping capability induced by renewable supply is **flexible ramping**, also referred to as **flexiramp**. Flexible ramp requirements interact with real-time dispatch and pricing due to the fact that they impose the reservation of a certain amount of headroom from a subset of technologies that qualify for responding rapidly to sustained ramp episodes resulting from the uncontrollable fluctuation of renewable resources.

Note that reserves are commonly one-way substitutable. This means that resources that are technically capable of offering frequency containment reserve can also offer frequency restoration reserve and replacement reserve, and resources that are technically capable of offering frequency restoration can also offer replacement reserve. The implications of one-way substitutability for the design of ancillary services markets are discussed later in this chapter.

---

**Example 6.1** *Interaction of restoration and replacement reserve.* Consider a power plant with the following characteristics:

- maximum contribution to upward and downward frequency restoration reserve of 20 MW,
- maximum contribution to replacement reserve of 10 MW,
- minimum capacity of 100 MW and maximum capacity of 170 MW, and
- planned production for hours 1, 2, 3, and 4 of 110, 120, 150, and 150 MW respectively.

The example is depicted in figure 6.2. The contribution of the plant to downward frequency restoration reserve is 10 MW for hour 1 (since more would violate the

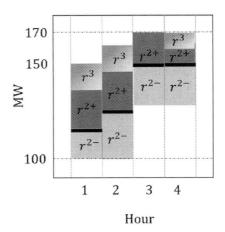

**Figure 6.2** Example of a generator offering energy and multiple types of reserve. Capacities are indicated by dashed lines, production by dark black lines, downward frequency restoration reserves contribution in the light gray area, upward frequency restoration reserves in the dark gray area, and replacement reserves in the light gray area with gradient.

minimum capacity), 20 MW for hour 2 (since more would violate both minimum capacity and maximum contribution to reserves), and 20 MW for hour 3 (since more would violate the maximum contribution to reserves). The *potential* contribution to upward frequency restoration reserve is 20 MW for all hours because there is always at least 20 MW available between maximum capacity and planned production. However, in hours 3 and 4 there are multiple options: the generator must decide if it will allocate its upward capacity to frequency restoration or replacement reserve. For example, in hour 3 the generator may contribute 20 MW to upward frequency restoration reserve (hence 0 MW to replacement reserve) and in hour 4 it may contribute 10 MW to both. This allocation is depicted in figure 6.2.

**Example 6.2** *Interaction of spinning and non-spinning reserve.* Consider a generator that can contribute up to 150 MW among spinning and non-spinning reserve, and has a minimum and maximum capacity of 100 MW and 170 MW, respectively. The planned production of the generator is 110 MW, 0 MW, and 0 MW for hours 1, 2, and 3 respectively (i.e. the plant is shut down in period 2). Since the plant is capable of contributing up to 150 MW to operating reserve, whether it is active or not, the effective contribution to spinning reserve is 60 MW for hour 1 (more would violate the maximum capacity of the plant) and 150 MW for hours 2 and 3. This example is depicted in figure 6.3.

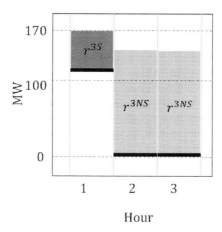

**Figure 6.3** Distinction between spinning (dark gray area) and non-spinning (light gray area) operating reserve.

## 6.2    Co-optimization of energy and reserves

The commitment of generation capacity for reserve increases the cost of operating the system, since capacity that is used for reserve cannot be used for producing power. This section presents an extension of the economic dispatch model in order to account for the requirement of the system to carry reserve capacity.

The simplest approach for committing reserve is to set reserve requirements exogenously. Reserve is typically offered by generators, although loads may also participate in the provision of reserve in certain markets. An auxiliary decision variable $r_g$ is introduced, which corresponds to the commitment of reserve capacity from each generator $g$. A reserve target $R$ is enforced through the following inequality constraint:

$$\sum_{g \in G} r_g \geq R,$$

where $G$ is the set of generators in the system.

In order to model the fact that generator capacity that is allocated towards reserve cannot be allocated for producing power, the following constraint is introduced:

$$p_g + r_g \leq P_g,$$

where $P_g$ is the maximum capacity of a generator $g$, and $p_g$ is the amount of power produced by the generator.

The amount of reserve that can be offered by a certain generator depends on how fast this reserve needs to be made available, as shown in figure 6.4. Frequency containment reserve is the highest quality reserve, followed by frequency restoration reserve and finally replacement reserve. A limit $R_g$, which depends on the response

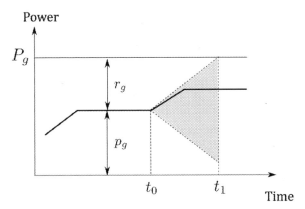

**Figure 6.4** The amount of available reserve depends not only on the technical constraints of a generator, but also on the type of reserve being offered. In this figure, the gray surface depicts the ramp rate constraints of a generator. Any reserve that needs to be made available at $t_0$ within $t_1 - t_0$ is limited by ramp rate constraints, whereas any slower type of reserve is limited by generator capacity.

time of the required reserve as well as the technical constraints of a generator (e.g. the ramp rate, or bid limits such as those that are applied in certain North European markets), is imposed on the amount of reserve that can be offered by a generator:

$$r_g \leq R_g.$$

The following model extends the economic dispatch model in order to account for reserves. Transmission constraints are ignored, and a single type of reserve is considered in order to simplify the exposition and the examples:

$$(EDR): \max_{p,d,r} \int_0^d MB(x)dx - \sum_{g \in G} \int_0^{p_g} MC_g(x)dx$$

$$(\lambda): \quad d - \sum_{g \in G} p_g = 0$$

$$(\mu): \quad R \leq \sum_{g \in G} r_g$$

$$r_g \leq R_g, g \in G$$

$$p_g + r_g \leq P_g, g \in G$$

$$p_g \geq 0, r_g \geq 0, g \in G$$

$$d \geq 0.$$

**Example 6.3** *Provision of reserve by the most expensive units.* Consider the system of example 3.9, with slightly modified characteristics, as shown in table 6.1. The demand

**Table 6.1  List of generators for example 6.3.**

| Generator | Marg. cost ($/MWh) | Max (MW) | Ramp (MW/min) |
|-----------|-------------------|----------|---------------|
| Cheap | 0 | 100 | $+\infty$ |
| Moderate | 10 | 100 | 1 |
| Expensive | 80 | 100 | 5 |

is 100 MW and is inelastic. Suppose that the system requires frequency restoration reserve equal to 100 MW, which needs to be available in 10 minutes. The optimal solution utilizes the most expensive generators to the greatest extent possible for providing reserve. The expensive generator provides reserve up to its ramp rate, 50 MW. The moderately expensive generator also provides reserve up to its capacity, 10 MW. The remaining reserve, 40 MW, is provided by the cheap generator. Once this capacity is allocated for replacement reserve, the remaining capacity is dispatched in merit order for providing power. The moderately expensive generator produces 40 MW, the expensive generator produces nothing, and the cheap generator produces 60 MW.

Note that there is no explicit economic cost for offering reserve in the objective function of $(EDR)$. Although actual market models sometimes include such costs for supplying reserve, and thus allow generators to submit information about economic costs that they incur for providing reserve, the opportunity cost of booking reserve capacity for the provision of reserve and not using this capacity in order to produce energy is accounted for endogenously in $(EDR)$ and should not appear as a separate term in the objective function of $(EDR)$.

Additional features can be added to the aforementioned model. In contrast to frequency restoration reserve, which is typically an upwards corrective action, frequency containment and frequency restoration reserve is often differentiated between upwards and downwards reserve.[1] In this case, separate variables need to be introduced for upwards and downwards reserve. In addition, reserve requirements can be imposed simultaneously in the model, with higher quality technologies being able to cover the needs of both higher and lower quality reserve. Denote $R1^+$ and $R1^-$ as the up and down requirement, respectively, for frequency containment reserve, and $R2, R3$ as the frequency containment and replacement reserve requirement, respectively. Denote $r1^+_{g,1}, r1^+_{g,2}, r1^+_{g,3}$ as the amount of high-quality capacity allocated towards satisfying frequency containment reserve (upward), frequency restoration reserve, and replacement reserve requirements, respectively. Denote $r1^-_g \geq 0$ as the amount of downwards frequency containment reserve capacity. Denote $r2_{g,2}, r2_{g,3}$ as the amount of medium quality reserve capacity that is allocated towards satisfying frequency restoration and replacement reserve requirements, respectively. Finally, denote $r3_g$ as the amount of low-quality reserve capacity that is allocated towards

---

[1] In certain systems, certain types of frequency restoration reserve are only used for upwards balancing, which is what will be assumed for the following exposition.

replacement reserve requirements. Recall that the amount of reserve that can be offered by a generator depends both on its technical constraints and the response time that is available for the reserve in question. Denote $R1_g, R2_g, R3_g$ as the amount of frequency containment, frequency restoration, and replacement reserve, respectively, that can be offered by a generator. The following constraints then need to be imposed:

$$\sum_{g \in G} r1_{g,1}^+ \geq R1^+, \sum_{g \in G} r1_g^- \geq R1^-,$$

$$\sum_{g \in G} (r1_{g,2}^+ + r2_{g,2}) \geq R2, \sum_{g \in G} (r1_{g,3}^+ + r2_{g,3} + r3_g) \geq R3,$$

$$p_g + \sum_{i=1}^{3} r1_{g,i}^+ + \sum_{i=2}^{3} r2_{g,i} + r3_g \leq P_g, p_g - r1_g^- \geq 0, g \in G,$$

$$\sum_{i=1}^{3} r1_{g,i}^+ \leq R1_g, r1_g^- \leq R1_g, \sum_{i=2}^{3} r2_{g,i} \leq R2_g, r3_g \leq R3_g, g \in G.$$

An alternative to fixed requirements for coping with contingencies is a **security-constrained economic dispatch (SCED)** model. Such a model dispatches the system by representing endogenously the reaction of the system operator to the occurrence of a contingency. This is achieved by introducing a two-stage model, where $p_g$ denotes first-stage decisions and $p_g(\omega)$ represents second-stage decisions, where $\omega$ corresponds to the realization of a contingency, i.e. the failure of a component in the system. The two stages can be linked by a ramp rate constraint:

$$-R_g \leq p_g(\omega) - p_g \leq R_g, g \in G.$$

The full model can then be described as:

$$(SCED): \min_p \sum_{g \in G} \int_0^{p_g} MC_g(x)dx$$

$$p_g \leq P_g, g \in G$$

$$\sum_{g \in G} p_g = D$$

$$p_g(\omega) \leq P_g \cdot 1_g(\omega), g \in G, \omega \in \Omega$$

$$\sum_{g \in G} p_g(\omega) = D, \omega \in \Omega$$

$$-R_g \leq p_g(\omega) - p_g \leq R_g, g \in G, \omega \in \Omega : 1_g(\omega) = 1$$

$$p_g \geq 0, g \in G$$

$$p_g(\omega) \geq 0, g \in G, \omega \in \Omega.$$

Here, $\Omega$ is the set of contingencies, and $D$ is the demand in the system. The expression $1_g(\omega) = 0$ indicates that generator $g$ is not available in contingency $\omega$. Note that the ramping constraints are relaxed for generators that are out of order in a given contingency. Demand is assumed inelastic, and the goal is to meet demand in the base case as well as in every contingency while minimizing the cost of the base case. A specific instance of this model is the case where each contingency corresponds to the failure of a single generator and the set of contingencies includes the entire set of generators. This is known as the $N - 1$ **criterion** because the system is required to be able to withstand the failure of 1 component and operate with its remaining $N - 1$ components without shedding load. This security criterion is commonly used in operations as well as planning. The $N - 1$ security criterion extends to transmission-constrained models, where the set of contingencies also includes the failure of transmission system elements, i.e. lines and transformers. A more stringent variation of the $N - 1$ security criterion is the $N - k$ security criterion, whereby the system is able to withstand the loss of $k$ components simultaneously without shedding load. Security-constrained models give rise to computationally challenging problems due to the large number of possible component failures.

---

**Example 6.4** *Security constrained economic dispatch.* Consider a security-constrained economic dispatch model for the system of example 6.3. Recall that, in this example, frequency restoration reserve needs to be made available within 10 minutes. Notice that the security-constrained model produces the same solution as example 6.3. The moderately expensive generator produces 40 MW, and the cheap generator produces 60 MW. In the case of an outage of the cheap generator, the moderately expensive generator can increase output to 50 MW and the expensive generator increases its output to 50 MW. Notice that, although (*EDR*) is simpler to solve, the specification of reserve requirements is a challenging task in more complex systems than the one studied in this example. At the same time, the choice of reserve requirements is crucial since it strongly influences costs and the reliability of the solution, as well as market prices.

---

In addition to security constraints and fixed reserve requirements, import constraints can be used in order to further secure the system. Import constraints set limits on the total flow of power over sensitive groups of lines, which are referred to as import groups.

---

**Example 6.5** *Import constraints.* The logic behind import constraints is explained in figure 6.5. In the example of the figure, $B1$ represents a load pocket with a substantial amount of local demand. Lines $K1$ and $K2$ link the load pocket to the remaining

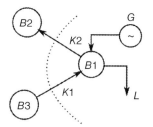

**Figure 6.5** An example of import constraints.

system. The arrows on the lines indicate the reference direction of each line. The lines connected to bus $B1$ constitute an import group, because if generator $G$ within $B1$ fails, it is necessary to import power from outside the load pocket. The set $\{K1, K2\}$ is defined as an import group, and a total power flow limit over the lines of the import group is determined. Suppose that an import limit of 100 MW is enforced in the example of the figure. Then, if $f_k$ denotes the flow of power over lines, the following constraint is added to the transmission-constrained economic dispatch problem:

$$f_{K1} - f_{K2} \leq 100 \text{ MW}.$$

Import constraints are determined based on operating experience instead of a systematic procedure, and are therefore subject to the same weaknesses as fixed reserve requirements. The lines that define an import group, the number of import groups, and the limit of total flow over the import group are a matter of judgment. Poor choices for any of these parameters may result in operating inefficiencies.

## 6.3    Markets for reserve

### 6.3.1    Single type of reserve

#### Simultaneous clearing of energy and reserves

The analysis of markets for reserves is driven by the fact that both reserve and energy use up the same finite resource, which is generator capacity. It can therefore be argued that both products should be priced, and that the auction should be conducted simultaneously, because the price of one product influences the other product.

**Definition 6.2**    A **simultaneous auction for energy and reserves** is conducted as follows:

- Sellers submit *ramp rates* and increasing bids: price–quantity pairs that represent the price at which they are willing to supply a certain quantity. Buyers submit decreasing bids: price–quantity pairs indicating the price that they are willing to pay for a specific quantity.
- The market operator solves ($EDR$) and announces $\lambda$ as the market clearing price for power and $\mu$ as the market clearing price for reserve.

One notable difference between the simultaneous auction and the economic dispatch auction presented in section 4.2 is that in the simultaneous auction generators need to submit, in addition to marginal cost information, certain details about their technical characteristics (ramp rates). In chapter 7, it is shown that additional detailed information about technical constraints is also required in day-ahead power pools, where unit commitment decisions are made.

Recall that the trading of energy in the economic dispatch auction results in a transfer from buyers to sellers, with no funds left to the market operator. The procurement of reserve from generators is typically financed by the system operator. These funds are ultimately charged to consumers and other market agents that benefit from reliability, since reserve is procured by the system operator on behalf of system users in order to ensure that they can enjoy reliable system operation.

---

**Example 6.6** *Energy and reserve prices in a co-optimization induce the optimal dispatch.* Consider again the system of example 6.3. The optimal prices produced from the model are $\lambda^\star = 10$ $/MWh for energy and $\mu^\star = 10$ $/MWh for reserve. It is worth examining the incentives of each generator in the market:

- The moderately expensive generator maximizes profits by offering as much reserve as possible, because each unit of reserve earns a profit of 10 $/MWh, whereas energy production earns zero profit. Note that this is exactly the optimal dispatch for this generator, which is asked to offer 10 MW of reserve. The generator is indifferent towards producing any amount of energy given the market clearing price for energy, including the optimal dispatch quantity of 40 MW.
- The expensive generator is not willing to produce any energy at a price of 10 $/MWh, since this results in losses of 70 $/MWh. In contrast, it is willing to offer as much reserve as possible, since this earns a profit of 10 $/MWh. This is indeed the case in the optimal dispatch.
- The cheap generator is indifferent between producing energy or offering reserve, since both activities earn the generator a profit of 10 $/MWh. Since both activities are profitable, the generator wishes to use up its entire capacity. This is indeed the case in the optimal dispatch, where the generator produces 60 MW of power and 40 MW of reserves.

Note that 1000 $/h are transferred from buyers to sellers for procuring energy. The system operator needs to pay generators an additional 1000 $/h for buying reserve on behalf of market participants.

---

## Competitive market model for energy and reserves

The simultaneous auction for energy and reserves allocates resources optimally, provided agents bid truthfully. Moreover, the outcome of the simultaneous auction coincides with the outcome of bilateral trading in a competitive market. To demonstrate this statement rigorously, it is necessary to model a competitive market. The first step towards a competitive market model is to determine the products and agents in the market. The following agents can be identified in such a model: (i) a system operator that procures reserves in order to ensure reliability, (ii) power producers that sell power and reserves, and (iii) power consumers that procure power.

Once the agents and products of the market are determined, the price and quantity adjustment of the market are described mathematically. Denote $\lambda$ as the price for energy and $\mu$ as the price for reserve. The market clearing condition (price adjustment) for power can be expressed as an equality constraint:

$$d - \sum_{g \in G} p_g = 0.$$

The market clearing condition (price adjustment) for reserve can be expressed as the following complementarity constraint:

$$0 \leq \mu \perp \sum_{g \in G} r_g - R \geq 0.$$

Profit maximization (quantity adjustment) for each producer $g \in G$ can be expressed as follows:

$$\max_{p,r} \lambda \cdot p_g + \mu \cdot r_g - \int_0^{p_g} MC_g(x)dx$$
$$p_g + r_g \leq P_g$$
$$p_g, r_g \geq 0.$$

Surplus maximization (quantity adjustment) can be expressed as follows:

$$\max_d \int_0^D MB(x)dx - \lambda \cdot d$$
$$d \geq 0.$$

Having defined the competitive market model, the following proposition can be established.

**Proposition 6.3** *Provided agents bid truthfully, the simultaneous auction for energy and reserves produces an allocation and prices for power and reserves that correspond to a competitive market equilibrium. The converse is also true.*

*Proof* The result is a special case of proposition 4.11.    □

## Sequential clearing of energy and reserve

An alternative to the simultaneous auction for energy and reserves is the sequential procurement of reserves and energy. One possible and common[2] configuration of sequential clearing is for reserve capacity to be auctioned first and, after reserve commitments have been finalized, a separate auction for energy is conducted. In the case of sequential trading of reserves and energy, it is up to the market agents to realize that there are opportunities for profit in the energy market and adjust their bids accordingly in the reserve market. The following example suggests that the sequential auctioning of energy and reserves produces the same outcome as a simultaneous auction.

If agents can perfectly anticipate the competitive equilibrium price of energy, then the sequential clearing of reserves and then energy can yield the same outcome as the co-optimization of energy and reserves. The analytical argument is not presented here, but relies on KKT analysis. Instead, the idea is illustrated through the following example. Note, however, that perfect anticipation of competitive equilibrium energy

---

[2] The auctioning of reserves before energy is related to the goal of the system operator to ensure security first. Alternative configurations where energy is auctioned before reserve also exist, but are less common.

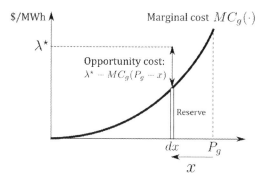

**Figure 6.6** The opportunity cost of allocating slice $dx$ for reserves, instead of using it to sell energy at a price $\lambda^\star$, is $\max(0, \lambda^\star - MC_g(P_g - x))$.

prices is unlikely in practice (due to the multitude of market clearing time periods and reserve products), which is a reason why sequential clearing can lead to inefficiencies in practice.

---

**Example 6.7** *Sequential clearing requires anticipation of prices.* Consider the system of example 6.3. Suppose that agents bid in a uniform auction for reserve first, and then in a uniform auction for energy. Suppose that agents share the same belief about the price of energy in the second stage, i.e. all agents believe that the price of energy will be $\lambda^\star$. This means that agents should bid the slice $dx$ at the maximum of $\lambda^\star - MC_g(P_g - x)$ and zero (see figure 6.6). Since the expensive generator has the highest marginal cost (80 $/MWh), this generator is cleared first for reserve because it submits the lowest bid for reserve. The cleared quantity is 50 MW, the maximum amount of reserve that the expensive generator can offer. The moderately expensive generator has the second-lowest bid, and is cleared for 10 MW of reserve. The cheap generator is cleared for the remaining amount of reserve, 40 MW. Given these market clearing quantities in the reserve market, and provided agents bid truthfully, the expensive generator offers 50 MW at 80 $/MWh in the energy market. The moderately expensive generator offers 90 MW at 10 $/MWh, and the cheap generator offers 60 MW at 0 $/MWh. The market clears with a price of 10 $/MWh when the demand is 100 MW. Returning to the reserve auction, it can be seen that the expensive and moderately expensive generators bid their reserve for a price of 0 $/MWh, while the cheap generator offers reserve at 10 $/MWh. This implies that the market clearing price for reserve is 10 $/MWh. The result is identical to the result of example 6.3.

---

Note that, in contrast to a simultaneous auction for energy and reserves, in a sequential auction producers bid supply functions for reserve, which reflect the **opportunity cost** of keeping capacity out of the energy market in order to offer it as reserve. In the context of this text, opportunity cost is understood as the profit that is foregone when allocating a scarce resource to one use, among a set of mutually exclusive uses, relative to the profit that the resource could have earned by allocating it to the most profitable alternative.

## 6.3.2   Multiple types of reserve

Example 6.7 demonstrates that the sequential clearing of energy and reserves for a single reserve product reproduces an identical outcome as the simultaneous clearing. The situation becomes more complex when multiple reserve products need to be auctioned due to the fact that capacities of higher quality (i.e. faster) can substitute for reserves of lower quality. The design of auctions for substitutable services poses fundamental market design dilemmas. Should the auctions be pay-as-bid or uniform price? If there are multiple reserve products that are substitutable (e.g. frequency containment reserve can substitute frequency restoration reserve, which can substitute replacement reserve), how should the auction be designed in order to clear these products? A number of designs can be envisioned, three of which are presented in this section. The relative advantages and weaknesses of these designs are compared through an example.

The first auction to be considered is a cascading multi-product auction. This is a sequence of uniform price auctions conducted as follows:

**Definition 6.4**   Multi-product reserve auction 1.

- Sellers submit increasing offer curves for each type of reserve.
- In order of higher- to lower-quality reserves:
  - Conduct a uniform price auction for the given reserve, and
  - if there are remaining bids that are out of the money, cascade them to the auction for the immediately inferior reserve.

The motivation for this auction is that, since higher-quality technologies can substitute for lower-quality technologies, out of the money bids should not be lost and should be cascaded to auctions for reserves of inferior quality. A major drawback of this auction is that it may lead to price reversal if agents bid truthfully, whereby lower-quality reserve products pay higher prices in the market. Price reversal sends the wrong signal to suppliers and creates incentives for misrepresenting the quality of the reserve that the suppliers can offer or the price at which they can offer it, even in the absence of market power.

---

**Example 6.8** *Price reversals in sequential reserve auctions.* Consider a cascading multi-product auction for frequency containment and frequency restoration reserves, as shown in figure 6.7. The demand for frequency containment reserve is 400 MW, and the demand for frequency restoration reserve is 350 MW. The frequency containment reserve market has one offer for 600 MW at 10 $/MWh and one offer for 50 MW at 15 $/MWh. The frequency restoration reserve market has one offer for 25 MW at 5 $/MWh and one offer for 400 MW at 20 $/MWh. The frequency containment reserve auction clears at 10 $/MWh, as indicated in the lower left panel.

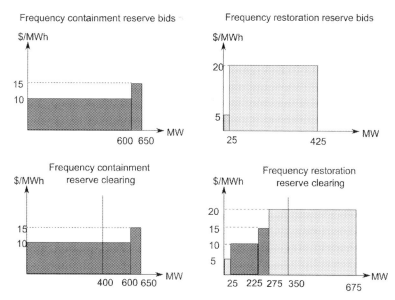

**Figure 6.7** The cascading multi-product auction of example 6.8. The upper graphs represent the bids for frequency containment (dark gray) and frequency restoration (light gray) reserve. The lower graphs represent the clearing of frequency containment and frequency restoration reserve.

The frequency containment reserve bids that are out of the money are cascaded to the frequency restoration reserve auction, as indicated in the lower right panel. The frequency restoration reserve auction clears at 20 $/MWh. Note that the price for frequency restoration reserve is double the price of frequency containment reserve. The quantities cleared by these cascading auctions are obviously not maximizing profit for producers. This indicates that the auction rules are expected to cause incentive problems, since producers will be tempted to submit bids that do not truthfully represent their cost to the market operator. The cost of reserves, as reflected by the submitted bids, is 8375 $/h. The total payment of the auction is 11000 $/h. The auction results in an efficient outcome provided agents bid truthfully. However, there is price reversal, which implies that eventually agents are expected to deviate from truthful bidding.

The price reversal observed in the cascading multi-products auction can be mitigated by having cascaded bids set the price, regardless of the auction in which they are cleared. This motivates the following alternative to a cascading multi-product auction.

**Definition 6.5**   Multi-product reserve auction 2.

• Sellers submit increasing offer curves for each type of reserve.
• In order of higher- to lower-quality reserves:
  – Clear bids in order of increasing cost, and

- if there are remaining bids that are out of the money, cascade them to the auction for the immediately inferior reserve.
- Set the market clearing price for each reserve product to the price of the highest accepted bid, *regardless* of the auction in which that bid was cleared.

---

**Example 6.9** *Price reversals in sequential reserve auctions (again).* Consider a cascading multi-product auction as described above for the bids of the previous example, which are shown in figure 6.7. The price paid for frequency containment reserve is 15 $/MWh because this is the price of the most expensive frequency containment reserve bid that is in the money, even though this bid is cleared for providing frequency restoration reserve. Frequency restoration reserve is priced at 20 $/MWh because the most expensive secondary reserve bid that is in the money costs 20 $/MWh, and it is cleared for offering frequency restoration reserve. The cost of the accepted bids is the same as in the previous example; therefore the allocation is efficient. The payment of the auction is higher, and amounts to 13000 $/h. Despite the fact that the allocation is efficient, price reversal occurs again. However, the extent of price reversal diminishes relative to the previous example.

---

An alternative to the cascading auctions is a multi-product auction that optimizes reserves simultaneously, by solving the following model:

$$(Res): \min_{r1,r2} \sum_{g \in G} \int_0^{r1_{g,1}+r1_{g,2}+r2_g} OC_g(x)dx$$

$$(\mu1): \sum_{g \in G} r1_{g,1} \geq R1$$

$$(\mu2): \sum_{g \in G} (r1_{g,2} + r2_g) \geq R2$$

$$(\rho1_g): r1_{g,1} + r1_{g,2} \leq R1_g, g \in G$$

$$(\rho2_g): r2_g \leq R2_g, g \in G$$

$$r1_{g,1} \geq 0, r1_{g,2} \geq 0, r2_g \geq 0, g \in G,$$

where $OC_g$ represents the marginal opportunity cost that a generator incurs by supplying reserve and keeping its capacity out of the energy market. An application of proposition 4.11 implies that using $\mu1$ and $\mu2$ as the market clearing prices for frequency containment and frequency restoration reserve, respectively, is compatible with the profit maximization incentives of producers. This motivates the following uniform price auction for reserves.

**Definition 6.6** Multi-product reserve auction 3.

- Suppliers submit increasing bids for reserves: price–quantity pairs that indicate the amount of reserves that they are willing to provide for a given price.

- The market operator solves (*Res*) and announces $\mu 1$ as the uniform price for frequency containment reserve and $\mu 2$ as the price for frequency restoration reserve.

The advantage of this auction design is that it prevents price reversal, as demonstrated formally in the following proposition.

**Proposition 6.7**   *In the uniform price auction of definition 6.6, the price for higher quality reserve is higher:* $\mu 1 \geq \mu 2$.

*Proof*   Consider the KKT conditions:

$$0 \leq r1_{g,1} \perp MC_g(r1_{g,1} + r1_{g,2} + r2_g) - \mu 1 + \rho 1_g \geq 0, g \in G,$$
$$0 \leq r1_{g,2} \perp MC_g(r1_{g,1} + r1_{g,2} + r2_g) - \mu 2 + \rho 1_g \geq 0, g \in G.$$

Provided $R1 > 0$, it must be the case that $r1_{g,1} > 0$ for some $g$. The first complementarity condition then implies that

$$\mu_1 = MC_g(r1_{g,1} + r1_{g,2} + r2_g) + \rho 1_g.$$

The conclusion then follows from the fact that $\mu_2 \leq MC_g(r1_{g,1} + r1_{g,2} + r2_g) + \rho 1_g$.
$\square$

---

**Example 6.10**   *Correction of price reversal problem in co-optimization.* Consider the system of example 6.8. The solution of (*Res*) results in equal frequency containment and frequency restoration reserve prices, $\mu 1^\star = \mu 2^\star = 20$ \$/MWh. The allocation is efficient and the prices are consistent with the profit maximization goal of producers. However, the payment to suppliers is 15000 \$/h, which is the most expensive solution relative to the two alternative auctions that were considered previously.

---

A criticism against marginal cost pricing as described above in the (*Res*) model is that, in order to achieve an efficient outcome while being consistent with the incentives of producers, the auction surrenders to suppliers the benefits that they could have gained by not bidding truthfully.

## 6.4    Operating reserve demand curves

Energy and reserve co-optimization models with fixed reserve requirements have been a standard approach for committing reserves in various worldwide markets, with US markets spearheading the effort. One practical drawback of fixed reserve requirements is that they may result in highly volatile energy prices when the system is strained.

---

**Example 6.11** *Price volatility in markets with fixed reserve requirements.* Consider a system with $D$ MW of inelastic demand, with a value of load that is equal to 1000 $/MWh, and 100 MW of elastic demand. The aggregate marginal benefit function $MB_L$ is given by the following expression:

$$MB_L(x) = \begin{cases} 1000 \text{ \$/MWh}, & 0 \text{ MW} \le x \le D \text{ MW}, \\ 1000 - 10 \cdot (x - D) \text{ \$/MWh}, & D \text{ MW} < x \le D + 100 \text{ MW}. \end{cases}$$

The system has an installed capacity of 10000 MW. The aggregate marginal cost function is

$$MC_G(x) = 0.015 \cdot x \text{ \$/MWh}.$$

Suppose that the system has a fixed reserve requirement of $R = 1000$ MW. This is equivalent to adding 1000 MW, with an arbitrarily high valuation, to the demand function. To see this, note from $(EDR)$ that, for large $D$, the system will curtail demand in order to ensure that enough generator capacity is available to satisfy the reserve requirement if necessary. This is equivalent to setting the reserve requirement at a valuation greater than the maximum valuation of the demand function. The energy price is given by (see problem 6.6):

$$\lambda^\star = \begin{cases} 0.015 \cdot (99.85 + 0.9985 \cdot D) \text{ \$/MWh}, & 0 \text{ MW} \le D \le 8913.5 \text{ MW}, \\ 1000 - 10 \cdot (9000 - D) \text{ \$/MWh}, & 8913.5 \text{ MW} < D \le 9000 \text{ MW}, \\ 1000 \text{ \$/MWh}, & D > 9000 \text{ MW}. \end{cases}$$

The evolution of the energy price as a function of inelastic demand in the system is presented in figure 6.8. Note that the equilibrium price jumps from 135 $/MWh to 1000 $/MWh when demand increases from to 8913.5 MW to 9000 MW. This is due to the limited elasticity of demand in the system.

---

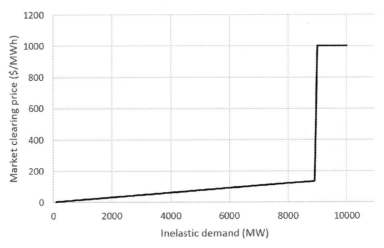

**Figure 6.8** The market clearing price as a function of inelastic demand in example 6.11.

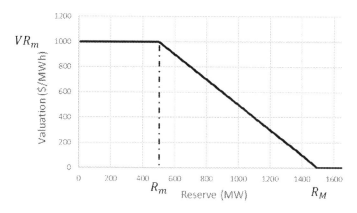

**Figure 6.9** Example of an ORDC. The system operator values increments of reserve up to $R_m$ at a high value $VR_m$, whereas increments above $R_M$ are valued at zero.

The previous example highlights that energy prices can be highly volatile in systems with limited demand response. In conditions of scarcity, a slight increase of demand can move the price from the marginal cost of the most expensive unit to the value of lost load.

High prices indicate scarcity and the need for investment in capacity. However, in systems with limited energy elasticity, price volatility can increase the risk of investment. **Operating reserve demand curves (ORDC)** have been proposed as an approach for achieving high energy prices in conditions of scarcity through price spikes that are more frequent but less elevated. This is achieved by expressing reserve requirements through a demand function, rather than a hard constraint that needs to be satisfied.

Consider, as an example, the ORDC in figure 6.9. The straight part of the ORDC to the left of $R_m$ corresponds to levels of leftover reserve in the system that are so low that reliable operation is threatened. At these levels of reserves, the system operator is willing to pay a high price in order to procure an additional increment of reserve capacity, because increments of reserve capacity go a long way in safeguarding system reliability. If this valuation is higher than the value of involuntary load curtailment, then in the context of an energy and reserves co-optimization this means that the system operator is willing to resort to involuntary load shedding in order to avoid depleting reserves to a level below $R_m$. The part of the curve to the right of $R_M$ corresponds to levels of available reserve for which the system operator is not willing to pay anything for additional reserve capacity, because at these levels of reserve the system is so reliable that additional increments of reserve would not make a noticeable difference in terms of reliability. The linear interpolation between $R_m$ and $R_M$ aims at providing an estimate for the valuation of the system operator for increments of reserve in between the tight conditions at $R_m$ and the comfortable conditions at $R_M$. This gives rise to an energy and reserves co-optimization model that can be described as follows:

$$(ORDC): \quad \max_{p,d,r,dr} \int_0^d MB(x)dx + \int_0^{dr} MR(x)dx - \sum_{g \in G} \int_0^{p_g} MC_g(x)dx$$

$$(\lambda): \quad d - \sum_{g \in G} p_g = 0$$

$$(\mu): \quad dr - \sum_{g \in G} r_g = 0$$

$$r_g \leq R_g, g \in G$$

$$p_g + r_g \leq P_g, g \in G$$

$$p, d, r, dr \geq 0.$$

Here, $MR(\cdot)$ is the marginal benefit for reserve, and is a decreasing function of the acquired reserve. This demand function is determined by the system operator.

**Definition 6.8**  A uniform pricing auction for reserves based on an ORDC is conducted as follows:

- Suppliers submit increasing bids for power; consumers submit decreasing bids for power. The system operator submits decreasing bids for reserve.
- The market operator solves $(ORDC)$ and announces $\lambda$ as the uniform price for power and $\mu$ as the uniform price for reserve.

The important effect of ORDCs is that they uplift energy prices under conditions of scarcity. This is driven by a *no-arbitrage condition* in the energy and reserve markets. Concretely, as the reader is asked to demonstrate in problem 6.1, since there is a marginal unit in the system that splits its available capacity between energy and reserve, this implies that the profit margin in the energy market should equal the profit margin in the reserve market for this unit. This means that energy prices and reserve prices track each other and are merely separated by the marginal cost of the marginal unit. Thus, even if the demand side of the market is inelastic, as long as there is price elasticity in the reserve market through an ORDC, the energy price behaves more smoothly as a function of system demand. This effect is illustrated in the following example.

---

**Example 6.12** *Reducing energy price volatility through an ORDC.* Consider the system of example 6.11, where the reserve requirement $R = 1000$ MW is replaced by an ORDC with $R_m = 500$ MW, $R_M = 1500$ MW, and $VR_m = 1000$ \$/MWh, i.e. the valuation of the ORDC to the left of $R_m$ is equal to value of lost load (VOLL). The energy price for the $(EDR)$ and $(ORDC)$ models is shown in figure 6.10. Note that, under tight system conditions, the energy price rises smoothly to the valuation of inelastic consumers.

---

The shape of the ORDC is an important determinant for the performance of the mechanism. Fixed reserve requirements are a special case of ORDC, whereby the system operator places an unlimited valuation for increments of reserve up to

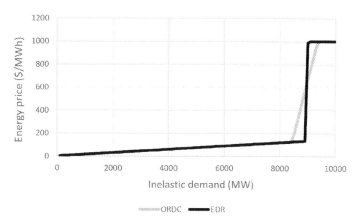

**Figure 6.10** Energy price as a function of demand for the ($EDR$) and ($ORDC$) models in example 6.12. An ORDC results in a smoother price profile as inelastic load varies in the market.

the level of the fixed reserve requirement, and zero valuation for increments beyond this level. Stepped ORDCs, which have been used in recent years by various system operators (such as Ireland, ISO-NE, MISO, CAISO, and SPP), are an evolution of fixed reserve requirements where the willingness of the system operator to pay for increments of reserve diminishes in steps, as the system becomes increasingly comfortable. There is an elegant theory that connects the shape of ORDCs to loss of load probability (LOLP) and the value of lost load (VOLL). Such ORDCs that are based on VOLL and LOLP are currently either employed or proposed in PJM, ERCOT, Belgium, Great Britain, Greece, and Poland.

---

**Example 6.13** *An ORDC based on LOLP and VOLL.* Consider a system with imbalances that obey a normal distribution, and an ORDC that is formulated according to the following formula:

$$MR(x) = (VOLL - \widehat{MC}) \cdot LOLP(x). \tag{6.1}$$

The formula is intended to capture the incremental value of an additional increment of reserve capacity in terms of keeping the probability of involuntary load shedding in check. Here, $VOLL$ corresponds to an estimate of the value of lost load in the system, $\widehat{MC}$ is an approximation of the incremental cost of producing additional energy in the system, and $LOLP(x)$ is the loss of load probability given that the system is carrying $x$ MW of reserve:

$$LOLP(x) = \mathbb{P}[Imb > x] = 1 - \mathbb{P}[Imb \le x].$$

For a system with a VOLL of 1000 $/MWh, an incremental cost of serving energy $\widehat{MC} = 50$ $/MWh, and a normal imbalance distribution with mean 0 MW and standard deviation 300 MW, the ORDC can be computed as follows:

$$MR(x) = 950 \cdot (1 - \Phi_{0,300}(x)),$$

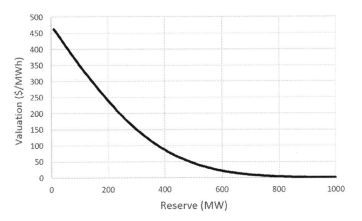

Figure 6.11 The ORDC of example 6.13.

where $\Phi_{\mu,\sigma}(\cdot)$ is the cumulative distribution function of a normal distribution with mean $\mu$ and standard deviation $\sigma$. The resulting ORDC is presented in figure 6.11.

---

The energy prices produced by an ORDC, such as those presented in figure 6.10, contribute towards *scarcity pricing*, i.e. uplifting energy prices above the marginal cost of the most expensive producer in the market. Scarcity pricing contributes towards remunerating the investment cost of all plants in the market, including peaking plants. Scarcity pricing based on ORDC is essentially a real-time mechanism. Although ORDCs can be introduced in the day-ahead market, and should be consistent with real-time ORDCs, the mechanism is essentially a real-time one. Concretely, the units that are rewarded most by scarcity pricing through ORDCs are those that can respond to real-time scarcity by increasing their output when needed. This is because those units that can increase their production at moments when the system needs them most also collect a real-time price that is augmented. This creates a so-called **pay-for-performance** incentive, meaning that the mechanism, by construction, rewards those units that not only promise to provide ancillary services to the system in real time, but actually deliver on their promise. Another way in which flexible technologies benefit from the mechanism is through the real-time price of reserve capacity, namely the multiplier $\mu$ in model $(ORDC)$. Concretely, technologies that are eligible for offering reserve capacity to the system are paid a price $\mu$ for each MW of reserve headroom that they have available in real time. This real-time market for reserve capacity generates, through back-propagation of real-time prices to forward reserve markets, a forward price of reserve that is intended to attract investment in flexible capacity. The process of back-propagation from real-time to forward markets is discussed through quantitative models in chapter 9, but the intuition is straightforward: in a market equilibrium with risk-neutral agents, forward prices of products or services cluster around their expected real-time values.

**Example 6.14** *Remunerating reserve capacity in real time.* Consider a unit that operates in a real-time market with an ORDC. Suppose that the $(ORDC)$ model produces an energy price $\lambda^\star = 60$ \$/MWh and a reserve price of $\mu^\star = 10$ \$/MWh. Suppose, furthermore, that the unit produces 10 MW of power in real time and has a technical maximum of 100 MW. Assuming that the unit has not sold any energy or reserve in the day-ahead or other forward markets, the total remuneration of the unit is:

- Energy payments: $60\frac{\$}{\text{MWh}} \times 10$ MWh $= \$600$
- Reserve payments: $10\frac{\$}{\text{MWh}} \times 90$ MWh $= \$900$

Thus, the 90 MW of the unit that are not used for producing energy are automatically assumed to be available as reserve, and are remunerated accordingly, as long as a real-time market for reserve is in place.

The fact that scarcity pricing based on ORDCs is essentially a real-time mechanism suggests that the implementation of the mechanism *requires* a real-time co-optimization of energy and reserves. This is not true: it is possible to introduce the mechanism in markets that do not co-optimize energy and reserve in real time. For instance, the majority of European markets do not co-optimize energy and reserve in real time. The way in which this can be done is inspired by the goal of avoiding arbitrage between energy and reserve markets (see problem 6.1). The idea is to compute an approximation of the dual multiplier $\mu$ of the $(ORDC)$ model in real time using telemetry information. The telemetry gives a real-time measure of the available reserve headroom $R$. Empirical estimates of LOLP and VOLL, and the telemetry data, can be used for estimating $\mu$ through formula (6.1), since the price of reserve is the evaluation of the ORDC at the measured value of reserve in real time. This approximation, which is referred to as an **ORDC adder** or **scarcity adder**, should be used

1. for settling real-time reserve and
2. as an uplift to the price that is used for settling real-time energy.

This ex-post approach for computing ORDC adders has been used in Texas and proposed in Belgium.

**Example 6.15** *Settlement with ORDC adders in a market without real-time co-optimization of energy and reserves.* Consider a system with the imbalance characteristics and VOLL of example 6.13, but where there is no real-time co-optimization of energy and reserves. Suppose that the real-time price of energy in this market is $\widehat{MC} = 50$ \$/MWh, that the available reserve in the system in real time is 600 MW, and that the unit is producing 10 MW and has a capacity of 100 MW. We can then use formula (6.1) in order to compute a scarcity adder, $\tilde{\mu}$:

$$\tilde{\mu} = (VOLL - \widehat{MC}) \cdot LOLP(R) = (1000 - 50) \cdot (1 - \Phi_{0,300}(600)) = 21.61 \$/\text{MWh}.$$

The settlement then proceeds as in example 6.14, where the energy price $\widehat{MC}$ of a market without co-optimization is uplifted by the adder $\tilde{\mu}$:

- Energy payments: $(50 + 21.61)\frac{\$}{\text{MWh}} \times 10\text{MWh} = \$716.10$
- Reserve payments: $21.61\frac{\$}{\text{MWh}} \times 90\text{MWh} = \$1944.90$.

Compare this to an energy-only market remuneration, $50\frac{\$}{\text{MWh}} \times 10\text{MWh} = \$500$. Essentially, the entire capacity of the unit, which is performing by being available in real time under tight system conditions, receives an adder of $\tilde{\mu}$. The difference in payments between the two designs is thus $(1944.9 + 716.1) - 500 = 21.61 \times 100 = \$2161$.

---

Example 6.15 describes how an ex-post scarcity adder can be computed based on ORDC, even in the absence of co-optimization in real time. One problematic interpretation of this implicit co-optimization is to use this adder only for uplifting energy prices, without also putting in place a real-time market for reserve. Such a setup fails to achieve many of the intended benefits of scarcity pricing, while also introducing potential distortions in competition and arbitrage opportunities. Such a design can emerge in systems that have put in place a real-time market for energy, but have omitted to put in place a real-time market for reserve. It is interesting, if not alarming, to note that *none* of the European markets trade reserve in real time, whereas *all* US markets trade reserve in real time.

Theory has been developed for the extension of ORDCs to multiple reserve products. Insights deriving from this theory have been applied in the Texas market. The complication of multiple reserve products relates to *one-way substitutability*: certain technologies are fast enough to qualify for both fast as well as slow reserves, whereas others are not technically capable to qualify for fast reserves, but only slow reserves. Adders for multiple reserve products should thus ensure that reserves of higher quality (i.e. reserves that need to be covered by fast technologies) should be remunerated at a price that is no less than that of reserves with lower quality (i.e. reserves that can be covered by both fast as well as slow technologies).

## 6.5    Balancing

Balancing is the task of adjusting power production and consumption in real time in order to ensure power consumption exactly equals power supply during an actual operating interval. This task is typically performed by resources that have committed to offer reserve, though other flexible resources can also contribute to balancing the system.

The **balancing market** is the **real-time market** for energy, which is presented in chapter 4, and which in principle should set the price for the settlement of forward contracts, as discussed in section 9. Nevertheless, it should be noted that there exist markets where the day-ahead price is used for settling forward contracts.

The balancing market consists of (i) collecting bids by resources that can adjust their production or consumption in real time, (ii) activating these resources in order to relieve any imbalances, and (iii) charging market participants who deviate from their forward positions. Resources that bid in the balancing market include reserves and other market participants who are prepared to adjust their position even though they have not committed to do so by selling reserve to the market. Reserves are obliged to bid in the balancing market an amount of power (in the upward or downward direction, depending on the type of reserve that has been sold) *at least* equal to the amount of reserve capacity that has been cleared by each of these resources. In certain markets, resources that are obliged to offer balancing energy due to the fact that they have been cleared in reserve markets are also referred to as **balancing service providers (BSPs)**. Instead, resources that offer balancing energy without being obliged to do so are sometimes referred to as **free bids**.

Balancing markets can either be conducted by collecting increment/decrement bids from market participants, or by re-running economic dispatch against the real-time supply and demand functions of the system. In this paragraph the first approach is explained, while the second approach was elaborated on in chapter 4.

Consider a forward market (e.g. a day-ahead or hour-ahead market) that has cleared in advance of real time at a certain price–quantity pair $(\lambda_0, Q_0)$, as indicated in figure 6.12. The logic of the balancing market is to allow resources to move upwards or downwards, relative to their position in the forward market, in exchange for a reward (in the case of upward changes) or payment (in the case of downward changes). The logic of increment bids in the balancing market is the following. (i) Generators that provide upward balancing energy need to produce extra power; therefore they require a payment in order to cover their additional fuel costs. (ii) Loads that provide upward balancing energy need to cut down on consumption; therefore they require a payment in order to cover the lost value from consuming less power. Thus, upward balancing generates a payment from the auctioneer (typically the system operator) to the agents. A similar logic applies to decremental bids. However, the direction of payments is reversed. (i) Generators that provide downward balancing energy are required to reduce their power output; therefore they achieve savings in their fuel costs. (ii) Loads that provide downward balancing energy are required to increase their consumption; therefore they gain additional value from the extra power that they consume. Thus, downward balancing energy generates a payment from agents to the auctioneer (typically the system operator). The logic of bid construction is demonstrated in the following example.

---

**Example 6.16** *Generating balancing market bids from earlier auction outcomes.* Consider the market of figure 6.12, with the offers indicated in table 6.2. The market clears at a price of 70 $/MWh at a quantity of 220 MW. The balancing market consists of upward and downward balancing energy bids, which are listed in table 6.3. For example, the bid Inc1 originates from the bid S4, which corresponds to a flexible resource and for which 30 MW have yet to be cleared. Similarly, bid Inc2 corresponds to bid S6 and bid Inc3 corresponds to bid D1. Bid Dec1 originates from

**Table 6.2 Supply and demand bids for the market of example 6.16.**

| Supply offer | Marg. cost ($/MWh) | Quantity (MW) | Flexible? |
|---|---|---|---|
| S1 | 25 | 40 | No |
| S2 | 40 | 80 | Yes |
| S3 | 60 | 80 | No |
| S4 | 70 | 50 | Yes |
| S5 | 75 | 40 | No |
| S6 | 100 | 50 | Yes |

| Demand offer | Valuation ($/MWh) | Quantity (MW) | Flexible? |
|---|---|---|---|
| D1 | 110 | 100 | Yes |
| D2 | 80 | 120 | No |
| D3 | 55 | 90 | No |
| D4 | 30 | 70 | Yes |

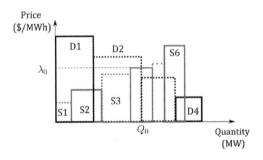

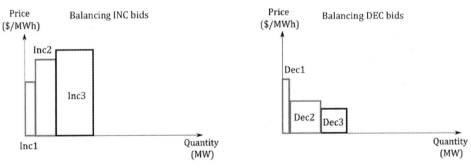

**Figure 6.12** A forward (e.g. day-ahead or hour-ahead) market followed by a balancing market. In the top figure, black bids indicate consumers and gray bids indicate producers. Bids with a dashed border indicate resources that are not flexible, and therefore cannot be offered in the real-time market. Bids with a solid border indicate resources that can move rapidly, and therefore can be offered in the real-time market as incremental or decremental bids. The lower left figure indicates marginal cost bids for upward balancing energy. The lower right figure indicates valuation bids for downward balancing energy.

**Table 6.3 Flexible bids for the balancing market of example 6.16.**

| Inc offer | Marg. cost ($/MWh) | Quantity (MW) |
| --- | --- | --- |
| Inc1 | 70 | 30 |
| Inc2 | 100 | 50 |
| Inc3 | 110 | 100 |

| Dec offer | Valuation ($/MWh) | Quantity (MW) |
| --- | --- | --- |
| Dec1 | 70 | 20 |
| Dec2 | 40 | 80 |
| Dec3 | 30 | 70 |

the 20 MW of bid S4 that have already been cleared, Dec2 corresponds to S2, and Dec3 corresponds to D4.

Having clarified the logic of bid construction, a model of the balancing market is now presented. Denote the set of downward balancing energy offers as $D$ and the set of upward balancing energy offers as $U$. Denote the marginal benefit of downward bid $d$ as $MB_d$ and the offered quantity as $\Delta_d$. Denote the marginal cost of upward bid $u$ as $MC_u$ and the offered quantity as $\Delta_u$. The variable $\delta^+$ (respectively $\delta^-$) indicates the amount of upward (respectively downward) activation that is cleared in the balancing market. The demand for upward or downward activation is indicated by $\Delta$ and corresponds to a random disturbance that occurs to the system in real time. A positive sign of $\Delta$ indicates the need for upward activation, a negative sign indicates the need for downward activation. Then the balancing market solves the following model:

$$\max_{\delta^+,\delta^-} \sum_{d \in D} MB_d \delta_d^- - \sum_{u \in U} MC_u \delta_u^+$$

$$\sum_{u \in U} \delta_u^+ - \sum_{d \in D} \delta_d^- = \Delta$$

$$\delta_u^+ \le \Delta_u, u \in U$$

$$\delta_d^- \le \Delta_d, d \in D$$

$$\delta_u^+, \delta_d^- \ge 0, u \in U, d \in D.$$

**Example 6.17** *Clearing the balancing market.* Consider again the market of example 6.16. Suppose that, after the forward (e.g. day-ahead or hour-ahead) market clears, the generator offering bid S3 fails. This results in a shortage of 80 MW, which needs to be made up through increment bids in the balancing market. The real-time price can be computed by referring to the increment offers in table 6.3. The shortage of 80 MW corresponds to an inelastic demand for 80 MW of upward balancing energy, which results in a market clearing price of 100 $/MWh. In fact, any price between 100 and 110 $/MWh clears the market, since bids Inc1 and Inc2 are fully accepted, while

Table 6.4 Real-time supply and demand bids for the market of example 6.17.

| Supply offer | Marg. cost ($/MWh) | Quantity (MW) |
| --- | --- | --- |
| S1 | $-\infty$ | 40 |
| S2 | 40 | 80 |
| S4 | 70 | 50 |
| S6 | 100 | 50 |

| Demand offer | Valuation ($/MWh) | Quantity (MW) |
| --- | --- | --- |
| D1 | 110 | 100 |
| D2 | $-\infty$ | 120 |
| D4 | 30 | 70 |

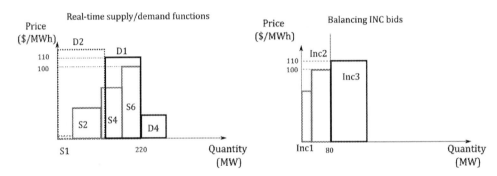

Figure 6.13 The clearing of the real-time market of example 6.17 using (left) the real-time supply and demand functions and (right) the activation of incremental bids.

bid Inc3 is fully rejected. Another way to obtain the same result is by considering the real-time supply and demand functions of the system, as indicated in table 6.4. Note that the bid S3 is absent in the real-time supply function because the unit has failed, and bid S5 is absent because the resource was not accepted in the forward market and is not flexible, so it cannot participate in the real-time market. Similarly, bid D3 is not present because it was not cleared in an earlier market and it corresponds to an inflexible resource that is not fast enough to be activated in real time. Figure 6.13 demonstrates the clearing of the real-time market using the real-time supply function and the incremental bids. Both processes result in an identical dispatch and an identical real-time market clearing price. We underline that the inflexible bids that have been accepted in the forward market are considered price-taking at the stage of real-time market clearing. Concretely, S1 has a marginal cost that is arbitrarily negative, which is equivalent to the bid being cleared by default. Likewise, D2 has a valuation that is arbitrarily positive, which is also equivalent to the bid being cleared by default. This is indicated in the left part of figure 6.13 by moving these bids to the left of their respective merit orders.

Certain markets differentiate the price awarded to resources for providing incremental or decremental adjustments, and the price charged to agents for causing the need for balancing. The latter is referred to in certain markets as the **imbalance charge**. The imbalance charge is charged to **balance responsible parties (BRPs)** for any deviation between their financial positions and the actual physical injection or withdrawal of power, and is only applied to the quantity of the deviation.

Note that the previous model not only resolves the imbalance $\Delta$, but also allows agents to trade electricity in real time in case new offers arrive in the balancing market that have not been introduced in previous forward markets. In case the market operator wishes to resolve an imbalance rather than enabling real-time trade of power, the model can be modified as follows. If $\Delta > 0$, then the balancing market model is written as

$$\min_{\delta} \sum_{u \in U} MC_u \delta_u^+$$

$$\sum_{u \in U} \delta_u^+ = \Delta$$

$$\delta_u^+ \leq \Delta_u, u \in U$$

$$\delta_u^+ \geq 0, u \in U.$$

If $\Delta < 0$, then the balancing market model is written as

$$\max_{\delta} \sum_{d \in D} MB_d \delta_d^-$$

$$\sum_{d \in D} \delta_d^- = |\Delta|$$

$$\delta_d^- \leq \Delta_d, d \in D$$

$$\delta_d^- \geq 0, d \in D.$$

## Problems

**6.1** *No-arbitrage condition in energy and reserves.* Use the KKT conditions of $(ORDC)$ to show that if $R_g$ is unbounded then the reserve price $\mu^\star$ and energy price $\lambda^\star$ satisfy the following relationship:

$$\mu^\star = \lambda^\star - MC_{g_m}(p_{g_m}^\star),$$

where $MC_{g_m}(p_{g_m})$ is the marginal cost of the marginal unit in the system. This is the unit that is providing both reserve and energy: $p_{g_m}^\star > 0$ and $r_{g_m}^\star > 0$. Interpret this result as a no-arbitrage condition.

**6.2** *Nonzero reserve prices.* Show that, if $R_g$ is unbounded in $(EDR)$, then either $\mu^\star = 0$ or

$$\mu^\star = MB(d^\star) - MC_{g_m}(p_{g_m}^\star).$$

Here, $g_m$ is a marginal generator (which we can assume exists) that provides both energy and reserve.

**6.3** *Security-constrained economic dispatch as a linear program.* Implement (*SCED*) and confirm the results of example 6.4.

**6.4** *Energy and reserve co-optimization as a linear program.* Implement (*EDR*) and confirm the results of example 6.3 and example 6.6 (i.e. reproduce the reserve allocation and the prices of energy and reserve).

**6.5** *Co-optimization of multiple types of reserve.* Implement (*Res*) and confirm the results of example 6.10.

**6.6** *Price volatility in markets with fixed reserve requirements.* Implement a mathematical programming model that approximates the prices that are derived in example 6.11. Consider the following approximations in your model:

- Discretize the supply function by considering 1000 units with identical generation capacity.
- Discretize the price-elastic demand by considering 100 consumers with identical load.
- Compute the market clearing price for 9100 different levels of inelastic demand by considering increments of inelastic demand of 100 MW.

**6.7** *Computing energy prices with fixed reserve requirements.* Prove the closed-form expression for the market clearing price in example 6.11. Compare the market clearing price to that of problem 6.6.

**6.8** *Co-optimization of energy and reserve with an ORDC.* Confirm the results of example 6.12 using a mathematical programming language. Use the same settings as in problem 6.6, and an increment of 10 MW in the ORDC.

# Bibliography

**Section 6.1**   The classification of ancillary services is based on the Federal Energy Regulatory Commission, although variations exist to the proposed classification of the FERC (Hirst & Kirby 1996, Stoft 2002). The classification of reserves is based on Callaway and Hiskens (2011), Wood, Wollenberg, and Sheblé (2013) and E-Bridge (2014). Notwithstanding, it is important to note that the terminology that is actually used in different systems may vary widely, both across geographies as well as over time. The implementation of flexible ramp in the California market is presented in CAISO (2015), while the efficiency of the mechanism relative to a scheduling of the system using stochastic programming is compared by Wang and Hobbs (2016).

**Section 6.2**   Energy and reserve co-optimization models with multiple types of reserve are abundant in the literature (Wu et al. 2004, Simoglou, Biskas, and Bakirtzis 2010, Baland 2014). Security-constrained economic dispatch is a widely studied topic, dating to early work by Monticelli, Pereira, and Granville (1987), and has been used as a framework for ensuring **reserve deliverability** (Zheng & Litvinov 2008, Chen, Gribik & Gardner 2014). Reserve deliverability refers to the ability of delivering balancing energy from committed reserves, while respecting network constraints.

Import constraints are discussed in Papavasiliou and Oren (2013) and are used in various systems, including California and Norway (Papavasiliou et al. 2022).

**Section 6.3**    The simultaneous pricing of energy and reserves is analyzed extensively in the literature (Wu et al. 2004, Hogan 2005, Baldick 2006), and has been widespread practice in US ISOs (CAISO 2013, PJM Interconnection 2017). The material and examples on markets for multiple types of reserve is based on Oren (2001). Although reserves are priced based on the co-optimization of energy and reserves in an **integrated scheduling process** in certain European markets such as Greece and Cyprus (Papavasiliou 2021a), the current prevailing practice in European markets is for reserves to be cleared in a reserves-only auction either before (e.g. Belgium, Germany) or after (e.g. Italy, Spain) the energy market (Papavasiliou, Smeers & de Maere-d'Aertrycke 2021). Nevertheless, Article 40 of the Electricity Balancing Guideline of the European Commission (European Commission 2017a) carves a path for the introduction of co-optimization of energy and reserves in the integrated day-ahead European energy market.

**Section 6.4**    The anchoring of the shape of ORDCs to LOLP and VOLL derives from theory developed by Hogan (2013). The effect of the shape of ORDCs on energy prices is investigated in the context of different systems, including Texas (Zarnikau et al. 2020), Illinois (Zhou & Botterud 2014), and Belgium (Cartuyvels & Papavasiliou 2022). The state of US ISOs in terms of implementing ORDCs is documented in New York ISO (2019), and the state of European markets is documented in Papavasiliou et al. (2023). Day-ahead ORDCs, and the need for consistency with real-time ORDCs, are analyzed by Hogan and Pope (2019). The process of the back-propagation of real-time reserve and energy prices in markets with ORDCs to forward energy and reserve markets can be represented quantitatively through coherent risk measures, which are defined and discussed in chapter 9. A model that explains this back-propagation of real-time prices to forward markets is developed by Papavasiliou, Smeers & de Maere-d'Aertrycke (2021). The implementation of scarcity pricing without an explicit co-optimization of energy and reserve in real time is described by Hogan (2005) and has been implemented in Texas (ERCOT 2015). Analogous proposals have been made for Europe (Papavasiliou et al. 2023), since the upcoming pan-European balancing platforms are foreseen to be energy-only platforms at the time of writing this book. The absence of a real-time market for reserve in European markets and some of the resulting adverse implications are first pointed out in Papavasiliou (2020), and subsequently analyzed in Papavasiliou and Bertrand (2021). The generalization of ORDCs to multiple reserve products with one-way substitutability is developed by Hogan (2013). Its implementation in Texas is documented in ERCOT (2015).

**Section 6.5**    There is a propensity to violate the law of one price (Jevons 1879, Cramton & Stoft 2006) in balancing market design by distinguishing BRPs and BSPs in terms of settlement. Although the two roles are different in operations, paying them differently in the real-time energy market (i.e. settling BRPs at imbalance charges that diverge from the balancing prices at which BSPs are paid) can distort competition and introduce arbitrage opportunities. This topic has been investigated in the literature in the context of the Japanese (Matsumoto, Bunn & Yamada 2021) and European (Papavasiliou & Bertrand 2021) balancing markets. This issue is becoming increasingly challenging in European market design, due to the push towards the harmonization of the pan-European balancing market.

# 7 Unit commitment

The scheduling of electric energy production is not independent across time periods. Generators are subject to various operating constraints that create dependencies in operations from one hour to the next. Costs may also depend on past decisions. This creates the need to represent the power scheduling problem across time periods. The **unit commitment** problem is the problem of scheduling the on–off status of units, while accounting for time-dependent operating constraints and costs.

Power system operations require a continuous effort to balance supply and demand. Certain decisions need to be made in advance due to operating constraints of generators (e.g. minimum up and down times, ramping rates) in order to ensure that supply can balance demand in real time. This day-ahead scheduling procedure is based on forecast demand and fixes the schedule of slow units. In real time, the system operator balances demand using whatever generation is available, *given* day-ahead scheduling decisions.

The distinction between day-ahead scheduling and real-time dispatch is universal across system operations. The unit commitment model is used for solving the day-ahead scheduling problem, while the economic dispatch model is used for real-time dispatch. A detailed example of day-ahead operations followed by real-time dispatch is presented in figure 3.12.

Section 7.1 presents the formulation of the unit commitment problem. The design of day-ahead markets that can handle the binary nature of on–off decisions and the non-convex costs of unit commitment is discussed in section 7.2. Various approaches for pricing despite the non-convexity of the unit commitment problem are discussed in section 7.3.

## 7.1 Optimization models of unit commitment

The unit commitment problem is defined over a scheduling horizon of $T$ time periods. In order to simplify the exposition of the model, transmission constraints are ignored and demand is assumed to be fixed. The focus is on the description of generator costs and constraints.

Since unit commitment is a scheduling problem, it is necessary to introduce binary variables $u_{gt}$ that indicate whether a unit $g$ is on (in which case $u_{gt} = 1$) or off (in which case $u_{gt} = 0$) in a given time period $t$. The amount of power production $p_{gt}$ and reserve $r_{gt}$ offered by generator $g$ at each time period $t$ are continuous decision variables.[1] The sequence of commitment, production, and reserve decisions over the entire horizon of the scheduling problem is represented as $u_g = (u_{g1}, \ldots, u_{gT}) \in \{0,1\}^T$, $p_g = (p_{g1}, \ldots, p_{gT}) \in \mathbb{R}^T$, and $r_g = (r_{g1}, \ldots, r_{gT}) \in \mathbb{R}^T$ for each generator $g \in G$, where $G$ represents the set of generators.

In addition to a production cost function $PC_g \colon \mathbb{R} \to \mathbb{R}$, which depends exclusively on production decisions, $p_{gt}$, at each time period, a generator $g$ incurs a commitment cost $UC_g \colon \{0,1\}^T \to \mathbb{R}$ that depends on commitment decisions, $u_g$, over the entire horizon. The commitment cost can include startup costs that are incurred whenever a unit transitions from $u_{g,t-1} = 0$ to $u_{gt} = 1$, as well as minimum load costs that are incurred at every time period that a unit is running, independently of how much it is producing. Total costs can then be represented as

$$TC_g(u_g, p_g) = UC_g(u_g) + \sum_{t=1}^{T} PC_g(p_{gt}),$$

where $TC_g \colon \{0,1\}^T \times \mathbb{R}^T \to \mathbb{R}$ is the total cost function of generator $g$.

---

**Example 7.1** *A total cost function with startup and min load costs.* Let $S_g$ represent the startup cost and $K_g$ represent the minimum load cost for a certain generator $g$. Let $MC_g \colon \mathbb{R} \to \mathbb{R}$ represent a marginal cost function. Then the total cost function can be expressed as

$$TC_g(u_g, p_g) = \sum_{t=1}^{T} \left( K_g u_{gt} + S_g v_{gt} + \int_0^{p_{gt}} MC_g(x)dx \right),$$

where $v_{gt}$ indicates whether a unit has started up in period $t$ or not:

$$v_{gt} = \begin{cases} 1 & \text{if } u_{g,t-1} = 0, u_{gt} = 1, \\ 0 & \text{otherwise.} \end{cases}$$

---

The unit commitment problem can be expressed as the following mixed integer program:

$$\begin{aligned}
(UC) \colon \min_{u,p,r} \ & \sum_{g \in G} TC_g(u_g, p_g) \\
& h_g(p_g, r_g, u_g) \le 0, g \in G \\
& \sum_{g \in G} p_{gt} = D_t, t = 1, \ldots, T \\
& \sum_{g \in G} r_{gt} = R_t, t = 1, \ldots, T \\
& u_{gt} \in \{0,1\}, g \in G, t = 1, \ldots, T,
\end{aligned} \tag{7.1}$$

---

[1] Multiple types of reserves can be modeled, as shown later in this chapter.

where $h_g \colon \mathbb{R}^{2T} \times \{0,1\}^T \rightarrow \mathbb{R}^{n_g}$ represent private operating constraints of unit $g$, $D_t$ represents hourly demand, and $R_t$ represents the hourly reserve requirement. The operating constraints that correspond to $h_g(p_g, r_g, u_g) \leq 0$ are analyzed in detail in the subsequent paragraphs.

**Initial conditions**   The scheduling of units over a certain horizon requires knowledge of the initial conditions of the units. In particular, the unit commitment history and production history are required (although the reserve provision history may be omitted) for a horizon of $T_0$ periods prior to the scheduling horizon $\{1, \dots, T\}$. The length of $T_0$ depends on the amount of time needed for decisions made in a certain time $t$ to not have an impact on present decisions. For example, if a unit needs to be kept off for 24 hours once it has been shut down, then $T_0$ cannot be less than 24 hours. The initial commitment and production decisions are denoted by $u0_g$ and $p0_g$, respectively, for each generator $g \in G$.

**Transitions**   In order to describe scheduling constraints, it is first necessary to define variables that indicate when a unit has been started up and shut down. In what follows, $u$ indicates on status, $v$ indicates a startup decision, and $z$ indicates a shutdown decision. The startup and shutdown decision variables can then be linked to the unit commitment decision variables as follows:

$$u_{gt} = u_{g,t-1} + v_{gt} - z_{gt}, g \in G, t = 2, \dots, T.$$

**Minimum up/down times**   Generators require a minimum of amount of time in order to cool off once they are shut down. Similarly, they cannot be shut down once they have been started up, before a minimum amount of time has elapsed. These constraints are demonstrated in figure 7.1. Denoting $UT_g$ and $DT_g$ as the minimum

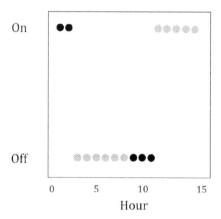

**Figure 7.1** Consider a generator with a minimum up time of five hours and a minimum down time of six hours. Gray marks correspond to forced states due to minimum up and down times. Black marks correspond to free choices.

up and minimum down time of a generator $g$, respectively, the minimum up and down time constraints can be expressed as follows:

$$\sum_{\tau=t-UT_g+1}^{t} v_{g\tau} \leq u_{gt}, g \in G, t = UT_g, \ldots, T, \tag{7.2}$$

$$\sum_{\tau=t-DT_g+1}^{t} z_{g\tau} \leq 1 - u_{gt}, g \in G, t = DT_g, \ldots, T. \tag{7.3}$$

**Generator temperature**   Generators require a certain amount of fuel in order to start up, which is not used as useful electric power, but is instead used to simply warm up the plant. The temperature of a generator therefore determines how much fuel is required in order to start it up. The temperature state of a generator can be discretized in order to be represented in a unit commitment model. The set of temperatures is denoted by $\Theta$.

---

**Example 7.2** *Dependency of startup cost on temperature.* Consider a generator with three generator states.

- Hot: 200 GJ is required in order to start the generator up 1–16 hours after it has been shut down.
- Warm: 220 GJ is required in order to start the generator up 17–24 hours after it has been shut down.
- Cold: 250 GJ is required in order to start the generator up 25 or more hours after the generator has been shut down.

The generator temperature set is then $\Theta = \{\text{Hot, Warm, Cold}\}$.

---

**Temperature-dependent startup**   Denote $v_{glt}$ as a binary variable that indicates whether a generator has been started up in temperature state $l$ at period $t$ or not. Since a generator can only be started up from a single temperature state, the following constraint is enforced:

$$v_{gt} = \sum_{l \in \Theta} v_{glt}, g \in G, t = 1, \ldots, T.$$

If temperature state $l$ occurs within $\underline{T}_{gl}$ to $\bar{T}_{gl}$ periods after shutdown, then a startup at temperature $l$ is possible only if the generator has been shut down within $\underline{T}_{gl}$ to $\bar{T}_{gl}$ periods before. Therefore,

$$v_{glt} \leq \sum_{\tau=t-\underline{T}_{gl}+1}^{t-\bar{T}_{gl}} z_{g\tau}, l \in \Theta, t = \underline{T}_{gl}, \ldots, T.$$

**Startup/shutdown profiles**   When generators are started up, their power output follows a predefined sequence for a certain number of periods. This sequence is called a startup profile. Similarly, in order for a generator to shut down, it is required

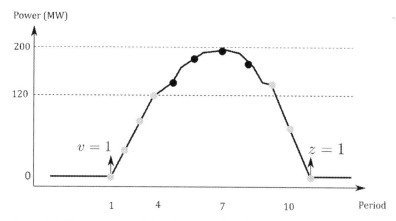

**Figure 7.2** The startup and shutdown profile of a generator. Gray points indicate $u_{gt}^{SU} = 1$ or $u_{gt}^{SD} = 1$, whereas black points indicate $u_{gt}^{DISP} = 1$.

to follow a predefined shutdown sequence, called a shutdown profile. Startup and shutdown profiles are illustrated in figure 7.2.

---

**Example 7.3** *Startup profile of a ramp-limited unit.* Consider a generator with a technical minimum of 120 MW and a ramp rate of 1 MW/min. When this generator is started up, it is required, on the one hand, to produce 120 MW or more as soon as possible. However, it cannot exceed its ramp rate of 60 MW/hour. A possible startup profile of the generator is therefore (60 MW, 120 MW).

---

Startup profiles may be dependent on the temperature of the generator, namely on the time that has elapsed since the generator was shut down. There is typically no such temperature dependency for shutdown profiles.

In order to describe startup and shutdown profiles, binary variables $u_{gt}^{SU}$, $u_{gt}^{SD}$, $u_{gt}^{DISP}$ are introduced. These binary variables indicate, respectively, whether a generator $g$ is following a startup profile, a shutdown profile, or no profile at all (in the latter case the generator is referred to as being *dispatchable*) at a given period $t$. The following constraint ensures that a generator is in one of the three states at any given period:

$$u_{gt} = u_{gt}^{SU} + u_{gt}^{DISP} + u_{gt}^{SD}, g \in G, t = 1, \ldots, T.$$

Whether or not a generator is following a startup profile depends on the temperature of the generator when it was started up and how long the startup profile lasts for the given temperature, which is denoted as $T_{gl}^{SU}$ (for example, a startup profile for a warm temperature state may last for a shorter amount of time than a startup profile of a cold temperature state). In order to determine if $u_{gt}^{SU}$ is active, it is therefore necessary to check $T_{gl}^{SU}$ periods in the past for all different temperature states:

$$u_{gt}^{SU} = \sum_{l \in \Theta} \sum_{\tau=t-T_{gl}^{SU}+1, \tau \geq 1}^{t} v_{gl\tau}, g \in G, t = \max_{l \in \Theta} T_{gl}^{SU}, \dots, T.$$

Denote $T_g^{SD}$ as the duration of the shutdown, and note that $T_g^{SD}$ is independent of temperature. In order to determine whether $u_{gt}^{SD}$ is active, it is necessary to check $T_g^{SD}$ periods into the future to see whether the unit is being shut down:

$$u_{gt}^{SD} = \sum_{\tau=t}^{t+T_g^{SD}-1} z_{g\tau}, g \in G, t = 1, \dots, T - T_g^{SD} + 1.$$

**Startup/shutdown production**    Denote $P_{gl\tau}^{SU}$ as the sequence of production levels for a startup profile when a unit is started up in temperature state $l$. Then the production of the generator, as it is following a startup profile, can be determined by fixing the production variable to the corresponding level within the profile:

$$p_{gt}^{SU} = \sum_{l \in \Theta} \sum_{\tau=t-T_{gl}^{SU}+1, \tau \geq 1}^{t} v_{gl\tau} P_{gl, t-\tau+1}^{SU}, g \in G, t = \max_{l \in \Theta} T_{gl}^{SU}, \dots, T.$$

Denote $P_{g\tau}^{SD}$ as the shutdown profile of a generator. The same rationale as before applies to determining shutdown production, the only differences being that it is necessary to look into future time periods and that there is no dependency in temperature state:

$$p_{gt}^{SD} = \sum_{\tau=t+1}^{t+T_{gl}^{SD}} z_{g\tau} P_{g, \tau-t}^{SD}, g \in G, t = 1, \dots, T_g^{SD}.$$

**Dispatchable production**    Denote $P_g^-$ and $P_g^+$ as the technical minimum and technical maximum of a generator $g$, respectively. When startup and shutdown profiles are ignored, production limits are represented as

$$P_g^- u_{gt} \leq p_{gt} \leq P_g^+ u_{gt}, g \in G, t = 1, \dots, T.$$

The existence of startup and shutdown profiles invalidates these constraints. Instead, it is necessary to express production limits such that $p_{gt}$ is equal to $p_{gt}^{SU}$ if the generator is following a startup profile, equal to $p_{gt}^{SD}$ if the generator is following a shutdown profile, and is in between its technical limits if the generator is in dispatchable mode. This can be accomplished through the following constraints:

$$p_{gt} \geq p_{gt}^{SU} + p_{gt}^{SD} + P_g^- u_{gt}^{DISP}, g \in G, t = 1, \dots, T, \tag{7.4}$$

$$p_{gt} \leq p_{gt}^{SU} + p_{gt}^{SD} + P_g^+ u_{gt}^{DISP}, g \in G, t = 1, \dots, T. \tag{7.5}$$

Note that these constraints are adequate because, on the one hand, if a generator is following a profile then the third term will be zero and only one of the two remaining terms will be nonzero, thereby fixing $p_{gt}$ to the profile production, whereas if no profile is active then the standard production limit constraints are obtained.

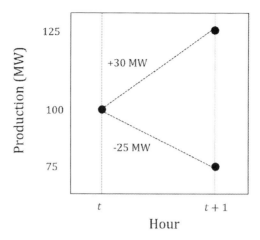

**Figure 7.3** A generator that produces 100 MW at period $t$ cannot produce less than 75 MW or more than 130 MW at hour $t + 1$ if it has a 25 MW ramp down and a 30 MW ramp up rate.

**Ramp rates**   Ramp rate limits require that the production of units from one period to the next cannot change too rapidly, either in the upward or downward direction. This is indicated graphically in figure 7.3. Denoting $R_g^-$ and $R_g^+$ as the ramp down and ramp up rate limits, respectively, ramp constraints in simple unit commitment models are commonly expressed as

$$-R_g^- \leq p_{gt} - p_{g,t-1} \leq R_g^+, g \in G, t = 2, \ldots, T.$$

The introduction of profiles requires a modification of these constraints. In particular, startup and shutdown profiles may violate ramp rate constraints. It is therefore desirable to deactivate ramp rate constraints during startup and shutdown, using a big-M parameter:

$$p_{gt} - p_{g,t-1} \leq R_g^+ + M \cdot u_{gt}^{SU}, g \in G, t = 2, \ldots, T,$$
$$p_{g,t-1} - p_{gt} \leq R_g^- + M \cdot u_{gt}^{SD}, g \in G, t = 2, \ldots, T.$$

**Commitment costs**   Commitment costs correspond to startup fuel consumption and minimum load consumption. Startup fuel consumption depends on the temperature state of the generator. Denote $SUC_{gl}$ as the startup cost for temperature state $l$. Also, denote $MLC_g$ as the minimum load cost that is incurred during every period that a generator is online. Then the total commitment cost of operating a power plant can be expressed as

$$UC_g(u_g) = \sum_{t=1}^{T} \left( \sum_{l \in (\cdot)} SUC_{gl} v_{glt} + MLC_g u_{gt} \right), g \in G.$$

Since a single temperature for the generator can be active at a time, at most one of the terms in the first sum is nonzero at any given period. Moreover, the startup cost is separate from the fuel cost resulting from startup profile power production, which is captured by the variable cost terms.

**Production costs**    The relationship between the amount of fuel injected in a power plant (measured commonly in MMBtu) and the electric power produced (measured commonly in MW) is typically nonlinear. The **average heat rate** of a plant, measured in MMBtu/MWh, is the ratio of *total* fuel consumption to *total* electric power production of a plant. Similarly, the **marginal heat rate**, also measured in MMBtu/MWh, measures the derivative of fuel consumption with respect to electric power production. Given a fuel price $FP$, commonly measured in \$/MMBtu, the marginal heat rate curve can be used to compute the production cost function $PC_g$ of a power plant. Denote $MHR_g$ as the marginal heat rate curve of a power plant. The variable cost is then given by

$$PC_{gt}(p_{gt}) = FP \cdot \int_0^{p_{gt}} MHR_g(x)dx, g \in G, t = 1, \ldots, T.$$

The marginal heat rate curve can be approximated by a finite number of segments. However, if the marginal heat rate is not increasing, binary variables must be introduced in the model in order to indicate which segment is currently active.

---

**Example 7.4** *Nonincreasing marginal heat rate curve.* Consider the marginal heat rate curve of figure 7.4. The marginal heat rate curve consists of three segments. These segments would be activated in order 2–1–3 unless binary variables are introduced in order to prevent this.

---

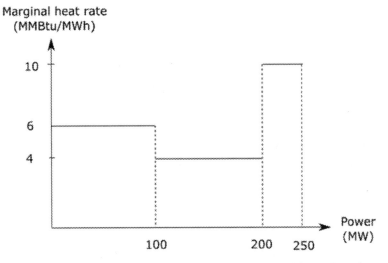

**Figure 7.4** The marginal heat rate curve of a generator, which consists of three segments. Binary variables must be introduced in order to ensure that segment 1 (0–100 MW) is used prior to segment 2 (100–200 MW).

Denote $S$ as the total number of segments that define a heat rate curve and $P_{gs}^{+}$ as the width of each segment $s$. The following constraint activates the first segment once a generator is started up:

$$u_{g,s_1,t} = u_{gt}, g \in G, t = 1, \ldots, T.$$

The following constraint ensures that a segment cannot be activated before the previous segment has been fully utilized:

$$u_{g,s+1,t} \leq \frac{p_{gst}}{P_{gs}^{+}}, g \in G, s = 1, \ldots, S-1, t = 1, \ldots, T.$$

The production within each segment is bounded by the width of the segment:

$$0 \leq p_{gst} \leq P_{gs}^{+} u_{gst}, g \in G, s = 1, \ldots, S, t = 1, \ldots, T.$$

The total power production of a plant is the sum of the power production at each segment:

$$p_{gt} = \sum_{s \in S} p_{gst}, g \in G, t = 1, \ldots, T.$$

Denote $MHR_{gs}$ as the marginal heat rate of the given segment. Then the total variable cost is computed as:

$$VC_{gt}(p_{gt}) = FC \cdot \sum_{s=1}^{S} MHR_{gs} \cdot p_{gst}, g \in G, t = 1, \ldots, T.$$

**Frequency restoration reserves**    Reserves influence unit commitment because they interact with the production of power. In what follows, capacity limit constraints are modified in order to account for reserves. It is possible to build more sophisticated models that account for the interplay of ramp rate constraints and reserves.

Frequency restoration reserves are often differentiated between upwards and downwards capacity, respectively denoted as $r2_{gt}^{+} \geq 0$ and $r2_{gt}^{-} \geq 0$. Since frequency restoration reserves occupy generator capacity, it is necessary to rewrite the minimum and maximum capacity constraints of equations (7.4) and (7.5) as follows:

$$p_{gt} - r2_{gt}^{-} \geq p_{gt}^{SU} + p_{gt}^{SD} + P_{g}^{-} \cdot u_{gt}^{DISP}, g \in G, t = 1, \ldots, T,$$
$$p_{gt} + r2_{gt}^{+} \leq p_{gt}^{SU} + p_{gt}^{SD} + P_{g}^{+} \cdot u_{gt}^{DISP}, g \in G, t = 1, \ldots, T.$$

Denote reserve contribution limits for upward and downward reserve as $MR2_{g}^{+}$, $MR2_{g}^{-}$, respectively. Reserve contribution constraints are then expressed as follows:

$$r2_{gt}^{-} \leq MR2_{g}^{-} \cdot u_{gt}^{DISP}, g \in G, t = 1, \ldots, T,$$
$$r2_{gt}^{+} \leq MR2_{g}^{+} \cdot u_{gt}^{DISP}, g \in G, t = 1, \ldots, T.$$

Aggregate replacement reserve requirements for upward and downward reserve are denoted as $RR2_{t}^{+}$ and $RR2_{t}^{-}$, respectively. The aggregate reserve requirement constraint is then imposed as:

$$\sum_{g \in G} r2_{gt}^- \geq RR2_t^-, t = 1, \dots, T,$$

$$\sum_{g \in G} r2_{gt}^+ \geq RR2_t^+, t = 1, \dots, T.$$

**Replacement reserves**   Consider replacement reserves that are differentiated between spinning reserve, which can only be offered by units that are online, and non-spinning reserve, which can be offered by units that are off but can be started up rapidly. Denoting $r3_{gt}^S \geq 0$ as spinning reserve, its modeling follows the same logic as the modeling of upwards frequency restoration reserve:

$$p_{gt} + r2_{gt}^+ + r3_{gt}^S \leq p_{gt}^{SU} + p_{gt}^{SD} + P_g^+ \cdot u_{gt}^{DISP}, g \in G, t = 1, \dots, T.$$

The modeling of non-spinning reserve $r3_{gt}^{NS} \geq 0$ is slightly different. Since there is no conflict of non-spinning reserve with production, the amount of non-spinning reserve that can be offered is only limited by the maximum capacity of the plant, and non-spinning reserve can only be offered if the plant is off:

$$r3_{gt}^{NS} \leq P_g^+ \cdot (1 - u_{gt}), g \in G, t = 1, \dots, T.$$

Denote $MR3_g$ as the replacement reserve limit that can be offered by a generator. Since a generator cannot offer spinning and non-spinning reserve simultaneously, the following constraint holds:

$$r3_{gt}^S + r3_{gt}^{NS} \leq MR3_g, g \in G, t = 1, \dots, T.$$

Aggregate reserve requirements are denoted as $RR3_t$. The replacement reserve requirement constraint is imposed as follows:

$$\sum_{g \in G} \left( r3_{gt}^S + r3_{gt}^{NS} \right) \geq RR3_t, t = 1, \dots, T.$$

**Stochastic unit commitment and scenario selection**   One drawback of the aforementioned model is the difficulty of determining reserve requirements. These reserve requirements strongly influence cost and reliability. However, determining their level is nonobvious. Matters are complicated in systems with large amounts of renewable energy sources, where operational uncertainty and the reliance of the system on reserves is increased. Various alternatives have been proposed to the unit commitment model proposed above, where uncertainty is represented endogenously within the unit commitment model.

The **stochastic unit commitment model** represents uncertainty in terms of a finite set of scenarios $\Omega$. Each scenario is weighed by a certain probability $P_\omega$ and the objective is to commit generators in such a way that the expected cost of operations is minimized.

In its simplest form, the problem can be cast as a two-stage optimization. Generators are partitioned between conventional units and renewable units. The output of conventional units can be controlled. The operation of renewable generators imposes no fuel cost to the system, but the output of these units is uncertain and can only be curtailed.

The sequence of events in the stochastic unit commitment model is as follows:

1. In the first stage, generators are committed.
2. Uncertainty is revealed. This uncertainty can involve the outage of certain generators (or transmission lines) as well as the forecast error of production from renewable resources or the forecast error of loads.
3. In the second stage, the system responds to uncertainty by dispatching generators and loads according to the scenario that materializes.

Second-stage decisions are now indexed by scenario because these decisions can be different depending on the scenario that materializes. These decisions include production and demand, as well as the resulting net injections and power flows in the network.

Note that the first-stage decision variables, unit commitment and startup, are not indexed by scenario. This means that units need to be committed without knowing which scenario will materialize. The purpose of the model is to select these commitments in such a way that minimizes the expected cost of operations. Two-stage stochastic models are discussed in further detail in section 8.1.1.

The trade-off in committing reserves is that an overly conservative commitment results in increased fuel costs due to the fixed startup and minimum load cost, as well as the curtailment of renewable energy due to the minimum run level of conventional units. Conversely, enough reserve resources need to be committed in order to ensure the reliable supply of power against a broad range of scenarios. Stochastic unit commitment strives to determine the optimal balance between economic efficiency and reliability in short-term operations.

The stochastic unit commitment model represents uncertainty explicitly, as opposed to the deterministic unit commitment model with reserve requirements. A drawback of the model is the fact that adding a large number of scenarios increases the size of the problem. Moreover, a detailed model of uncertainty is required in order to generate scenarios. Finally, the selection of a finite set of scenarios that accurately represent the full range of possible outcomes is crucial for the performance of the model and is also nontrivial. The bibliography section discusses literature on these topics that can serve as a point of departure for the interested reader.

**Security-constrained unit commitment**    Security-constrained unit commitment is an alternative approach for committing reserves that follows the logic of security-constrained economic dispatch. This model attempts to commit units such that no load is shed in real time given the failure of any *individual* component (i.e. generator, line, transformer) in the network. The objective is to minimize the cost of operations for the outcome where there is no contingency, while being able to fulfill the entire demand in the second stage, regardless of which component may fail.

The model can be formulated as a two-stage decision process, similar to the stochastic unit commitment model. "Scenarios" now represent single-component contingencies and renewable supply is replaced by its forecast value. The objective function accounts exclusively for the cost of the base case scenario, in which no component fails. There is one set of constraints for each scenario (or contingency), and the requirement is that the entire demand is satisfied.

Due to the fact that load shedding is not an option in the model, the model may be infeasible in the second stage given a certain first-stage unit commitment decision. This has implications for the algorithms that can be used for solving the model.

Similar to the stochastic unit commitment model, the security-constrained model entails more detail about uncertainty than the deterministic unit commitment model with reserve requirements. Also, the size of the security-constrained unit commitment problem makes it difficult to solve.

## 7.2    Market design for unit commitment

This text has progressively covered complex market designs for real-time dispatch. The market design for day-ahead scheduling relies on the same principles. However, the non-convexity of costs and constraints complicates the design of day-ahead markets.

Prior to discussing the design of day-ahead markets, it is worth reflecting on the role of day-ahead markets. As shown in figure 3.12, at some point close enough to real time, the system operator must ensure that enough units will be online when they will be needed. This point is typically the day ahead, when thermal units need to be notified about whether or not they should be started up in order to be available in the following day. The day-ahead market is therefore situated at the border between decentralized forward trading and centralized real-time control.

### 7.2.1    The two-settlement system

The day-ahead market is a forward market for trading electricity. In order to provide financial certainty to market agents without distorting their real-time incentives to produce and consume efficiently, the day-ahead forward market and real-time market are designed as a **two-settlement system**, where the day-ahead market functions as a forward market for buying or selling electricity that is traded in the real-time market.

Forward contracts are analyzed in detail in section 9.1. In short, the two-settlement system functions as follows. When a generator sells an amount $Q_1$ of power in the day-ahead market at a price $\lambda_1$, the generator receives a total payment of $\lambda_1 \cdot Q_1$. However, having sold the energy in the day-ahead market, the generator is obliged to produce it in real time. Any deviation from the quantity $Q_1$ is settled at the real-time energy price $\lambda_0$. If the generator produces $Q_0 > Q_1$, then the generator is paid $\lambda_0$ for the extra power. In the opposite case, the generator pays $\lambda_0$ for the shortage. Therefore, the payoff that the generator receives by selling $Q_1$ in the day-ahead market and producing $Q_0$ in the real time market is

$$R = \lambda_1 \cdot Q_1 + \lambda_0 \cdot (Q_0 - Q_1).$$

This is the two-settlement system for trading power in the day-ahead market and balancing the position in the real-time market. For loads, the two-settlement system applies identically. Loads that buy $Q_1$ in the day-ahead market and consume $Q_0$ in the real-time market *pay* $\lambda_1 \cdot Q_1 + \lambda_0 \cdot (Q_0 - Q_1)$.

In section 9.1 we demonstrate that the two-settlement system provides risk hedging to market participants in the day ahead, without removing their incentives to behave efficiently in real time. Therefore, real-time markets can be designed using the principles that have been outlined in the previous chapters.

## 7.2.2 Design dilemmas

Moving away from real-time markets to earlier short-term markets[2] differs in two notable ways. Physical assets do not need to be monitored individually, and costs and constraints are no longer convex. This leads to alternative options for organizing the market at the day-ahead time frame. A central dilemma in these alternatives is whether or not system and market operation are separated. We discuss some of these alternatives in the present section, and deep-dive into the issue of pricing in non-convex market models in section 7.3.

### Portfolio-based versus unit-based designs

Whereas real-time operations require the system operator to be able to identify and control individual generation assets so as to ensure that the system is operated securely, earlier markets (such as the intraday and day-ahead market) are not required to adhere to such a requirement. The degree to which decisions about commitment need to be centralized or left up to market participants in the day-ahead time frame is a topic of intense debate between experts.

In portfolio-based designs, assets are aggregated at the stage of market clearing. Aggregate offers are constructed by the owners of the portfolios, without specifying to the auctioneer how these assets are expected to deliver the promised offer physically. The market is cleared, and it is then the responsibility of the portfolio owners to disaggregate the market clearing outcome to schedules of individual assets. These schedules are announced to the system operator after day-ahead market operations, but before real time, in the **nomination** stage.

Note that the day-ahead market clearing function in this setting can be entirely separated from the nomination function. In Europe, which mostly operates a portfolio-based design, the day-ahead market is operated by exchanges, whereas the nominations are received by the system operators, which can check whether the nominated schedules are compatible with network constraints, ancillary service requirements, and other system needs.

---

[2] The subsequent discussion refers to the day-ahead market. Note, however, that similar dilemmas apply to forward markets that take place later in time, such as intraday auctions.

In unit-based designs, separate physical assets are bid separately to the market. The representation of generation assets at the physical layer means that the network can also be represented at the physical layer, since the precise location of different assets can be specified. By contrast, in portfolio-based designs the information about the precise location of a physical asset is lost if two assets that are bundled into a portfolio are connected in different nodes of the physical network.

Unit-based designs are often associated with an integration of market and system operation, whereby the auctioneer is also the system operator and no nomination (i.e. disaggregation) of physical asset schedules is required after the day-ahead market. This is typical of US designs. Nevertheless, it is also possible to separate auctions from system operation in a unit-based design. For instance, the day-ahead market in Greece is operated by a power exchange that is a separate entity from the system operator.

Unit-based designs do not offer as much flexibility to traders. However, they allow for closer monitoring of market power, since every market offer can be associated to a specific technology. Portfolio-based designs prevent nodal pricing when assets from different locations are aggregated into a portfolio.

**Exchanges versus pools**

The non-convexity of the day-ahead market clearing model means that proposition 4.11 no longer holds. Therefore, it is no longer guaranteed that we can find a price that can induce all market participants to follow a schedule that maximizes social welfare, given their stated market offers.

---

**Example 7.5** *Inexistence of a market clearing price.* Consider a system with an inelastic demand of 360 MW. Three (or more) identical generators are available in the system. The generators have a technical capacity of 200 MW, and incur a startup cost of $1000 and a marginal cost of 5 $/MWh. The generators require a price of 10 $/MWh in order to recover their startup cost. At that price, each generator earns 5 $/MWh above its marginal cost, which is exactly enough to cover its startup cost when it is producing at full output (200 MW). For any price below 10 $/MWh, the generator is not willing to produce. At 10 $/MWh, the generator is willing to produce, but *exactly* at its maximum capacity. The supply and demand "curves" are shown in figure 7.5. It can be observed that there is no price that equates supply to demand. For any price below 10 $/MWh demand exceeds supply, for any price equal to or above 10 $/MWh supply exceeds demand.

---

The complication of the previous example arises from the fact that the non-convex cost of the generators induces them to either produce nothing, or produce at their technical capacity. In contrast, convex costs imply that supply changes smoothly as the market price changes.

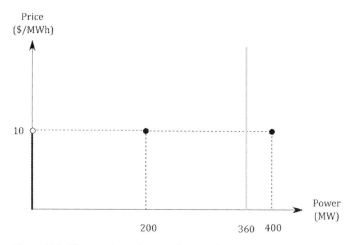

**Figure 7.5** The auction of example 7.5. The demand curve is in gray; the (discontinuous) supply curve is in black.

There are two different philosophies for dealing with non-convexities, and they relate to the separation of market and system operation and who foots the bill for dealing with non-convexities:

- Keeping the optimal schedule and using side payments in order to induce assets to follow the optimal schedule. These side payments are financed by the auctioneer. This is the approach that is currently adopted in US designs, and the specific way in which the price is defined gives rise to various alternatives: IP pricing, pricing using a linear relaxation, convex hull pricing, and approximations of convex hull pricing.
- Sacrificing the optimal schedule if there is no price that can induce it and continuing to reject candidate schedules until one can be found that can be supported by a market clearing price. There are no side payments in such a setting,[3] thus the auctioneer is not financially liable for such side payments. This is the predominant design in Europe.

The first alternative described above corresponds to **power pools**. Power pools are characterized by the use of side payments. There are various other features that are typical of power pools, although they are not necessary. Power pools are often associated with joint system and market operation, meaning that the system operator is also the operator of the pool. Side payments related to non-convexities are bundled with various other operation costs. Power pools often involve unit-based offers and allow bidders to submit multipart offers that are used for populating a detailed unit commitment model, such as the one described in section 7.1.

The second alternative described above corresponds to **exchanges**. Exchanges are characterized by the fact that they do not use side payments, and they are

---

[3] The meaning of this statement is formalized carefully in section 7.3, after some definitions are introduced.

often associated with portfolio-based designs. Exchanges are often separated from system operation. Exchanges typically do not offer multipart bids, but rather specify standardized products. Nevertheless, these standardized products can become quite complex in practice in terms of specifications, since they attempt to achieve the same flexibility as unit commitment models implemented in power pools (i.e. allowing the portfolio owners to specify on–off constraints, startup and minimum load costs, minimum up and down times, ramp rates, and so on).

## 7.3    Pricing with non-convexities

### 7.3.1    Inexistence of a clearing price

The design of day-ahead markets is complicated by the non-convexity of the unit commitment problem. According to proposition 4.11, in a convex market model the dual multiplier of the market clearing constraint is guaranteed to induce market participants to follow the optimal solution of the auction model. What if the market clearing model is not convex?

---

**Example 7.6** *MILP without a market clearing price.* Consider a market with the offers indicated in table 7.1. The offers are depicted graphically in figure 7.6. The market model can be described mathematically as follows:

**Table 7.1  Data for example 7.6.**

| Bid | Quantity (MWh) | Price ($/MWh) | Min. acceptance (MWh) |
|---|---|---|---|
| A (buy) | 10 | 300 | 0 |
| B (buy) | 14 | 10 | 0 |
| C (sell) | 12 | 40 | 11 |
| D (sell) | 13 | 100 | 0 |

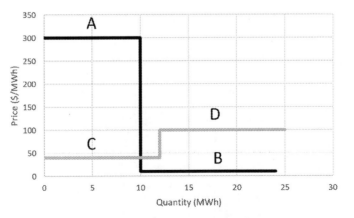

**Figure 7.6**  The production and consumption offers of example 7.6.

$$\max_{p,d,u} 300 \cdot d_A + 10 \cdot d_B - 40 \cdot p_C - 100 \cdot p_D$$

$$d_A + d_B - p_C - p_D = 0$$

$$11 \cdot u_C \le p_C \le 12 \cdot u_C$$

$$d_A \le 10, d_B \le 14, p_D \le 13$$

$$d_A, d_B, p_D \ge 0$$

$$u_C \in \{0, 1\}.$$

The optimal solution for this example is for order C to be accepted, and for it to produce 11 MWh: $p_C^\star = 11$ MWh, $u_C^\star = 1$. Order A is fully covered, $d_A^\star = 10$ MWh, and order B consumes 1 MWh, $d_B^\star = 1$ MWh. Order D produces nothing at the optimal solution: $p_D^\star = 0$ MWh. Although it would be preferable to only use 10 MWh of order C and avoid matching the extra MWh to order B, the fact that order C needs to produce at least 11 MWh implies that we need to tolerate the inefficient matching of its 11th MWh with order B in order to reap the gains of its low cost. The welfare of the resulting solution is equal to $2570. The model presented above is a mixed integer linear program (MILP), and is thus non-convex. Proposition 4.11 therefore no longer applies, which implies that we have no guarantee that we can find a price that induces the optimal reactions described previously. Indeed, we can verify that no price works:

- At a price below 40 $/MWh, profit-maximizing behavior results in under-supply because order C is not willing to produce, whereas order A is willing to consume its entire quantity.
- At a price of 40 $/MWh, order A is the only one willing to buy. But order C, although indifferent about producing any quantity, needs to produce at least 11 MWh, which implies that there is an oversupply.
- At a price above 40 $/MWh, order C wishes to produce its entire quantity, whereas order B is not willing to buy; thus there is oversupply.

In order to cope with the inexistence of market clearing prices, current European market design "paradoxically rejects" non-convex bids. To understand the motivation of this approach, note that, in the extreme, if we were to remove all non-convex bids from the market, we would be left with a convex model. And since a convex model is guaranteed to have a market clearing price, gradually removing orders from the market and subsequently checking if the remaining orders can be induced to follow the centralized scheduling solution is a process that is guaranteed to terminate finitely.

The side effect of doing this is that **paradoxically rejected orders** emerge. These are orders that would prefer to be matched given the prevailing market price.

It is interesting to underline the psychological dimension of this arrangement. Institutionally, this approach has been able to persist because it is deemed acceptable to prevent orders from being matched and thus reap their maximum possible gain, whereas matching orders that suffer a loss at the prevailing price has been deemed unacceptable.

An important disadvantage of this approach is the fact that it prioritizes the pricing rule over efficiency. Removing orders from the order book means that the welfare of the resulting market clearing solution will be no greater than that of the full order book, and may be strictly less.

---

**Example 7.7** *Coping with inexistence of price through paradoxically rejected orders.* We return to example 7.6 and consider a new order book, where order C is no longer present. The market clearing model for the remaining order book can be expressed as follows:

$$\max_{p,d} 300 \cdot d_A + 10 \cdot d_B - 100 \cdot p_D$$
$$(\lambda): \quad d_A + d_B - p_D = 0$$
$$d_A \leq 10, d_B \leq 14, p_D \leq 13$$
$$d_A, d_B, p_D \geq 0.$$

This is a convex model, and the market clearing price is indicated as $\lambda$. The clearing price is equal to 100 \$/MWh. The welfare of the new solution is \$2000. Welfare is reduced by \$570 compared to the matching of example 7.6. This loss of welfare is due to the fact that order C, which is more than two times cheaper than order D, has been removed from the order book. We confirm that order C is paradoxically rejected: at the prevailing market price of 100 \$/MWh, it would strictly prefer to produce 12 MWh than to not be matched at all.

---

The alternative paradigm, which is implemented in US market design, is to maintain the welfare-maximizing solution and to try to find a price that comes close to being aligned with agents' incentives. "Comes close" is formalized quantitatively in section 7.3.2. The process is illustrated graphically in figure 7.7.

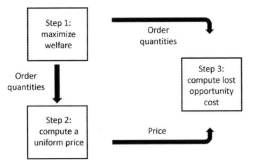

**Figure 7.7** US markets maintain the welfare-maximizing primal solution. This is used as fixed input for computing prices in a separate pricing run. Given prices and order matches, it is then possible to compute lost opportunity cost.

## 7.3.2   Measuring deviation from equilibrium

Even if an equilibrium market clearing price is not guaranteed to exist, it is still meaningful to issue a price that "comes close" to achieving this goal. But how does one quantify "comes close"?

The answer to this question is inspired by the quantity adjustment process that drives agents' behavior in a market equilibrium. Since the agents' goal is to maximize surplus, a metric of interest is the surplus that they can achieve when they are allowed to react freely to the market price signal (or **self-schedule**), versus the surplus that they can achieve when they follow the dispatch instructed by the market operator. This gives rise to the definition of lost opportunity cost.

Concretely, given a price signal $\lambda$, we define **lost opportunity cost (LOC)** as the difference between the maximum surplus that an agent can achieve by self-scheduling, minus the surplus that the agent can achieve by following the market instructions. LOC is zero in convex market models. If the auctioneer arranges to pay LOC to market participants *only* if they follow the market schedule, then LOC can act as an incentive for inducing agents to follow the market schedule voluntarily.

---

**Example 7.8** *Lost opportunity cost as a function of price.* The LOC of different agents in the market of example 7.6 is presented in figure 7.8. We note that LOCs are large for low prices, and large again for high prices. They become the lowest for intermediate prices (and they become zero in convex markets, at the level of the equilibrium price). In order to appreciate why LOC behaves in this way, we consider two cases: a price of 10 $/MWh (see example 7.12), and a price of 100 $/MWh (see example 7.7).

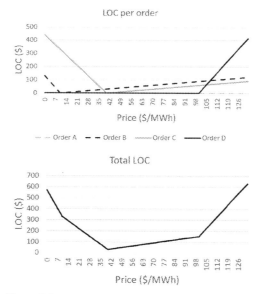

**Figure 7.8** (Top) Lost opportunity cost decomposition and (Bottom) total lost opportunity cost in example 7.8.

For the case of 10 $/MWh, lost opportunity costs are computed as follows:

- Order A would like to buy 10 MWh, and does exactly that, so the LOC is zero.
- Order B is indifferent for any quantity bought, thus the LOC is zero.
- Order C would like to sell 0 MWh, and sells 11 MWh at a loss of $330, thus the LOC is $330.
- Order D would like to sell 0 MWh, and does exactly that, thus the LOC is zero.

The total LOC is therefore $330. For the case of 100 $/MWh, lost opportunity costs are computed as follows:

- Order A would like to buy 10 MWh, and does exactly that, so the LOC is zero.
- Order B would like to buy 0 MWh and does exactly that, so the LOC is zero.
- Order C would like to sell 12 MWh (at a profit of $720) and sells 0 MWh, thus the LOC is $720.
- Order D is indifferent for any quantity sold, thus the LOC is zero.

The total LOC in the case of a price of 100 $/MWh is thus $720. We notice that this LOC comes in the form of a foregone opportunity to sell at a profit, which is to be contrasted to the opportunity cost that results from being forced to produce at a loss.

The example further illustrates the U shape of the total opportunity cost function in the lower panel of figure 7.8. When the market price is too low, large opportunity costs originate from producers who produce at a loss and from buyers who are not matched. When prices are too high, LOCs emerge from producers who are not matched from producing, and for inducing buyers who are matched to consume. The balance is struck at the equilibrium price, which results in the LOC being equal to zero. In non-convex market models, such an equilibrium price is not guaranteed to exist, but opportunity cost can still be defined, and one pricing method in particular (convex hull pricing, which is presented later) aims at minimizing it.

---

The previous example highlights the decomposition of LOC between a part that relates to foregone opportunity and a part that relates to forced losses. The latter is referred to as **make-whole payments**.

In order to avoid ambiguity, we provide mathematical definitions of LOC and make-whole payments below, using the nomenclature of section 4.3. We specifically consider a generic agent with decisions $(x, q)$, a benefit function $f(x)$, and a set of constraints $g(x) \leq 0$ and $h(x) = q$. This can be used to represent both consumers and producers. In the case of producers, $q$ is negative and $f(x)$ is equal to minus cost.

Given a market clearing price $\lambda^\star$, the surplus maximization problem of the agent can be expressed as:

$$\max_{x,q}(f(x) - (\lambda^\star)^T q)$$
$$g(x) \leq 0$$
$$h(x) = q.$$

Denote by $(\bar{x}, \bar{q})$ the market instruction,[4] and by $(x^\star, q^\star)$ the optimal self-dispatch decision that solves the surplus maximization problem above. Denote by $\bar{\Pi}$ and $\Pi^\star$ the market surplus and the self-dispatch surplus, respectively.

**Definition 7.1    Lost opportunity cost (LOC)** is defined as

$$LOC = \Pi^\star - \bar{\Pi}.$$

**Definition 7.2    Make-whole payment (MWP)** is defined as the amount of refunds that an agent needs in case it makes a negative surplus by following the market schedule $(\bar{x}, \bar{q})$ given the market price $\lambda^\star$:

$$MWP = \max(0, -(f(\bar{x}) - (\lambda^\star)^T \bar{q})).$$

---

**Example 7.9** *Make-whole payments.* We return to example 7.8 and consider the case where price $\lambda^\star$ is equal to 10 \$/MWh. We demonstrate in example 7.8 that the total LOC in this case is equal to \$330, and is due to order C producing at a loss. This LOC thus consists entirely of make-whole payments. In the case where the market clearing price $\lambda^\star$ is equal to 100 \$/MWh, we demonstrate in example 7.8 that the lost opportunity cost is \$720, due entirely to foregone opportunities of order C. Thus, make-whole payments at this price are equal to zero.

---

Some clarifications about nomenclature are in order. LOC is referred to in certain papers as **uplift**, whereas other papers reserve the term uplift for make-whole payments. Uplift is a more general term that refers to side payments required in market operations for certain out of market actions and other purposes. Side payments resulting from non-convexities constitute part of uplift. Using the term uplift for LOC or MWP reflects a deeper debate about what is deemed desirable when confronted with non-convexities in a market model: whereas some experts advocate for minimizing make-whole payments (which also implies a lower financial exposure for the market operator), others argue for minimizing lost opportunity cost (which is founded on a market equilibrium argument).

It is interesting to point out that one of the agents that maximizes surplus in the market is the network operator, and LOC can be defined meaningfully for the case of network operators, as in the case of generators and consumers. The LOC of network operators is referred to in the literature as **potential congestion revenue shortfall**. The naming of the term reflects the fact that it measures the possible inadequacy of revenues that are collected by network operators for paying back holders of financial transmission rights.

## 7.3.3    Alternative pricing proposals

As indicated in figure 7.7, the first step in market clearing in US markets is to compute the primal optimal commitment and dispatch decisions in a primal computation step.

---

[4] The market instructions are essentially limited to $\bar{q}$, but we can represent as $\bar{x}$ the internal decisions of the agent that achieve $\bar{q}$ while maximizing surplus.

Prices are then computed *given* this primal optimal solution in a second step. Side payments[5] are computed in step 3 of figure 7.7, given the schedule computed in step 1 and the prices computed in step 2.

We explore a number of alternatives in this section that have also seen their way to practical implementation (either exactly, or approximately). Nevertheless, we underline that numerous other alternatives have been proposed and analyzed in the literature; therefore, the exposition does not attempt to be exhaustive. We specifically focus on the following alternatives:

- Convex hull pricing: A proxy of this approach is currently adopted in MISO.
- Integer programming (IP) pricing: This approach is currently employed in CAISO.
- Linear programming relaxation: For reasons that are explained below, this can be interpreted as an approximation of convex hull pricing, and is currently adopted in PJM.

In the following discussion, we use the unit commitment problem of equation (7.1) as the generic representation of the market clearing model employed in figure 7.7. Step 1 of figure 7.7 amounts to computing primal optimal solutions $(u, p, r)$ by solving problem $(UC)$.

### Convex hull pricing

**Convex hull pricing** computes prices $(\lambda, \lambda^R)$ in step 2 of figure 7.7 by solving the following problem:

$$(CHP): \min_{p,u,r} \sum_{g \in G} TC_g^{\star\star}(u_g, p_g)$$

$$(p_g, r_g, u_g) \in conv(X_g), g \in G$$

$$(\lambda): \qquad \sum_{g \in G} p_{gt} = D_t, t = 1, \ldots, T$$

$$(\lambda R): \qquad \sum_{g \in G} r_{gt} = R_t, t = 1, \ldots, T.$$

Here, $X_g = \{(p_g, r_g, u_g) : h_g(p_g, r_g, u_g) \leq 0, u_g \in \{0,1\}^T\}$, $conv(X)$ corresponds to the convex hull of set $X$, and $f^{\star\star}$ corresponds to the tightest convex approximation of function $f$. The convex hull price of energy is obtained as the dual multiplier $\lambda$ of the market clearing constraint for energy, and the convex hull price of reserve is obtained as the dual multiplier $(\lambda^R)$ of the market clearing constraint for reserve.

The intuition of convex hull pricing is that it corresponds to the equilibrium price that would have emerged from a convex economy that is the "closest possible" to the non-convex economy that we are confronted with. This concept is illustrated in example 7.10.

---

[5] The use of side payments in step 3 essentially amounts to discriminatory pricing, and is a key differentiation between pools and exchanges. In European jargon, these discriminatory payments are termed **nonuniform pricing**.

**Figure 7.9** Convex hull of the aggregate cost function of orders C and D in example 7.10.

---

**Example 7.10** *Convex approximation of a non-convex economy.* Figure 7.9 presents the aggregate cost function of orders C and D in example 7.6. This is the closest convex function that lies underneath the aggregate cost function that measures the lowest cost at which orders C and D can make a given quantity available to the market. The original cost function is non-convex, because at the 11th MWh there is a drop in total cost due to the fact that we can switch from a mix of orders C and D to simply relying on order C, since the minimum output limit of order C is no longer binding. The convex approximation of this cost function increases steadily, at a rate of 40 \$/MWh up to the 12th MWh, and increases at a rate of 100 \$/MWh thereafter. The derivative of this convex function is exactly the gray function in figure 7.6. Crossing this gray supply function with the black demand function of figure 7.6 gives an equilibrium convex hull price of 40 \$/MWh.

---

One very interesting and attractive aspect of convex hull pricing is the fact that it minimizes lost opportunity cost. As we argue in section 7.3.2, since lost opportunity cost is a measure of how far off equilibrium a market finds itself, the ability of convex hull pricing to minimize LOC certainly counts as an important advantage of this pricing proposal. In fact, LOC turns out to be the duality gap of the $(UC)$ model of equation (7.1) when the complicating constraints (supply equals demand in the energy and reserves market) are relaxed.

**Proposition 7.3** *Lost opportunity cost is the duality gap of $(UC)$ when we relax the market clearing constraints, and convex hull pricing minimizes lost opportunity cost.*

*Proof* Consider the Lagrange relaxation of the $(UC)$ problem:

$$g(\lambda, \lambda^R) = \min_{u,p,r} \left\{ \sum_{g \in G}(TC_g(u_g, p_g) - \sum_{t=1}^{T} \lambda_t \cdot p_{gt} - \sum_{t=1}^{T} \lambda_t^R \cdot r_{gt}) + \right.$$
$$\left. \sum_{t=1}^{T} \lambda_t \cdot D_t + \sum_{t=1}^{T} \lambda_t^R \cdot R_t \right\}$$
$$h_g(p_g, r_g, u_g) \leq 0, g \in G$$
$$u_{gt} \in \{0,1\}, g \in G, t = 1, \ldots, T.$$

But this dual function is, by definition, equal to the maximum profit $\Pi^\star(\lambda, \lambda R)$ that agents can achieve when they are self-scheduling in response to the market prices $(\lambda, \lambda R)$:

$$g\left(\lambda, \lambda^R\right) = \sum_{g \in G} \Pi_g^\star(\lambda, \lambda R) + \sum_{t=1}^{T} \lambda_t \cdot D_t + \sum_{t=1}^{T} \lambda_t^R \cdot R_t.$$

The primal optimal solution $(\bar{u}, \bar{p}, \bar{r})$ of $(UC)$ results in the lowest possible primal objective function $p^\star$, which can also be expressed as a sum of agents' profits:

$$p^\star = \sum_{g \in G} TC_g(\bar{u}_g, \bar{p}_g)$$

$$= \sum_{g \in G} \left( TC_g(\bar{u}_g, \bar{p}_g) - \sum_{t=1}^{T} \lambda_t \cdot \bar{p}_{gt} - \sum_{t=1}^{T} \lambda R_t \cdot \bar{r}_{gt} \right) +$$

$$\sum_{t=1}^{T} \lambda_t \cdot D_t + \sum_{t=1}^{T} \lambda R_t \cdot R_t$$

$$= \sum_{g \in G} \bar{\Pi}_g(\lambda, \lambda R) + \sum_{t=1}^{T} \lambda_t \cdot D_t + \sum_{t=1}^{T} \lambda R_t \cdot R_t.$$

The LOC for any choice of market clearing price $(\lambda, \lambda R)$ can thus be seen as the difference between the dual function $g(\lambda, \lambda R)$ and the primal optimal objective function value $p^\star$:

$$LOC(\lambda, \lambda R) = \sum_{g \in G} \Pi_g^\star(\lambda, \lambda R) - \sum_{g \in G} \bar{\Pi}_g(\lambda, \lambda R)$$

$$= g(\lambda, \lambda R) - p^\star.$$

This establishes the second part of the proposition. It remains to show that convex hull prices minimize the dual function $g(\lambda, \lambda R)$. This follows from theorem 6.2[6] of Wolsey and Nemhauser (1999), from which we can conclude that convex hull prices $(\lambda^\star, \lambda R^\star)$ are exactly the prices that maximize the dual function, $d^\star = g(\lambda^\star, \lambda R^\star)$:

$$p^\star - d^\star = \sum_{g \in G} \Pi^\star(\lambda^\star, \lambda R^\star) - \sum_{g \in G} \bar{\Pi}(\lambda^\star, \lambda R^\star)$$

$$= LOC(\lambda^\star, \lambda R^\star). \qquad \square$$

---

**Example 7.11** *Convex hull pricing minimizes lost opportunity cost.* Notice in the lower panel of figure 7.8 that the LOC is minimized at a price of 40 $/MWh. This is exactly the convex hull price that is computed in example 7.10. The minimum LOC at this price is due to order B being instructed to consume 1 MWh at a loss (it is willing to pay

---

[6] The theorem states that maximizing the dual function of a Lagrange relaxation is equivalent to solving a relaxation of the primal problem where the constraints that are not relaxed are replaced by their convex hull.

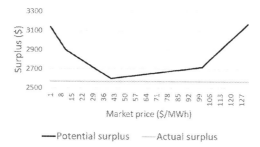

**Figure 7.10** Convex hull prices are the prices at which the potential surplus of agents is minimized. The lost opportunity cost is the difference between this potential surplus and the actual realized surplus given the market schedule.

10 $/MWh, whereas the market price is 30 $/MWh), and amounts to $30. This is the minimum LOC, and is, in particular, less than the LOC that is computed in example 7.8 for the case of EU pricing (which sets a price of 100 $/MWh, resulting in an LOC of $330) and IP pricing (which sets a price of 10 $/MWh, resulting in an LOC of $720). This minimum LOC is the difference between the dual function $d^\star = g(\lambda^\star)$ and the actual surplus of the agents in the market $p^\star$. The dual function $g(\lambda)$ can be interpreted as the maximum surplus that agents can achieve by self-scheduling given a price $\lambda$. This interpretation is indicated in figure 7.10.

Despite the aforementioned appeal of convex hull pricing, it has been criticized for some counterintuitive pricing outcomes (such as allowing units that are not online to affect price formation, or for resulting in price separation between nodes that are not congested).

The computation of convex hull prices can be nontrivial, since it requires describing the convex hull of the feasible set of generator constraints, as well as the tightest convex approximation of the cost function. Although the characterization of convex hulls is generally a hard problem, in certain cases (which arise often in power system operations) it can be easy. This issue is picked up again subsequently in this section, when we discuss linear programming relaxations.

An altogether different strategy for computing convex hull prices relies on the important observation of proposition 7.3 that convex hull prices maximize the dual function of $(UC)$. This opens up the possibility of using various convex non-differentiable optimization algorithms, which can also rely on decomposition and parallel computing, for solving the dual function optimization problem.

## IP pricing

**Integer programming (IP) pricing** computes prices $\left(\lambda, \lambda^R\right)$ in step 2 of figure 7.7 by solving the following problem:

$$(IP): \min_{p,r} \sum_{g \in G} TC_g(u_g^\star, p_g)$$

$$h_g(p_g, r_g, u_g^\star) \leq 0, g \in G$$

$$(\lambda): \quad \sum_{g \in G} p_{gt} = D_t, t = 1, \ldots, T$$

$$(\lambda R): \quad \sum_{g \in G} r_{gt} = R_t, t = 1, \ldots, T,$$

where $u^\star$ is the optimal commitment schedule computed in step 1 of figure 7.7.

IP pricing is attractive due to the simplicity of its implementation and its computational tractability, as well as the fact that it maintains some familiar pricing business rules that apply to convex models. For instance, agents that are characterized by convex models (such as network operators) face the same pricing business rules as they would under marginal pricing in a convex setting. However, IP pricing can result in non-negligible LOC, and this LOC is not guaranteed to decrease relative to the size of the market when the market size increases. By contrast, convex hull pricing is guaranteed to result in an elimination of LOC relative to the size of the market, as the size of the market increases, and the relative importance of non-convexities diminishes.

---

**Example 7.12** *IP pricing in the illustrative example.* As we show in example 7.6, the optimal decision $u_C^\star$ is to commit the order: $u_C^\star = 1$. We then arrive at the following market clearing model in step 2 of figure 7.7 for computing IP prices:

$$\max_{p,d} 300 \cdot d_A + 10 \cdot d_B - 40 \cdot p_C - 100 \cdot p_D$$

$$(\lambda): \quad d_A + d_B - p_C - p_D = 0$$

$$11 \leq p_C \leq 12$$

$$d_A \leq 10, d_B \leq 14, p_D \leq 13$$

$$d_A, d_B, p_D \geq 0.$$

As we show in exercise 7.3, the IP price is equal to 10 $/MWh. This can be understood intuitively by crossing supply and demand curves. Concretely, due to the constraint that $p_C \geq 11$, the supply curve of orders C and D essentially "starts" at 11 MWh (meaning that, at that point, it shoots from minus infinity to 40 $/MWh), and thus crosses the demand curve of the market at an equilibrium price of 10 $/MWh, which is the valuation of buy order B. This is illustrated graphically in figure 7.11.

---

### Linear programming relaxation

Pricing using the linear programming relaxation computes prices $(\lambda, \lambda^R)$ in step 2 of figure 7.7 by solving the following problem:

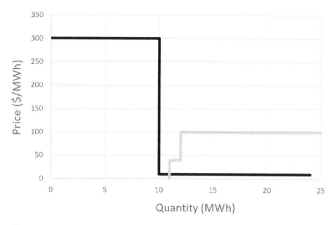

**Figure 7.11** The graphical interpretation of IP pricing that is provided in example 7.12.

$$(LPR):\ \min_{u,p,r} \sum_{g \in G} TC_g(u_g, p_g)$$

$$h_g(p_g, r_g, u_g) \leq 0, g \in G$$

$$(\lambda):\qquad \sum_{g \in G} p_{gt} = D_t, t = 1, \dots, T$$

$$(\lambda R):\qquad \sum_{g \in G} r_{gt} = R_t, t = 1, \dots, T$$

$$0 \leq u_{gt} \leq 1, g \in G, t = 1, \dots, T.$$

Thus, we compute prices by considering the linear programming relaxation of the original ($UC$) problem.

An important appeal of this method is the simplicity of its implementation. Interestingly, this approach can also be interpreted as an approximation of convex hull pricing. To see why this is the case, we first state theorem 2.1 from Balas (1998).

**Proposition 7.4 (Balas 1998, thm. 2.1)**   *Consider the set* $F = \cup_{h \in Q} F_h$, *where* $F_h = \{A^h x \geq d_0^h, x \geq 0\}$ *and* $Q$ *is a finite set. If every* $F_h$ *is nonempty, we can characterize the convex hull of* $F$ *as*

$$conv(F) = \left\{ x \in \mathbb{R}^n \mid \begin{array}{c} x = \sum_{h \in Q} \xi^h \\ A^h \xi^h - d_0^h \xi_0^h \geq 0, h \in Q \\ \sum_{h \in Q} \xi_0^h = 1 \\ (\xi^h, \xi_0^h) \geq 0, h \in Q \end{array} \right\}.$$

In order to relate this result to convex hull pricing, we can apply the theorem to the convex hull of the constraint set of order C in example 7.6. In exercise 7.4, the reader is asked to verify that the linear relaxation is exactly the convex hull of order C. The graphical interpretation is provided in figure 7.12.

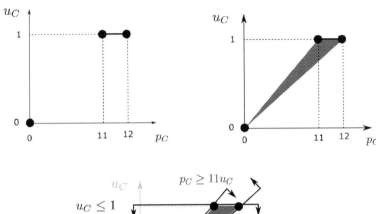

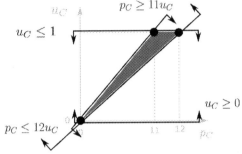

**Figure 7.12** The convex hull of the feasible set of order C in example 7.6 is simply the linear relaxation of the constraints of the order. (Top left) The feasible sets of $(p, u)$ for the case where the unit is off and on. (Top right) The convex hull of these two sets. (Bottom) The convex hull can be expressed as a set of linear inequalities, following the result of Balas.

**Example 7.13** *Linear programming relaxation in the illustrative example.* Applying the linear relaxation model to the illustrative example, we arrive at the following model:

$$\max_{p, u, d} 300 \cdot d_A + 10 \cdot d_B - 40 \cdot p_C - 100 \cdot p_D$$

$$(\lambda): \quad d_A + d_B - p_C - p_D = 0$$

$$11 \cdot u_C \leq p_C \leq 12 \cdot u_C$$

$$d_A \leq 10, d_B \leq 14, p_D \leq 13$$

$$d_A, d_B, p_D \geq 0$$

$$0 \leq u_C \leq 1.$$

As we can confirm in exercise 7.5, the market clearing price is 40 \$/MWh, which is exactly the convex hull price (as we have computed it in example 7.10).

### 7.3.4    The European exchange and EUPHEMIA

**Market products in the European power exchange**

The philosophy of the European power exchange is to use products that emulate the characteristics of thermal resources that are encountered in unit commitment models. These products include:

- Aggregated hourly orders (which include linear orders and stepwise orders).
- Complex orders (which include minimum income conditions and load gradients).
- Block orders (which include linked block orders, block orders in an exclusive group, and flexible hourly block orders).
- Merit orders and unique national price (PUN) orders.

A detailed exposition of these different products is out of scope for this textbook, though the products are described in publicly available documentation. Nevertheless, it is worth emphasizing that (some of) these products attempt to emulate the cost structure and constraints of thermal resources in unit commitment models. For instance, block orders are take-it-or-leave-it products that correspond to the on–off nature of unit commitment decisions. Linked blocks capture the dependency between startup and the ability of a unit to produce. Concretely, a child order in a linked block order cannot be accepted "more" than the parent, whereas a parent order can be accepted alone. This corresponds to a dependency whereby a unit can produce (the dispatch corresponding to the child order) as long as it is started up (the commitment corresponding to a parent order). Exclusive groups are another interesting example. Orders in an exclusive group are mutually exclusive, in the same way that different on–off trajectories are mutually exclusive in a unit commitment problem. A key difference is that the full set of trajectories can be endogenously optimized in a unit commitment model, whereas a subset of trajectories needs to be enumerated in an exclusive group.

**Formulation of the European power market clearing problem**

The European day-ahead market clearing model is a quadratic integer program subject to complementarity constraints. The quadratic objective function is the economic welfare produced by trading energy. The integer variables relate to the acceptance or rejection of block orders and other market products. The complementarity constraints relate to the pricing business rules that are deemed acceptable by European market stakeholders.

In order to formalize the European day-ahead market model, we return to our illustrative example. Stated in words, the goal is to maximize economic welfare, subject to the following constraints:

- Continuous orders (orders A, B, and D) are maximizing surplus given the clearing prices.
- Block orders (order C) are not paradoxically accepted, but may be paradoxically rejected.

The conditions that pertain to continuous orders can be expressed straightforwardly from the KKT conditions of the surplus maximization problems of these orders. For instance, the surplus maximization problem of order A can be written out as follows:

$$\max_{d_A \geq 0} (300 - \lambda) \cdot d_A$$
$$(\mu_A): \quad d_A \leq 10.$$

This surplus maximization can be expressed equivalently through the following set of KKT conditions:

$$0 \le d_A \perp \lambda - 300 + \mu_A \ge 0,$$
$$0 \le \mu_A \perp 10 - d_A \ge 0.$$

This set of conditions can be expressed equivalently as the following set of pricing business rules:

- If the order is fully executed ($d_A = 10$), then it is in the money or at the money ($300 \ge \lambda$).
- If the order is fractionally executed ($0 < d_A < 10$) then the order is at the money ($300 = \lambda$).
- If the order is rejected ($d_A = 0$), then it is at the money or out of the money ($300 \le \lambda$).

We now turn our attention to the case of block orders, and specifically consider order C. The tricky thing about expressing the pricing business rules pertaining to block orders is that we cannot fall back to KKT conditions, as we did for continuous orders, because the profit maximization of the block orders is a mixed integer linear program, and thus does not have corresponding KKT conditions. But its linear programming relaxation *does* correspond to an equivalent KKT system, and we focus on this KKT system.

The linear relaxation of the profit maximization of order C is expressed as follows:

$$\max_{u_C, p_C} (\lambda - 40) \cdot p_C$$
$$(\mu_C): \quad p_C - 12 \cdot u_C \le 0$$
$$(\nu_C): \quad 11 \cdot u_C - p_C \le 0$$
$$(s_C): \quad u_C \le 1$$
$$p_C, u_C \ge 0.$$

By strong duality (since the dual objective function is equal to $s_C$), we can conclude that $s_C$ is the profit of the order (which is exactly the primal objective function). The KKT conditions of this profit maximization problem can be written out as follows:

$$0 \le p_C \perp 40 - \lambda + \mu_C - \nu_C \ge 0,$$
$$0 \le u_C \perp -12 \cdot \mu_C + 11\nu_C + s_C \ge 0,$$
$$0 \le \mu_C \perp 12u_C - p_C \ge 0,$$
$$0 \le \nu_C \perp p_C - 11u_C \ge 0,$$
$$0 \le s_C \perp 1 - u_C \ge 0.$$

The key observation for formulating the pricing business rules that allow block orders to be paradoxically rejected is that this amounts to removing the complementarity operator in condition $0 \le s_C \perp 1 - u_C \ge 0$, and simply replacing it by the two inequalities $s_C \ge 0, u_C \le 1$. To see why this is the case, let us observe that the complementarity condition precludes that the order can have both a positive surplus *and* not be fully accepted: $s_C > 0$ and $u_C < 1$. This is exactly the essence of paradoxically rejected bids.

The overall market clearing model can thus be expressed as follows:

$$(EUDA): \quad \max_{p,d,u,\lambda,\mu,v,s} \quad 300 \cdot d_A + 10 \cdot d_B - 40 \cdot p_C - 100 \cdot p_D$$

$$0 \le d_A \perp \lambda - 300 + \mu_A \ge 0, 0 \le \mu_A \perp 10 - d_A \ge 0$$

$$0 \le d_B \perp \lambda - 10 + \mu_B \ge 0, 0 \le \mu_B \perp 14 - d_B \ge 0$$

$$0 \le p_D \perp 100 - \lambda + \mu_D \ge 0, 0 \le \mu_D \perp 13 - p_D \ge 0$$

$$0 \le p_C \perp 40 - \lambda + \mu_C - v_C \ge 0$$

$$0 \le u_C \perp -12\mu_C + 11v_C + s_C \ge 0$$

$$0 \le \mu_C \perp 12 \cdot u_C - p_C \ge 0$$

$$0 \le v_C \perp p_C - 11 \cdot u_C \ge 0$$

$$s_C \ge 0, u_C \in \{0, 1\}$$

$$dA + dB - pC - pD = 0.$$

The objective function aims to maximize economic welfare. The first three lines of constraints represent the pricing business rules of continuous orders. The remaining constraints, save for the last, correspond to the pricing business rules of blocks.

Note that this model is formulated in both primal and dual variables. Note, furthermore, that it is mixed integer, because it includes the binary variable $u_C$ in addition to the continuous primal and dual variables. Finally, the problem is subject to complementarity constraints, which encode the pricing business rules. This last element raises formidable computational challenges. Next, we sketch a branch and cut strategy for tackling the problem.

## EUPHEMIA

**EUPHEMIA** is the algorithm that is used for solving the pan-European day-ahead electricity market clearing model. The algorithm relies on a **branch-and-cut** backbone, which includes cuts for discarding solutions that do not admit prices that support the chosen matching of orders. The general structure of a branch-and-cut algorithm is described in figure 7.13.

We comment briefly on each of the functions (boxes) and checks (diamonds) in figure 7.13:

• The *primal problem* function amounts to a convex quadratic program, which is computationally easy. Integer variables are replaced by their linear relaxations.
• The *integer solution* check is a trivial inspection of whether or not the solution of the primal problem results in integer values for binary variables $u$, as in the standard branch-and-bound algorithm.
• The *feasible pricing problem* check can be formulated as a linear program, which is the complementarity system that enforces the pricing business rules described above for *fixed* values of the primal variables $(u, p, d)$. Since these primal variables

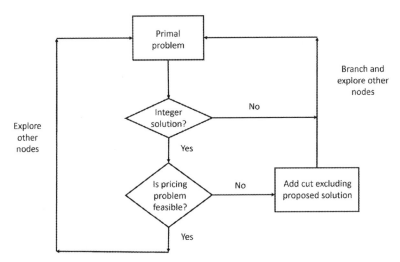

**Figure 7.13** A logical diagram of the branch-and-cut algorithm that forms the backbone of EUPHEMIA.

are fixed to the optimal solution of the *primal problem* box, this step is computationally easy and amounts to checking if a system of inequalities is feasible or not. The variables that are free in this computation are the dual variables of the problem, namely $(\mu, \nu, \lambda, s)$.

- The *add cut excluding proposed solution* function is where much of the "magic" in the EUPHEMIA algorithm happens, since problem-specific cuts are added to the problem that exclude not only the current solution from further consideration but also numerous other alternatives that never need to be explored in the branch-and-bound tree. Such "deep" cuts exploit the specific structure of the problem that EUPHEMIA is tackling.

## Problems

**7.1** *Validity of min up/down time constraints.* Verify that constraints (7.2) enforce the following implication: if $v_{gt} = 1$, then $u_{g\tau} = 1$ for all $t \le \tau \le t + UT_g - 1$. Similarly, verify that constraints (7.3) enforce the following implication: if $z_{gt} = 1$, then $u_{g\tau} = 0$ for all $t \le \tau \le t + DT_g - 1$.

**7.2** *A mixed integer linear program without a clearing price.* Implement example 7.6 as a mixed integer linear program in a mathematical programming language, and confirm the optimal solution and its objective function value.

**7.3** *Computing the IP price.* Implement example 7.12 as a linear program in a mathematical programming language, and confirm the market clearing price.

**7.4** *Linear relaxations and convex hulls.* Use proposition 7.4 to prove that the convex hull of order C in example 7.6 is given by the linear relaxation of the constraints of the order.

**7.5** *Computing the linear programming relaxation price.* Implement example 7.13 as a linear program in a mathematical programming language, and confirm the market clearing price.

**7.6** *EU day-ahead market model.* Formulate the (*EUDA*) model in a mathematical programming language, and confirm that the results correspond to the solution derived in example 7.7.

# Bibliography

**Section 7.1**  The unit commitment model presented in this section, which includes thermal states, startup/shutdown profiles, and dispatchable operation, is based on Simoglou et al. (2010). Startup and shutdown profiles are an important modeling aspect for explaining price formation in systems with considerable shares of thermal units, since units that are not in dispatchable mode should not set market clearing prices (Camelo et al. 2018). The complexity of the unit commitment model further escalates in the case of multistage combined cycle generation units, where different yet interdependent modes of operation need to be accounted for (Cohen & Ostrowski 1995, Simoglou et al. 2010, Papavasiliou, He & Svoboda 2015).

Since unit commitment models are large-scale mixed integer programs, the characterization of the convex hulls of the feasible set of units' operation in these models can significantly accelerate branch-and-bound solvers for solving these problems. The research in this area was pioneered by Rajan and Takriti (2005), who aimed to prove that the minimum up and down time constraints and transition equation provided in section 7.1 define the convex hull of a minimum up/down time polytope under certain conditions. Ramp constraints and other operational constraints complicate the characterization of the convex hull of the unit commitment problem, which motivated subsequent work to that of Rajan and Takriti (2005); see for instance Morales-España, Latorre, and Ramos (2013), Morales-España, Gentile, and Ramos (2015), Silbernagl, Huber, and Brandenberg (2015), Damcı-Kurt et al. (2016), Gentile, Morales-Espana, and Ramos (2017), and Queyranne and Wolsey (2017). A survey of the results from this literature is provided in table 2.1 of Stevens (2016).

The endogenous representation of uncertainty in unit commitment models was pioneered by Takriti, Birge, and Long (1996), shortly thereafter followed by Carpentier et al. (1996). Stochastic unit commitment returned to the forefront a few years after these publications due to the emergence of ambitious renewable energy integration policies. In this context, stochastic unit commitment became a natural tool for quantifying the locational reserve requirements in the system and simulating the operation of the system under uncertainty (Ruiz et al. 2009, Tuohy et al. 2009, Meibom et al. 2010, Papavasiliou, Oren & O'Neill 2011, Papavasiliou and Oren 2013). Due to the large scale of the resulting models, considerable research has focused on the development of decomposition methods for solving these problems (Nowak & Römisch 2000, Watson & Woodruff 2011, Papavasiliou, Oren & Rountree 2015, Aravena et al. 2021). Robust optimization (Bertsimas et al. 2013) and other frameworks of optimization under uncertainty have also been considered for supporting the operation of the system under deep penetration levels of renewable supply.

Scenario selection/reduction is a nontrivial challenge in the formulation of stochastic programming models. A classic approach, which is inspired by the idea of minimizing the distance of the probability distribution of the reduced scenario set to the original set of scenarios, relies on the theory developed by Dupacova, Gröwe-Kuska and Römisch (2003) and the faster variants proposed by Heitsch and Römisch (2003). The effectiveness of these methods in the context of stochastic unit commitment is demonstrated by Gröwe-Kuska et al. (2002). Approaches that aim at accounting for the interplay between selected scenarios and the objective function that is being optimized include the work of Morales et al. (2009) and an approach inspired by importance sampling by Papavasiliou and Oren (2013).

**Section 7.2**   Example 7.5 is based on Stoft (2002).

**Section 7.3**   Example 7.6 is sourced from Madani et al. (2018). Convex hull pricing appears in the context of general equilibrium models in the work of Starr (1969) and Arrow and Hahn (1971). It is advocated for in the context of short-term electricity markets and unit commitment by Hogan and Ring (2003); see also Gribik, Hogan, and Pope (2007). These works also refer to lost opportunity cost interchangeably as uplift. The fact that lost opportunity cost diminishes as market size increases in convex hull pricing is established in Chao (2019). A critical review of convex hull pricing is provided by Schiro et al. (2015), where the authors also refer to make-whole payments as uplift. The term potential congestion revenue shortfall, and its connection to revenue inadequacy in FTR payments, is used in Garcia, Nagarajan, and Baldick (2020). The fact that convex hull pricing minimizes lost opportunity cost is proven in Hogan and Ring (2003). Integer programming pricing is proposed by O'Neill et al. (2005). The reference to this pricing proposal as IP pricing is used in Hogan and Ring (2003). Primal methods for computing convex hull prices that rely on Dantzig–Wolfe decomposition have been recently proposed by Andrianesis et al. (2021), while a dual approach relying on the level method is presented by Stevens and Papavasiliou (2022). A linearization of the European power exchange model is presented in Madani and Van Vyve (2015).

# 8     Hydrothermal planning

This chapter focuses on medium-term hydrothermal planning. The general notation of multi-stage stochastic linear programs is introduced in section 8.1. We then focus on the formulation of the medium-term hydrothermal planning problem in section 8.2, and specifically analyze the dynamic programming value functions that are used for both solving the problem in practice and also deriving economically meaningful information from the model. Section 8.3 discusses the evaluation of performance in stochastic programming. This is a topic that extends beyond the scope of hydrothermal planning and is useful for determining the added value of stochastic programming formulations.

## 8.1   Multi-stage stochastic linear programming

In this section we introduce multi-stage stochastic linear programming. We do so gradually, by first describing two-stage stochastic programs in section 8.1.1. The representation of uncertainty in multiple stages is then taken on in section 8.1.2. Multi-stage stochastic linear programs are finally introduced in section 8.1.3.

### 8.1.1   Two-stage stochastic linear programs

Two-stage stochastic linear programs exhibit a two-stage structure that captures the interaction between decisions and information that is revealed over time. The typical sequence of events in two-stage stochastic programs is depicted in figure 8.1, and can be summarized as follows:

1. **First-stage decisions** $x$ are taken before any uncertainty is revealed.
2. Uncertainty is revealed in the form of an **outcome** $\omega \in \Omega$, which is drawn from a **sample space** $\Omega$.
3. The decision-maker reacts with **second-stage decisions** $y(\omega)$. Note that these second-stage decisions are indexed by $\omega$, meaning that they are adapted to the uncertainty that has been revealed. The notation $y(\omega)$ and $y_\omega$ is used interchangeably in what follows.

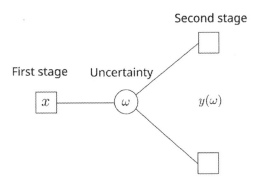

**Figure 8.1** Sequence of events in two-stage stochastic programs.

Decisions in figure 8.1 are indicated by squares, while the revelation of uncertainty is indicated by circles. The goal of the agent in this uncertain environment is to minimize expected cost, or maximize expected benefit. In order to define expectation, we must assume a certain probability measure on the events of the sample space. The expectation corresponding to this probability measure is indicated as $\mathbb{E}$. The overall problem can then be expressed as follows:

$$
\begin{aligned}
(TSLP): \quad &\min_{x,y} c^T x + \mathbb{E}\left[ q(\omega)^T y(\omega) \right] \\
&\text{s.t. } Ax = b \\
&\quad T(\omega)x + W(\omega)y(\omega) = h(\omega), \omega \in \Omega \\
&\quad x \geq 0, y \geq 0.
\end{aligned}
$$

We assume that there are $n_1$ first-stage decisions, thus $x \in \mathbb{R}^{n_1}$. For a given realization $\omega$, there are $n_2$ second-stage decisions, thus $y(\omega) \in \mathbb{R}^{n_2}$. The deterministic parameters of the first stage are $c \in \mathbb{R}^{n_1}$, $b \in \mathbb{R}^{m_1}$, $A \in \mathbb{R}^{m_1 \times n_1}$. Second-stage parameters are random, since they depend on the outcome $\omega$. Their dimensions are $q(\omega) \in \mathbb{R}^{n_2}$, $h(\omega) \in \mathbb{R}^{m_2}$, $T(\omega) \in \mathbb{R}^{m_2 \times n_1}$, and $W(\omega) \in \mathbb{R}^{m_2 \times n_2}$.

---

**Example 8.1** *A two-stage deterministic hydro planning model.* This example is based on example A.13 and problem A.2. Consider a system with two thermal plants. The first plant, G1, has a marginal cost of 10 \$/MWh and can produce up to 60 MW. The second plant, G2, has a marginal cost of 50 \$/MWh, and can produce up to 100 MW. The system also has a hydro unit that can store energy with an efficiency factor $\eta = 0.8$, as in problem A.2. The load is equal to 50 MW in period 1, and 100 MW in period 2. We are interested in bringing the model into the format of a generic two-stage model by introducing a variable that indicates the amount of energy that is stored in the hydro reservoir. In this specific example, we have $\Omega = \{1\}$, i.e. there is no uncertainty since there is a unique outcome. Naturally, its probability is equal to 1. The overall model can be formulated as follows:

$$\min_{p.dH,pH,e} \sum_{g \in G} MC_g \cdot p_{g1} + \sum_{g \in G} MC_g \cdot p_{g2}$$

$$p_{g1} \leq P_g, g \in G$$

$$e_1 = dH_1 - pH_1/\eta$$

$$D_1 + dH_1 - \sum_{g \in G} p_{g1} - pH_1 = 0$$

$$p_{g2} \leq P_g, g \in G$$

$$e_2 = dH_2 - pH_2/\eta + e_1$$

$$D_2 + dH_2 - \sum_{g \in G} p_{g2} - pH_2 = 0$$

$$e_2 = 0$$

$$p, pH, dH, e \geq 0.$$

The variable $p_{gt}$ corresponds to the production of thermal unit $g \in G$ in period $t$. The variable $dH_t$ corresponds to the electricity demand of the hydro unit in period $t$, $pH_t$ corresponds to the production of the hydro unit in period $t$, and $e_t$ corresponds to the amount of hydro energy in the hydro reservoir at the end of period $t$. In order to arrive at a model that is equivalent to that of example A.13 and problem A.2, we assume that the amount of energy in the hydro reservoir at the beginning of period 1 is zero, and require that the amount of energy at the end of period 2, $e_2$, is also zero. The first-stage decision vector $x$ consists of $p_{g1}$ for all $g \in G, pH_1, dH_1$, and $e_1$. The second-stage decision vector $y$ consists of $p_{g2}$ for all $g \in G, pH_2, dH_2$, and $e_2$. We assume that load $D_t$ needs to be fully satisfied at every time period. Note that the constraint that links the first to the second stage is the storage balance constraint, where we see that the storage level in period 2, $e_2$, depends on the amount of stored energy in period 1, $e_1$. All other constraints are either implicating first-stage variables alone, or second-stage variables alone. Solving the model, we find that the optimal solution is for unit G1 to produce at its maximum capacity in both periods. The excess energy in period 1 is stored in the hydro reservoir. Even if only a fraction of this energy is recovered in period 2 (only $pH_2 = 8$ MWh of electricity are produced out of the $e_1 = 10$ MWh that are stored), this is preferable to using unit G2 in period 2.

---

The basic hydrothermal planning model presented in example 8.1 typically unfolds with a monthly time step. The model can be extended to account for rainfall, uncertainty in various parameters (e.g. uncertainty in rainfall), hydro storage limits, time-varying fuel costs, and load shedding:

$$\max_{p.d.dH,pH,e} V \cdot d_1 - \sum_{g \in G} MC_{g1} \cdot p_{g1} + \sum_{\omega \in \Omega} P_\omega \cdot \left( V \cdot d_2(\omega) - \sum_{g \in G} MC_{g2} \cdot p_{g2}(\omega) \right)$$

$$p_{g1} \leq P_g, g \in G$$

$$e_1 = R_1 + dH_1 - pH_1/\eta$$

$$e_1 \leq E$$

$$d_1 + dH_1 - \sum_{g \in G} p_{g1} - pH_1 = 0$$

$$d_1 \leq D_1$$

$$p_{g2}(\omega) \leq P_g, g \in G, \omega \in \Omega$$

$$e_2(\omega) = R_2(\omega) + dH_2(\omega) - pH_2(\omega)/\eta + e_1, \omega \in \Omega$$

$$d_2(\omega) + dH_2(\omega) - \sum_{g \in G} p_{g2}(\omega) - pH_2(\omega) = 0, \omega \in \Omega$$

$$e_2(\omega) \leq E, \omega \in \Omega$$

$$d_2(\omega) \leq D_2, \omega \in \Omega$$

$$p, d, pH, dH, e \geq 0.$$

The probability of $\omega$ is indicated by $P_\omega$. Note that we only assume rainfall uncertainty in this formulation, although demand uncertainty (i.e. $D_2(\omega)$), uncertainty in component availability (i.e. $P_g(\omega)$), or uncertainty in fuel cost[1] (i.e. $MC_{gt}(\omega)$), can be added straightforwardly. Rainfall in period 1 is indicated by $R_1$, whereas rainfall in period 2 is indicated by $R_2(\omega)$. The valuation of consumers is indicated by $V$. The energy storage limit of the hydro reservoir is indicated by $E$. The marginal cost of technology $g$ in time step $t$ is indicated by $MC_{gt}$.

---

**Example 8.2** *A two-stage stochastic hydro planning model.* We return to the system of example 8.1, where we assume that $V = 1000$ \$/MWh, and that the reservoir can hold up to $E = 10$ MWh. Let us further assume that the rainfall in period 1 is $R_1 = 5$ MWh, and that the uncertainty in period 2 is described by two equally likely scenarios, $\Omega = \{1, 2\}$, with $P_1 = P_2 = 0.5$, where $R_2(1) = 0$ MWh and $R_2(2) = 10$ MWh. The optimal solution is depicted in figure 8.2. Note that the hydro management policy is fairly obvious. In period 1, unit G1 is used to the greatest extent possible, i.e. until the hydro reservoir is entirely full. In period 2, plant G2 is used to the extent that the low-cost unit G1 and the hydro plant (including the second-period rainfall) cannot cover demand. The resulting optimal objective function value is equal to \$147450. The optimal hydro management policy becomes less obvious in multi-stage models, where the trade-off is between over-filling reservoirs (with the risk of dumping excess water in high-rainfall scenarios) and under-filling reservoirs (with the risk of using high-cost thermal units or even resorting to load shedding). Stochastic programming models allow decision-makers to reach optimal decisions in such nonobvious settings.

---

[1] Uncertainty in right-hand side parameters such as rainfall, demand, or component availability are easier to handle computationally than uncertainty in fuel cost. The reason for this relates to the workings of the decomposition algorithms that are used for tackling the general form of the hydrothermal planning problem, and a precise explanation of the difficulty that is encountered when there is uncertainty in the parameters of the objective function of the problem exceeds the scope of this book. Hydrothermal planning problems are thus typically formulated with possibly time-varying fuel costs, which are, however, not uncertain.

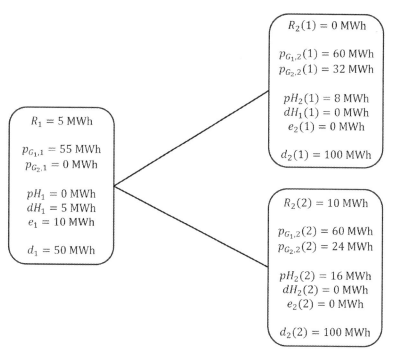

$R_2(1) = 0$ MWh

$p_{G_1,2}(1) = 60$ MWh
$p_{G_2,2}(1) = 32$ MWh

$pH_2(1) = 8$ MWh
$dH_1(1) = 0$ MWh
$e_2(1) = 0$ MWh

$d_2(1) = 100$ MWh

$R_1 = 5$ MWh

$p_{G_1,1} = 55$ MWh
$p_{G_2,1} = 0$ MWh

$pH_1 = 0$ MWh
$dH_1 = 5$ MWh
$e_1 = 10$ MWh

$d_1 = 50$ MWh

$R_2(2) = 10$ MWh

$p_{G_1,2}(2) = 60$ MWh
$p_{G_2,2}(2) = 24$ MWh

$pH_2(2) = 16$ MWh
$dH_2(2) = 0$ MWh
$e_2(2) = 0$ MWh

$d_2(2) = 100$ MWh

**Figure 8.2** Graphical illustration of the optimal solution of the two-stage hydrothermal planning problem under uncertainty in example 8.2.

If the optimal policy in example 8.2 is fairly obvious, there are two-stage stochastic programs where the interaction between time and uncertainty is already nontrivial, even if the decision problem unfolds in only two stages. One such famous example is the **newsboy problem** (or **newsvendor problem**), which is a classic two-stage stochastic program. In this problem, a newsvendor needs to decide how many newspapers to buy for the following day, while facing an uncertain demand. Newspapers can be bought today at a low cost, and resold tomorrow at a profit. Excess orders may result in waste if the demand that actually materializes is not high enough to absorb the ordered newspapers. A low number of orders today is also not favorable, because the newsvendor loses the opportunity to sell at a profit tomorrow. This highlights the nontrivial interplay between time and uncertainty in a two-stage setting. The reader is asked to implement the model and analyze the solution in problem 8.3.

## 8.1.2 Modeling multi-period uncertainty

Before generalizing the two-stage stochastic programming model to multiple stages, we discuss various options for representing uncertainty in multiple stages. The modeling alternatives presented here are not the only options, but they are chosen because they are encountered quite commonly in the stochastic programming literature.

When moving from two to multiple stages, it is important to represent the gradual revelation of information in the form of a stochastic process $\xi_t$ that unfolds over time.

We consider three models for representing this gradual revelation of information, where each is a special case of the preceding: **scenario trees**, **lattices**, and processes with **stagewise independence**.

A **scenario tree** is a graphical representation of a stochastic process $\{\xi_t\}, t = 1, \ldots, H$, i.e. a graph consisting of a set of nodes $N$ and a set of directed edges $E$, where:

- each node of the scenario tree corresponds to a history of realizations of the process up to a certain stage $t$: $\xi_{[t]} = (\xi_1, \ldots, \xi_t)$, and
- each edge of the scenario tree corresponds to transitions from $\xi_{[t]}$ to $\xi_{[t+1]}$.

The root of the tree corresponds to the first stage, $t = 1$. The **ancestor** of a node $\xi_{[t]}$, denoted $A\left(\xi_{[t]}\right)$, is the *unique* adjacent node on the scenario tree that precedes $\xi_{[t]}$. Mathematically,

$$A\left(\xi_{[t]}\right) = \left\{\xi_{[t-1]} : \left(\xi_{[t-1]}, \xi_{[t]}\right) \in E\right\}.$$

The **children**, or **descendants**, of a node, denoted $C\left(\xi_{[t]}\right)$, is the set of nodes of the scenario tree that are adjacent to $\xi_{[t]}$ and occur at stage $t + 1$:

$$C\left(\xi_{[t]}\right) = \left\{\xi_{[t+1]} : \left(\xi_{[t]}, \xi_{[t+1]}\right) \in E\right\}.$$

Note that each node is effectively tagged by a certain stage index $t$. If a node corresponds to a history of two realizations, it effectively corresponds to stage 2. If it corresponds to three realizations, it effectively corresponds to stage 3. In terms of actually implementing a scenario tree model in code, it is useful to explicitly associate a stage parameter to each node of the scenario tree.

In order to fully specify the multi-stage model of uncertainty, we need to assign a history $\xi_{[t]}$ to every node of the scenario tree as well as a transition probability $P\left(\xi_{[t+1]}|\xi_{[t]}\right)$ to every edge. Figure 8.3 depicts a four-stage scenario tree.

---

**Example 8.3** *A four-stage scenario tree of rainfall.* Returning to example 8.2, let us consider adding two additional stages to the model. The scenario tree is shown in figure 8.4. Each value indicated within the node corresponds to the realization $\xi_t$ of rainfall for the given stage, but essentially each node represents the history of all realizations leading up to that node. Each value indicated on the arc of the scenario tree corresponds to the transition probability from the origin to the destination node. Taking the top node of stage 2 as an example (i.e. the one where $\xi_2 = 10$), its children are the two top nodes of stage 3 (i.e. the ones with realizations $\xi_3 = 15$ and $\xi_3 = 5$), and its ancestor is the root node.

---

In probability theory, there is a distinction between outcomes (also referred to as **scenarios** in multistage stochastic programming) and **events**. Events are subsets of the set of outcomes. The set of outcomes $\Omega_H$ of a scenario tree is the set of trajectories from period 1 to the end of the horizon, $H$. It corresponds to the set of nodes of the scenario tree in the last stage $H$. Correspondingly, $\Omega_t$ corresponds to the set of scenarios from period 1 to period $t$.

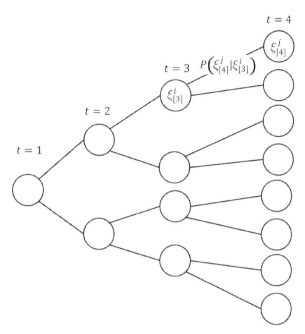

**Figure 8.3** A graphical representation of a scenario tree. Each node of the scenario tree $\xi_{[t]} \in N$ is associated with a history of realizations of the stochastic input. Each edge $(\xi_{[t]}, \xi_{[t+1]}) \in E$ is associated with a transition probability $P(\xi_{[t+1]}|\xi_{[t]})$, which implies a probability of realization for each node.

As mentioned previously, events are subsets of the set of scenarios. For instance, the top and next-to-top node of the last stage of the scenario tree of figure 8.4 correspond to the nodes $(5, 10, 15, 20)$ and $(5, 10, 15, 10)$, respectively. Their union corresponds to an event of $\Omega_4$, and is specifically the event in which there was a lot of rain in periods 1, 2, and 3. Events encode information in the sense that we can only tell them apart if we have enough information. The event $(5, 10, 15, 20)$ is the event in which there was a lot of rain in all periods of the horizon of the scenario tree in figure 8.4. Although this is an event that corresponds to the set of outcomes $\Omega_4$, it is *not* an event in the set of outcomes $\Omega_3$. Even if we know that we are in outcome $(5, 10, 15)$ of $\Omega_3$ (i.e. there was a lot of rain in periods 1, 2, and 3), this does *not* guarantee to us what outcome of $\Omega_4$ we are in (i.e. we do not know if there will be more rain or less rain in stage 4).

The probability of landing in a given node $\xi_{[t]}^i = (\xi_1^i, \xi_2^i, \dots, \xi_t^i)$ of the scenario tree is equal to the probability of traversing the trajectory that leads to that node:

$$P\left(\xi_{[t]}^i\right) = P\left(\xi_{[1]}^i\right) \cdot P\left(\xi_2^i | \xi_{[1]}^1\right) \cdot \dots \cdot P\left(\xi_t^i | \xi_{[t-1]}^i\right). \tag{8.1}$$

Scenario trees are fairly general in terms of the processes that they represent. A specific type of stochastic processes that are often used in hydrothermal planning models are **Markov processes**, i.e. processes where the conditional distribution of $\xi_t$

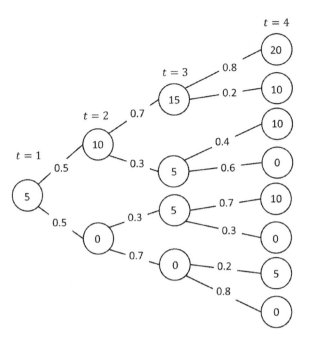

**Figure 8.4** The scenario tree of the four-stage model of rainfall uncertainty in example 8.3. The numbers in the nodes correspond to the realization $\xi_t$ of the process at stage $t$, while the numbers on the edges correspond to transition probabilities $P\left(\xi_{[t]}|\xi_{[t-1]}\right)$.

only depends on the last realization of the process, $\xi_{t-1}$, and not on the history of the process, $\xi_{[t-1]} = (\xi_1, \ldots, \xi_{t-1})$. Mathematically:

$$P\left(\xi_t|\xi_{[t-1]}\right) = P(\xi_t|\xi_{t-1}),$$

for all stages $t$ and possible histories $\xi_{[t]}$.

Markov processes can be represented as scenario trees, as long as the transition probabilities respect the Markov property. Alternatively, Markov processes can be represented as lattices.

A **lattice** is a special case of a scenario tree that can be used for representing a Markov process $\{\xi_t\}, t = 1, \ldots, H$, where

- each node of the lattice corresponds to a realization $\xi_t$, and
- each edge corresponds to a transition from $\xi_t$ to $\xi_{t+1}$ and a probability $P(\xi_{t+1}|\xi_t)$.

A lattice is represented graphically in figure 8.5.

Given the transition probabilities on the edges of the lattice, it is possible to associate a probability to each node of the lattice. This is the probability of landing at the respective node, conditional on starting at the root node. This probability can be computed recursively as follows, where $\xi_1^1$ is the unique value of $\xi_1$ at the root node and $\Xi_t$ is the set of possible values of $\xi_t$ in period $t$:

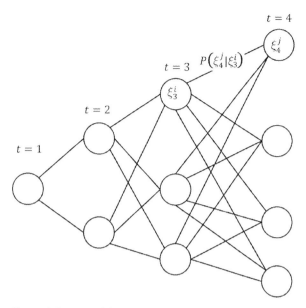

**Figure 8.5** A graphical representation of a lattice. Each node $\xi_t \in N$ is associated with a realization of the stochastic input. Each edge $(\xi_t, \xi_{t+1}) \in E$ is associated with a transition probability $P(\xi_{t+1}|\xi_t)$.

$$P(\xi_1 = \xi_1^1) = 1,$$

$$P(\xi_2 = n) = P(\xi_2 = n|\xi_1^1), n \in \Xi_2,$$

$$\vdots$$

$$P(\xi_t = n) = \sum_{m \in \Xi_{t-1}} P(\xi_t = n|\xi_{t-1} = m) \cdot P(\xi_{t-1} = m), n \in \Xi_t,$$

$$\vdots$$

$$P(\xi_H = n) = \sum_{m \in \Xi_{H-1}} P(\xi_H = n|\xi_{H-1} = m) \cdot P(\xi_{H-1} = m), n \in \Xi_H.$$

We can unfold lattices into scenario trees (in other words, we can represent Markov processes by scenario trees). On the contrary, we cannot always fold scenario trees into lattices.

---

**Example 8.4** *A scenario tree that is not a Markov process.* Consider the scenario tree of example 8.3. This is not a Markov process. To see why this is the case, note that:

$$P\left(\xi_4 = 10|\xi_{[3]} = (5, 10, 5)\right) = 0.4,$$

whereas

$$
\begin{aligned}
P(\xi_4 = 10|\xi_3 = 5) &= \frac{P(\xi_4 = 10, \xi_3 = 5)}{P(\xi_3 = 5)} \\
&= \frac{(0.5 \cdot 0.3) \cdot 0.4 + (0.5 \cdot 0.3) \cdot 0.7}{0.5 \cdot 0.3 + 0.5 \cdot 0.3} = 0.55.
\end{aligned}
$$

Since the two values are different, the Markov property is violated.

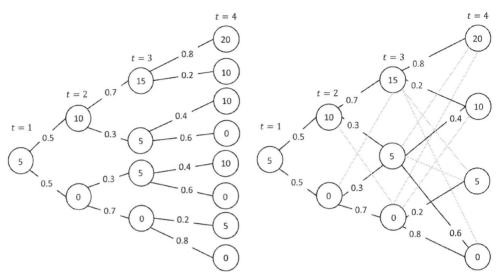

**Figure 8.6** (Left) The scenario tree of the four-stage model of Markovian rainfall uncertainty in example 8.5, as well as (Right) the corresponding lattice. The numbers in the nodes correspond to the realization $\xi_t$ of the process at stage $t$. The gray lines in the lattice correspond to transitions with zero probability.

**Example 8.5** *A four-stage scenario tree of rainfall that is Markov.* We return to the scenario tree of example 8.3, and modify it slightly so that it satisfies the Markov property. Specifically, the condition that needs to hold is that whenever $\xi_t = \xi_t^i$ in two different nodes of the same stage of the scenario tree, the transition probabilities from that node to each node of the next stage must be identical. Concretely, in stage 3 we have two possible paths that lead to the realization $\xi_3 = 5$. In order for the scenario tree to satisfy the Markov property, it is necessary for the transition probabilities of the scenario tree from $\xi_3 = 5$ to all nodes of stage 4 to be equal, regardless of whether the history that led to $\xi_3 = 5$ is $\xi_{[3]} = (5, 10, 5)$ or $\xi_{[3]} = (5, 0, 5)$. The scenario tree and the corresponding lattice are presented in figure 8.6.

**Stagewise independent processes** are stochastic processes $\{\xi_t\}, t = 1, \ldots, H$, in which the distribution of $\xi_t$ only depends on the stage $t$, i.e. $\xi_t$ is independent of $\xi_{[t-1]}$. Stagewise independent processes are thus Markovian and can be represented both as scenario trees as well as lattices. In the same way that not all scenario trees correspond to lattices, it is also true that not all lattices correspond to stagewise independent processes.

**Example 8.6** *A lattice that is not stagewise independent.* Consider the lattice in the right part of figure 8.6. The lattice is not stagewise independent. For instance, the probability of $\xi_4 = 10$ depends on $\xi_3$. Concretely,

$$P(\xi_4 = 10 | \xi_3 = 5) = 0.4,$$

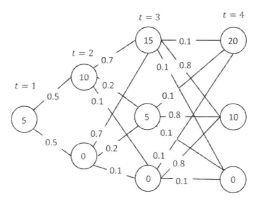

**Figure 8.7** The lattice of the four-stage model of stagewise independent rainfall uncertainty in example 8.7. The values in the nodes correspond to the realizations of $\xi_t$, while the values on the edges correspond to transition probabilities $P(\xi_{t+1}|\xi_t)$.

while

$$P(\xi_4 = 10|\xi_3 = 15) = 0.2.$$

Since the two probabilities are different, the condition of stagewise independence is violated.

**Example 8.7** *A four-stage lattice that is stagewise independent.* We return to the lattice of example 8.5 and modify it so that it satisfies stagewise independence. Specifically, we require that the probability leading into a node $\xi_t$ in stage $t$ is identical, no matter which node it is adjacent to in stage $t - 1$. For instance, $\xi_4 = 20$ occurs with a probability of 0.1, no matter whether the process is equal to $\xi_3 = 15$, 5, or 10 in the previous stage. This leads to the lattice that is depicted in figure 8.7.

In the following sections we discuss the formulation of multi-stage stochastic programs, as well as the notion of dynamic programming value functions for multi-stage problems. We specifically show how the different models of uncertainty affect the definition of value functions.

### 8.1.3  Multi-stage stochastic linear programs

We now combine the ideas of section 8.1.1 and section 8.1.2 in order to formulate multi-stage stochastic linear programs. A multi-stage stochastic linear program is a linear program that unfolds over multiple stages, where the decision-maker is required to reach decisions at each stage. The constraints of the model are linear, and so is the objective function. One possible way to formulate the model on a scenario tree is the following:

$(MSLP - ST)$:

$$\min_x \sum_{t=1}^{H} \sum_{n \in \Omega_t} P(n) \cdot c_t(n)^T x_t(n)$$

s.t. $W_1 x_1 = h_1$

$T_1(n)x_1 + W_2(n)x_2(n) = h_2(n), n \in \Omega_2$

$\vdots$

$T_{t-1}(n)x_{t-1}(A(n)) + W_t(n)x_t(n) = h_t(n), n \in \Omega_t$

$\vdots$

$T_{H-1}(n)x_{H-1}(A(n)) + W_H(n)x_H(n) = h_H(n), n \in \Omega_H$

$x_1 \geq 0, x_t(n) \geq 0, t = 2, \ldots, H, n \in \Omega_t.$

This formulation, which describes the overall problem as a monolithic massive linear program, is called the **extended form** of the multi-stage stochastic linear program. As we discuss subsequently, it is computationally intractable. But it is interesting to study in order to understand the nuances of different uncertainty models and what these nuances imply in terms of the dynamic programming value functions that are actually used for solving the problem.

In this model, $N = \cup_{t=1}^{H} \Omega_t$ corresponds to the set of nodes of the scenario tree. The probability of occurrence of node $n \in N$ is denoted as $P(n)$ and given by equation (8.1). The parameters $c_t$, $T_t$, and $W_t$ are sets of random variables, in the sense that they are indexed over the set of scenarios $\Omega_t$ for every stage $t = 2, \ldots, H$ of the model. It is crucial to point out that the decisions $x_t$ are *also* indexed by scenario $n \in \Omega_t$. This implies that the model determines a **policy**, i.e. a decision $x_t\left(\xi_{[t]}\right)$ that depends both on the stage $t$ and the available information $\xi_{[t]} \in \Omega_t$ that has been revealed so far to the decision-maker.

By construction, the model respects **non-anticipativity**. This means that the decisions in stage $t$ are not clairvoyant, i.e. they can only depend on information $n \in \Omega_t$ that has been revealed so far, and not on information $m \in \Omega_{t'}, t' > t$ that is not available to the decision-maker. In other words, the decisions in stage $t$ are identical for all $m \in \Omega_{t+1}$. For instance, in a two-stage model the decision-maker must reach a unique decision in the first stage, and cannot adapt its decision to different scenarios $\omega \in \Omega$, since the decision-maker does not know which scenario will actually transpire when deciding on its first-stage decision.

We further note that the model is formulated in a way that each stage only depends on the preceding stage. For instance, in a hydro planning model, we do not describe the amount of water held in period $t$ as the inflows minus outflows of all previous periods, but rather as the inflows of the current period minus outflows of current period plus the level of water up to now.

Solving a multi-stage stochastic linear program is generally intractable because the scenario tree can easily become immense, even in models where the uncertain parameters are discretized. This is where special models of uncertainty (such as the

Markov property or stagewise independence), combined with the special structure of linear programs, come into play. The special structure of linear programs can be exploited, and combined with the principle of dynamic programming, to produce computational techniques that can yield operationally useful results for power system models of large scale. Although we do not expand on the algorithmic techniques, we focus on the dynamic programming principle and, specifically, on the definition of value functions, because the value functions themselves (in particular the value of water) correspond to input that is used widely in short-term hydrothermal models.

The dynamic programming principle states that the optimal solution to a dynamic decision-making problem under uncertainty is such that the derived policy is also optimal for the subproblem that considers only a segment of the full horizon of the problem.

Concretely, working backwards in time, we can define the following **value function** for stage $H$ as a function of the decisions $x_{H-1}$ and information $\xi_{[H]}$ (which is all that we need to know in order to decide what to do in the last stage $H$ of the horizon):

$$
\begin{aligned}
Q_H(x_{H-1}, \xi_{[H]}) = \ &\min_{x_H} c_H(\xi_{[H]})^T x_H \\
& W_H(\xi_{[H]}) x_H = h_H(\xi_{[H]}) - T_{H-1}(\xi_{[H]}) x_{H-1} \\
& x_H \geq 0.
\end{aligned}
$$

If this function $Q_H$ can be computed, then it is possible to inject it into the decision problem of stage $H - 1$ in order to decide how to proceed in stage $H - 1$:

$$
\begin{aligned}
Q_{H-1}\left(x_{H-2}, \xi_{[H-1]}\right) = \ & \\
\min_{x_{H-1}} c_{H-1}\left(\xi_{[H-1]}\right)^T & x_{H-1} + \sum_{\xi_{[H]} \in \Omega_H} P\left(\xi_{[H]} | \xi_{[H-1]}\right) \cdot Q_H\left(x_{H-1}, \xi_{[H]}\right) \\
W_{H-1}\left(\xi_{[H-1]}\right) x_{H-1} & = h_{H-1}\left(\xi_{[H-1]}\right) - T_{H-2}\left(\xi_{[H-1]}\right) x_{H-2} \\
x_{H-1} & \geq 0.
\end{aligned}
$$

Note that the decision-making in stage $H - 1$ balances out the present cost of $x_{H-1}$, as expressed in the first term of the objective function, $c_{H-1}\left(\xi_{[H-1]}\right)^T x_{H-1}$, with the expected future cost of deciding $x_{H-1}$, as expressed in the second term of the objective function, $\sum_{\xi_{[H]} \in \Omega_H} P\left(\xi_{[H]} | \xi_{[H-1]}\right) \cdot Q_H\left(x_{H-1}, \xi_{[H]}\right)$.

Proceeding recursively, we arrive at the following **dynamic programming equation** for any stage $t = 1, \ldots, H - 1$:

$$
\begin{aligned}
Q_t\left(x_{t-1}, \xi_{[t]}\right) = \ &\min_{x_t} c_t(\xi_{[t]})^T x_t + \sum_{\xi_{[t+1]} \in \Omega_{t+1}} P\left(\xi_{[t+1]} | \xi_{[t]}\right) \cdot Q_{t+1}\left(x_t, \xi_{[t+1]}\right) \\
& W_t\left(\xi_{[t]}\right) x_t = h_t\left(\xi_{[t]}\right) - T_{t-1}\left(\xi_{[t]}\right) x_{t-1} \\
& x_t \geq 0.
\end{aligned}
$$

In intuitive terms, $Q_t\left(x_{t-1}, \xi_{[t]}\right)$ quantifies the expected future cost of deciding $x_{t-1}$ in stage $t - 1$, assuming that the available information in stage $t$ is $\xi_{[t]}$, and further assuming that the decision-maker will act optimally from stage $t$ onward. According to the dynamic programming equation, the decision process in stage $t$ takes into consideration both the current-period costs, as well as the future costs of present decisions.

## 8.2   The hydrothermal planning problem

Having formulated a general multi-stage stochastic linear program, we now proceed to the specific formulation of the multi-period hydrothermal planning problem.

### 8.2.1   Model formulation

A simplified version of the multi-period hydrothermal planning problem under uncertainty can be formulated as follows:

$$(Hydro - ST):$$

$$\max_{p,d,pH,dH,e} \sum_{t=1}^{H} \sum_{n \in \Omega_t} P(n) \cdot (V \cdot d_t(n) - \sum_{g \in G} MC_{gt} \cdot p_{gt}(n))$$

$$p_{gt}(n) \leq P_g, g \in G, t = 1, \ldots, H, n \in \Omega_t$$

$$e_t(n) = R_t(n) + dH_t(n) - pH_t(n)/\eta + e_{t-1}(A(n)),$$

$$t = 1, \ldots, H, n \in \Omega_t$$

$$d_t(n) + dH_t(n) - \sum_{g \in G} p_{gt}(n) - pH_t(n) = 0, t = 1, \ldots, H, n \in \Omega_t$$

$$e_t(n) \leq E, t = 1, \ldots, H, n \in \Omega_t$$

$$d_t(n) \leq D_t, t = 1, \ldots, H, n \in \Omega_t$$

$$p, d, pH, dH, e \geq 0.$$

For the first stage of the model, an initial level of hydro energy $e_0$ needs to be assumed.

An industry and academic standard for tackling the hydrothermal planning problem is the **stochastic dual dynamic programming (SDDP)** algorithm. The algorithm exploits the linear programming structure of the problem, as well as the information structure of specific uncertainty models (e.g. Markov uncertainty or stagewise independence). The algorithm combines Monte Carlo simulation with ideas from cutting plane methods in order to derive policies that are useful in system operation.

The above formulation simplifies various aspects of the real problem. In real applications, the model can include a complex representation of hydroelectric production as a function of outflow and head. Furthermore, hydrothermal systems often cover broad geographic regions where the management of upstream river flow affects downstream flow and thus downstream hydro storage. This network interaction is not represented in the simplified model of this section, but can be included in the linear model and handled by the algorithmic techniques discussed previously.

The horizon of the model is commonly a few years. Assuming monthly time steps, a ten-year planning horizon translates to a model with 120 time steps. In practical applications, the horizon of the hydrothermal planning problem can extend well beyond 120 stages.

For reasons that are discussed further in section 8.2.2, it is common to adopt the Markov assumption in the hydrothermal planning problem (which means that the problem can be formulated on a lattice). State-of-the-art formulations at the time of writing of this book involve 100 uncertainty realizations per stage. This is also commonly the number of uncertainty realizations that are adopted in practical industry-scale models.

At the time of writing, the dimensions[2] of the state vector $x_t$ count in the few tens (e.g. 50) in the state-of-the-art academic literature. In practical applications, the dimension of the state space can be much greater.

There are extensions of proposition 4.11 that establish the equivalence of the multi-stage stochastic hydrothermal planning program to a decentralized economic equilibrium, where risk-neutral agents maximize expected profits. The result can be extended to risk-averse agents, under certain assumptions on the availability of financial instruments in the market.

The information structure of the scenario tree suggests how the scenario tree, and the underlying linear program, can be represented in code.[3] Effectively, one needs to define a set of scenario tree nodes $N$. The uncertain parameters of the model are indexed with respect to the nodes of the scenario tree. The probability of occurrence of any given node in the scenario tree is the probability of realization of its history. Each node of the scenario tree is also associated with a time period.

All uncertain parameters and decisions are indexed over the nodes of the scenario tree. This means that it is redundant to additionally index stochastic parameters and decisions over time stages, since the stage is implied by the node of the scenario tree. Similarly, it is sufficient to index constraints over the nodes of the scenario tree, since the notation $t = 1, \ldots, H$ and $n \in \Omega_t$ is equivalent to $n \in N$, i.e. the two are equivalent to scanning all nodes of the scenario tree.

---

**Example 8.8** *A four-stage hydrothermal planning problem on a scenario tree.* Consider the scenario tree[4] of example 8.3, which is depicted in figure 8.4. Assume that the hydro reservoir is initially empty, and recall that the system consists of two thermal units. Unit G1 has a marginal cost of 10 $/MWh and a capacity of 60 MW, while unit G2 has a marginal cost of 50 $/MWh and a capacity of 100 MW. Load can be curtailed at a cost of 1000 $/MWh. Suppose that the reservoir can store 50 MWh,

[2] The state vector dimension in practical applications is commonly equal to the number of lags in time series models of rainfall plus the number of hydro reservoirs that are represented in the model. The relation between hydro reservoirs and the state vector is discussed in section 8.2.2.
[3] This is not the only way of representing a scenario tree. Alternative forms of modeling are discussed in the bibliographical survey of the present chapter.
[4] We assume an hourly time step in this example, which is not representative of real models, where stages are often monthly, but is sufficient for the needs of illustrating the concept of value functions in this chapter. Note, however, that the time resolution within each stage of realistic models can be hourly. The difference is that each stage typically lasts for a month instead of an hour.

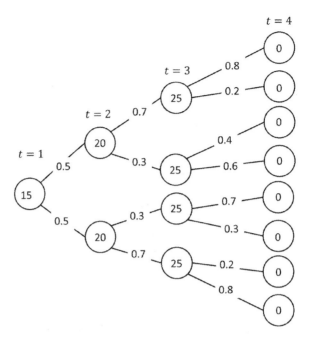

**Figure 8.8** The level of stored hydro energy in the four-stage model of rainfall uncertainty in example 8.8. The numbers in the nodes correspond to the level of stored hydro energy $e_t\left(\xi_{[t]}\right)$, while the numbers on the edges correspond to transition probabilities $P\left(\xi_{[t]}|\xi_{[t-1]}\right)$.

and its efficiency is 0.8. We additionally specify that the demand in period 3 is equal to 120 MW, and the demand in period 4 is 180 MW. In problem 8.4, the reader is asked to implement the hydrothermal planning model in code. The resulting model manages the hydro reservoir as indicated in figure 8.8, where we depict the hydro level at each scenario of each stage.

## 8.2.2   Value functions

In this section we discuss the application of dynamic programming in the context of the hydrothermal planning problem. The dynamic programming algorithm applied to multi-stage stochastic linear programming has interesting connections to linear programming duality. The output of the algorithm, which is the dynamic programming value functions, provides one of the most useful inputs to short-term scheduling models in hydrothermal systems. The **value of water**, which is the slope of these value functions, represents the opportunity cost of using water. From the point of view of short-term models, this translates to a "marginal cost" of water that allows us to decouple short-term operations from medium-term planning.

The practical problem that dynamic programming solves is that merely storing $(Hydro-ST)$ on a computer, let alone solving it, is impossible. This is due to the fact that scenario trees blow up in size once we start considering an interesting number of stages and random outcomes. For instance, if we were to formulate a hydrothermal planning problem with 121 stages and 5 realizations of uncertainty per stage, the last stage of the scenario tree would have $5^{120}$ nodes. As the usual comparison for convincing people about computational intractability goes, this is more than the number of atoms in the universe.

Recall that the dynamic programming algorithm commences from the last stage and proceeds backwards. Applying the general algorithm to the specific case of a multi-stage stochastic program on a scenario tree, the value function of the last stage can be described mathematically as follows:

$$Q_H\left(e_{H-1}, \xi_{[H]}\right) = \max_{p,d,pH,dH,e} V \cdot d - \sum_{g \in G} MC_{gH} \cdot p_g$$

$$(\mu_g): \quad p_g \leq P_g, g \in G$$

$$(\lambda H): \quad e = R_H\left(\xi_{[H]}\right) + dH - pH/\eta + e_{H-1}$$

$$(\lambda): \quad d + dH - \sum_{g \in G} p_g - pH = 0$$

$$(\delta): \quad e \leq E$$

$$(\nu): \quad d \leq D_H$$

$$p, d, pH, dH, e \geq 0.$$

Note that the minimization, which is how the generic $(MSLP\text{–}ST)$ model has been formulated, is replaced by a maximization in the specific case of the hydrothermal planning problem. The logic of the dynamic programming equations applies identically, with minimization replaced by maximization.

In contrast to $(Hydro-ST)$, which is an intractable linear program that cannot even be stored in computer memory, the above is a fairly light linear program since a single stage (the last stage $H$) and a single realization of uncertainty (the history $\xi_{[t]}$) is considered. The objective function value of this linear program is the value function $Q_H$. This value function depends on rainfall realizations and the level of hydro energy that is available in the second-to-last stage, $H-1$. We are interested in the value of $Q_H$ for different values of $\xi_{[H]}$ and $e_{H-1}$, because this information will be used for informing us how we should manage hydro reservoirs in the second-to-last stage, $H-1$.

Note that the value function $Q_H$ is indeed a function of $\xi_{[H]}$, because the rainfall in the second constraint depends on the node of the scenario tree $\xi_{[H]}$ that we find ourselves in during the final stage. It is also a function of $e_{H-1}$, because the chosen water level also affects the right-hand side of the second constraint of the above linear program.

It is interesting to note what the value function $Q_H$ is *not* a function of. In particular, the value function is *not* a direct function of the hydrothermal management decisions, $p, pH, dH$, or the served demand, $d$, in period $H-1$. The only thing that matters as far as decisions in stage $H$ are concerned is how these decisions in stage $H-1$ affect the hydro level $e_{H-1}$, which is the only decision that directly affects the value function $Q_H$.

**Proposition 8.1** *The value function $Q_H\left(e_{H-1}, \xi_{[H]}\right)$ can be expressed as follows:*

- *If $D_H > \eta \cdot \left(R_H(\xi_{[H]}) + e_{H-1}\right) + \sum_{g \in G} P_g$, then*

$$Q_H\left(e_{H-1}, \xi_{[H]}\right) = V \cdot \left(\sum_{g \in G} P_g + \eta \cdot \left(R_H(\xi_{[H]}) + e_{H-1}\right)\right) - \sum_{g \in G} MC_{gH} \cdot P_g.$$

- *Denote $\bar{g}^-$ as the unit before $\bar{g}$ in the merit order. If*

$$\eta \cdot \left(R_H(\xi_{[H]}) + e_{H-1}\right) + \sum_{g \in G: MC_{gH} < MC_{\bar{g}^- H}} P_g \leq D_H$$

$$\leq \eta \cdot \left(R_H(\xi_{[H]}) + e_{H-1}\right) + \sum_{g \in G: MC_{gH} < MC_{\bar{g}H}} P_g,$$

*then*

$$Q_H\left(e_{H-1}, \xi_{[H]}\right) = V \cdot D_H - \sum_{g \in G: MC_{gH} < MC_{\bar{g}H}} MC_{gH} \cdot P_g -$$

$$-MC_{\bar{g}H} \cdot \left(D_H - \sum_{g \in G: MC_{gH} < MC_{\bar{g}H}} P_g - \eta \cdot \left(R_H(\xi_{[H]}) + e_{H-1}\right)\right).$$

- *If $D_H < \eta \cdot \left(R_H(\xi_{[H]}) + e_{H-1}\right)$, then*

$$Q_H\left(e_{H-1}, \xi_{[H]}\right) = V \cdot D_H.$$

*Proof*   Using the mnemonic table 2.1, we can express the dual linear program that defines $Q_H\left(e_{H-1}, \xi_{[H]}\right)$ as follows:

$$\min_{\mu, \lambda H, \lambda, \delta, \nu} \sum_{g \in G} \mu_g \cdot P_g - \lambda H \cdot \left(R_H(\xi_{[H]}) + e_{H-1}\right) + \delta \cdot E + \nu \cdot D_H$$

$(p_g)$:  $\mu_g - \lambda \geq -MC_{gH}, g \in G$

$(d)$:  $\nu + \lambda \geq V$

$(pH)$:  $-\lambda H / \eta - \lambda \geq 0$

$(dH)$:  $\lambda H + \lambda \geq 0$

$(e)$:  $-\lambda H + \delta \geq 0$

$\mu \geq 0, \nu \geq 0, \delta \geq 0.$

We first show that $\lambda H \leq 0$. Suppose, for the sake of contradiction, that this is not the case, i.e. that $\lambda H > 0$. Then $\lambda \geq -\lambda H$ and $\lambda \leq -\lambda H / \eta$. Since $0 < \eta < 1$, these inequalities cannot hold simultaneously. This further implies that $\lambda \geq 0$.

We can already observe that $\delta = 0$ at the optimal solution. To show this, we can argue by contradiction. If $\delta > 0$ at the claimed optimal solution, we can improve the objective function by setting $\delta = 0$. This would improve the term $\delta \cdot E$ in the objective function. And it would continue to respect the last constraint of the dual problem, $\delta \geq \lambda H$, since we have argued in the previous paragraph that $\lambda H \leq 0$.

We can further observe that the optimal value of $\mu_g$ is $\max(\lambda - MC_{gH}, 0)$. Again, we can argue by contradiction. The dual variable cannot be lower, because it would violate the constraints of the model. And if it is greater, we can improve the objective function by decreasing it, without violating any of the constraints of the model.

Similarly, we can argue that $\nu$ is equal to $\max(V - \lambda, 0)$ at the optimal solution. Again, this can be argued by contradiction: the value of $\nu$ cannot be lower, otherwise it would violate one of the constraints of the dual problem. And it cannot be larger, because then we can lower its value and improve the objective function of the dual problem without violating any of the dual constraints.

Finally, we can observe that the optimal value of $\lambda H$ is $-\lambda \cdot \eta$. Once again we argue by contradiction: if it were greater than this value it would violate the third constraint of the dual problem. And if it were less then we could lower its value without violating any of the problem constraints, but improve the objective function of the problem.

Given the above, the optimal objective function value $d^\star$ of the dual problem can be expressed as follows:

$$d^\star = \sum_{g \in G} \max(\lambda - MC_{gH}, 0) \cdot P_g + \lambda \cdot \eta \cdot \big(R_H(\xi_{[H]}) + e_{H-1}\big) + \max(V - \lambda, 0) \cdot D_H.$$

We can now distinguish different cases, based on the different values that $\lambda$ can assume, which yield different expressions for the dual function.

- For $\lambda = 0$, the value of $d^\star$ is[5]

$$V \cdot D_H.$$

 This corresponds to a case where none of the thermal units are producing, and the demand is fully covered by the hydro plant.[6] The value of water (i.e. the subgradient of $Q_H$) with respect to $e_{H-1}$ is zero, because any additional water in the reservoir will not help. The system is already able to fully cover demand at zero cost.

- For $\lambda = MC_{\bar{g}H}$ for some unit $\bar{g}$,[7] we have a situation where unit $\bar{g}$ is marginal.[8] All units with a lower marginal cost are producing at their capacity, while all units with a higher marginal cost are producing zero. In this case, the value of $d^\star$ is

---

[5] The argument is identical for any $0 < \lambda < MC_{g^0 H}$, where $g^0$ is the unit with the lowest marginal cost.

[6] We know this because $\nu \geq V - \lambda > 0$, which means that $d = D_H$ since $\nu$ is complementary to the constraint $d \leq D_H$. Moreover, $\mu_g = \max(\lambda - MC_{gH}, 0) = 0$ for all $g \in G$, which means that $p_g = 0$, since $p_g$ is complementary to the first dual constraint, which is not tight. Since $\lambda H = -\eta \cdot \lambda$, the fourth dual constraint is not binding, which means that the complementary primal variable is zero, i.e. $dH = 0$. The power balance constraint then implies that the hydro production fully covers the demand, i.e. $pH = D_H$.

[7] The argument is identical for any $MC_{\bar{g}H} < \lambda < MC_{\bar{g}^+ H}$, where $\bar{g}^+$ is the unit that is immediately more expensive than $\bar{g}$, or for $MC_{\bar{g}H} < \lambda < V$ if $\bar{g}$ is the most expensive unit in the merit order.

[8] As in the first case, we can use the primal and dual constraints as well as the complementarity conditions in order to derive the optimal value of the primal variables.

$$V \cdot D_H - \sum_{g \in G: MC_{gH} < MC_{\bar{g}H}} MC_{gH} \cdot P_g -$$

$$-MC_{\bar{g}H} \cdot \left( D_H - \sum_{g \in G: MC_{gH} < MC_{\bar{g}H}} P_g - \eta \cdot \left( R_H(\xi_{[H]}) + e_{H-1} \right) \right).$$

The entire demand is satisfied, there is fuel cost due to the fact that all cheap units are producing at their technical limit, and the marginal unit is producing the energy needed to cover any leftover demand that is not covered by the cheap thermal plants and by the hydro reservoir. The value of water in this case is $\eta \cdot MC_{\bar{g}H}$, because 1 MWh of additional hydro energy contributes towards producing $\eta$ MWh less from unit $\bar{g}$.

- For $\lambda = V$, we have[9] all units producing at their max and the load being served up to the level of $\sum_{g \in G} P_g$. The value of $d^\star$ is thus

$$V \cdot \left( \sum_{g \in G} P_g + \eta \cdot \left( R_H(\xi_{[H]}) + e_{H-1} \right) \right) - \sum_{g \in G} MC_{gH} \cdot P_g,$$

where the first term corresponds to the consumer benefit up to the level of load that can be served, and the second term corresponds to the production cost of the thermal units that are operating at their full capacity. The value of water in this case is $\eta \cdot V$, because every extra MWh of water can contribute to a decrease of load shedding by $\eta$ MWh.

Note that values of $\lambda$ that are greater than $V$ cannot be dual optimal, because we can always do better by choosing $\lambda = V$.

$\square$

We observe that the value function is a piecewise linear concave function of $e_{H-1}$. This is already anticipated from linear programming; see the discussion in the end of section A.10. Moreover, from proposition 2.8 we know that the value of water, i.e. the slope of $Q_{H-1}$, is equal to the dual optimal multiplier $\lambda H$. We now have both the geometric as well as the physical intuition to understand why this slope changes. From a geometric standpoint, what happens is that the optimal basis of the last-stage problem changes as we vary the amount of water $e_{H-1}$ that is carried over to the last stage. This implies that the dual optimal multiplier of the last stage changes (again, see the discussion in the end of section A.10). From the standpoint of physical intuition, these changes of optimal basis correspond to "phase shifts," where the system moves from using only hydro energy under very comfortable conditions, to using thermal units as the system becomes increasingly tight, to ultimately shedding load when neither the hydro nor the thermal resources are sufficient to fully cover demand. As these phase shifts occur, the value of water increases.

---

[9] As in the first case, we can use the primal and dual constraints as well as the complementarity conditions in order to derive the optimal value of the primal variables.

Working recursively, the value function in stage $t$ is:

$$Q_t\left(e_{t-1}, \xi_{[t]}\right) =$$

$$\max_{p,d,pH,dH,e} \left[ V \cdot d - \sum_{g \in G} MC_{gt} \cdot p_g + \sum_{n \in \Omega_{t+1}} P\left(\xi_{[t+1]} = n | \xi_{[t]}\right) \cdot Q_{t+1}\left(e, \xi_{[t+1]}\right) \right]$$

$$(\mu_g): \quad p_g \leq P_g, g \in G$$

$$(\lambda H): \quad e = R_t\left(\xi_{[t]}\right) + dH - pH/\eta + e_{t-1}$$

$$(\lambda): \quad d + dH - \sum_{g \in G} p_g - pH = 0$$

$$(\delta): \quad e \leq E$$

$$(\nu): \quad d \leq D_t$$

$$p,d,dH,pH,e \geq 0.$$

The results that are obtained for the last-stage value function $Q_H\left(e_{H-1}, \xi_{[H]}\right)$ actually carry over to the general value function of stage $t$, $Q_t\left(e_{t-1}, \xi_{[t]}\right)$. In particular, the value function $Q_t\left(e_{t-1}, \xi_{[t]}\right)$ is a piecewise linear concave function of $e_{t-1}$. To gain an intuition about why this is the case, note that a piecewise linear concave function can be modeled through linear inequalities, as discussed in section A.9. And $Q_t\left(e_{t-1}, \xi_{[t]}\right)$ is the objective function of a linear program that is parametrized with respect to $e_{t-1}$, which is a right-hand side parameter of the linear program that defines $Q_t\left(e_{t-1}, \xi_{[t]}\right)$. As indicated in the discussion in the end of section A.10, we can then conclude that $Q_t\left(e_{t-1}, \xi_{[t]}\right)$ is indeed a piecewise linear concave function of $e_{t-1}$.

---

**Example 8.9** *Value functions on a scenario tree.* Consider the scenario tree of example 8.8, which is depicted in figure 8.4. The value function for the top node of stage 4 is computed using proposition 8.1. Concretely, we need to consider four cases:

- There is demand curtailment:

$$D_H > \eta \cdot \left(R_H(\xi_{[H]}) + e_{H-1}\right) + \sum_{g \in G} P_g$$

$$\Rightarrow 180 > 0.8 \cdot (20 + e_3) + (60 + 100)$$

$$\Rightarrow e_3 < 5.$$

In this case, the value function is:

$$Q_H(e_{H-1}, \xi_{[H]}) = V \cdot \left( \sum_{g \in G} P_g + \eta \cdot \left(R_H(\xi_{[H]} + e_{H-1})\right) \right) - \sum_{g \in G} MC_{gH} \cdot P_g$$

$$= 1000 \cdot (60 + 100 + 0.8 \cdot (20 + e_3)) - 10 \cdot 60 - 50 \cdot 100$$

$$= 170400 + 800 \cdot e_3.$$

At $e_3 = 5$, this evaluates to \$174400.

- The expensive thermal resource $G_2$ is marginal:

$$\eta \cdot (R_H(\xi_{[H]}) + e_{H-1}) + P_{G_1} \leq D_H$$
$$\leq \eta \cdot (R_H(\xi_{[H]}) + e_{H-1}) + P_{G_1} + P_{G_2}$$
$$\Rightarrow 0.8 \cdot (20 + e_3) + 60 \leq 180 \leq 0.8 \cdot (20 + e_3) + 60 + 100$$
$$\Rightarrow 5 \leq e_3 \leq 130.$$

In this case, the value function is:

$$Q_H(e_{H-1}, \xi_{[H]}) = V \cdot D_H - MC_{G_1H} \cdot P_{G_1}$$
$$- MC_{G_2H} \cdot (D_H - P_{G_1} - \eta \cdot (R_H(\xi_{[H]}) + e_{H-1})$$
$$= 1000 \cdot 180 - 10 \cdot 60 - 50 \cdot (180 - 60 - 0.8 \cdot (20 + e_3))$$
$$= 174200 + 40 \cdot e_3.$$

At $e_3 = 5$, this evaluates to \$174400, so the value function is indeed continuous at the break-point of 5 MWh. At $e_3 = 130$, the value function evaluates to \$179400.

- The cheap thermal resource $G_1$ is marginal:

$$\eta \cdot (R_H(\xi_{[H]}) + e_{H-1}) \leq D_H \leq \eta \cdot (R_H(\xi_{[H]}) + e_{H-1}) + P_{G_1}$$
$$\Rightarrow 0.8 \cdot (20 + e_3) \leq 180 \leq 0.8 \cdot (20 + e_3) + 60$$
$$\Rightarrow 130 \leq e_3 \leq 205.$$

In this case, the value function is:

$$Q_H(e_{H-1}, \xi_{[H]}) = V \cdot D_H - MC_{G_1H} \cdot (D_H - \eta \cdot (R_H(\xi_{[H]}) + e_{H-1}))$$
$$= 1000 \cdot 180 - 10 \cdot (180 - 0.8 \cdot (20 + e_3))$$
$$= 178360 + 8 \cdot e_3.$$

At $e_3 = 130$, this evaluates to \$179400, so the value function is indeed continuous at the break-point of 130 MWh. At $e_3 = 205$, the value function evaluates to \$180000.

- Hydro is the marginal resource:

$$D_H < \eta \cdot (R_H(\xi_{[H]}) + e_{H-1})$$
$$\Rightarrow 180 < 0.8 \cdot (20 + e_3)$$
$$\Rightarrow e_3 > 205.$$

In this case, the value function is:

$$Q_H(e_{H-1}, \xi_{[H]}) = V \cdot D_H$$
$$= 1000 \cdot 180 = 180000.$$

At $e_3 = 205$, this evaluates to \$180000, so the value function is indeed continuous at the break-point of 205 MWh. The resulting value function is depicted in figure 8.9.

In the case of Markov processes, the value functions coincide whenever the value of $\xi_t$ coincides in a given stage. In other words, the $Q_t$ functions are no longer

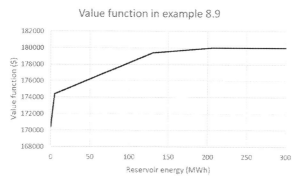

**Figure 8.9** The value function $Q_4 \left(e_3, \xi_{[4]}\right)$ for $\xi_{[4]} = (5, 10, 15, 20)$ in example 8.9.

dependent on $\xi_{[t]}$, but rather on $\xi_t$. This implies significant computational savings, because algorithms such as SDDP work by estimating value functions. It makes a material computational difference if these value functions need to be evaluated for every node of a scenario tree, versus every node of a lattice. The intuition behind this result is that the expected future cost of acting optimally in a Markov process only depends on the state $(e_t, \xi_t)$ that we find ourselves in at stage $t$, no matter how we got there.

---

**Example 8.10** *Value functions on a lattice.* Consider the scenario tree of figure 8.6. Using the procedure of example 8.9, we can estimate a value function (which is a piecewise linear concave function) for all nodes of the last stage. Referring to the left part of figure 8.6, the value functions in nodes $(5, 0, 5, 10)$ and $(5, 10, 5, 10)$ are identical, because the rainfall in these nodes is equal to $R_4 = 10$ MWh. By the same token, the value functions are identical in nodes $(5, 10, 5, 0)$ and $(5, 0, 5, 0)$ of the scenario tree, since $R_4 = 0$ in both nodes. According to the dynamic programming equation for stage $t$, the value function $Q_3 \left(e_2, \xi_{[3]}\right)$ for $\xi_{[3]} = (5, 10, 5)$ depends on the conditional expectation of the value functions $Q_4$. This conditional expectation is identical, because $0.4 \cdot Q_4(e_3, (5, 10, 5, 10)) + 0.6 \cdot Q_4(e_3, (5, 10, 5, 0)) = 0.4 \cdot Q_4(e_3, (5, 0, 5, 10)) + 0.6 \cdot Q_4(e_3, (5, 0, 5, 0))$. In other words, it is enough to define value functions on the nodes of the lattice in the right of figure 8.6 in order to determine the optimal hydro management policy, as opposed to defining these value functions for every node of the corresponding scenario tree.

---

In the case of stagewise independent processes, the value functions coincide for every node of a given stage $t$. In other words, the value functions are only functions of the hydro level $e_{t-1}$, $Q_t(e_{t-1})$, as opposed to being functions of the hydro level and realized uncertainty, $Q_t(e_{t-1}, \xi_t)$.

---

**Example 8.11** *Value functions for stagewise independent processes.* Consider the lattice of figure 8.7. In problem 8.5 the reader is asked to unfold this lattice into a scenario tree. Note that, in node $(5, 10, 15)$ of this scenario tree, the conditional expectation of the fourth-stage value functions is $0.1 \cdot Q_4(e_3, (5, 10, 15, 20)) + 0.8 \cdot Q_4(e_3, (5, 10, 15, 10)) + 0.1 \cdot Q_4(e_3, (5, 10, 15, 0))$. This is identical to the conditional expectation of all nodes in stage 3, exactly due to the stagewise independence assumption.

## 8.3    Performance of stochastic programs

Hydrothermal planning is a computationally challenging problem when formulated as a multi-stage decision-making model under uncertainty. There are alternative ways to estimate the cost that a decision-maker is faced with when confronted with uncertainty, which are easier to compute, and which provide interesting benchmarks for comparison against the performance of the stochastic programming solution. We focus on two specific benchmarks in this section. The first benchmark is the performance that we would attain if we could perfectly anticipate the future, which we refer to as perfect foresight. The second benchmark is the performance that we would attain if we would replace the random parameters in the model with their expected value, which gives rise to the expected value problem.

The ensuing discussion focuses on the two-stage stochastic linear program, (*TSLP*), which was introduced in section 8.1.1. The idea of perfect foresight generalizes straightforwardly to multi-stage models. For the case of the expected value problem, the generalization is not as straightforward.

Referring to the generic two-stage stochastic linear program, we can define the value function for a given realization $\xi$ of the second stage as follows:

$$z(x, \xi) = c^T x + Q(x, \xi) + I(x|\{Ax = b, x \geq 0\}),$$
$$Q(x, \xi) = \min_y \{q(\omega)^T y \mid W(\omega)y = h(\omega) - T(\omega)x\}.$$

The function $I(x|K)$ is equal to zero if $x \in K$ and equal to infinity otherwise. The function $z(x, \xi)$ can be interpreted as the cost that would be incurred given that we have decided $x$ in the first stage, given that realization $\xi$ occurs in the second stage, and given that we react optimally to this realization. It is an easy function to compute for a given argument $(x, \xi)$, because it involves solving a manageable linear program.

The **wait-and-see value** is the expected value of reacting with perfect foresight $x^\star(\xi)$ to $\xi$:

$$WS = \mathbb{E}[\min_x z(x, \xi)] = \mathbb{E}[z(x^\star(\xi), \xi)].$$

The **here-and-now value** is the expected value of the two-stage stochastic program:

$$SP = \min_x \mathbb{E}[z(x, \xi)].$$

The min and $\mathbb{E}$ operators are essentially swapped in the two definitions. The crucial difference is that, in the case of $WS$, we get to observe the uncertainty before deciding on our first-stage decision. This is unrealistic from the point of view of access to information, since we cannot adapt our decision $x$ to the realized uncertainty $\xi$: the whole point of a decision being first stage is that it has to be the same decision, no matter which $\xi$ materializes. But it provides an interesting performance benchmark, because we cannot hope to do better than $WS$. It also quantifies our willingness to pay for access to perfect information. Concretely, the **expected value of perfect information** is defined as:

$$EVPI = SP - WS.$$

It can be interpreted as the value of the ability to perfectly forecast the future. For maximization problems, the definition is the other way around, i.e. $EVPI = WS - SP$, since the wait-and-see approach corresponds to an upper bound on the performance that we can attain with the optimal solution of the stochastic program.

The wait-and-see value $WS$ is far easier to compute than $SP$, because each scenario $\omega$ can be tackled independently of other scenarios. Instead, $SP$ has a first-stage decision $x$ that creates interdependencies between scenarios, and thus results in a large-scale linear program, as discussed in sections 8.1 and 8.2.

---

**Example 8.12** *Expected value of perfect information in the hydrothermal planning problem.* Consider the hydrothermal planning problem of example 8.2, with the only difference being that the amount of rainfall in scenario 2 of period 2 is assumed to be equal to 55 MWh, i.e. $R_2(2) = 55$. The here-and-now value can be computed using the code of problem 8.2. It is equal to:

$$SP = \$148110.$$

The expected value of perfect information is computed using the code of problem 8.6:

$$WS = \$148115.$$

The EVPI is thus equal to:

$$EVPI = WS - SP = 148115 - 148110 = \$5.$$

In the case of perfect foresight, the decision is to carry less water in the reservoir. This is due to the fact that, in scenario 2, the perfect foresight policy does not have an interest in producing at the full capacity of unit 1: the demand of period 2 can be covered by production from G1 in period 2 (without efficiency losses) and hydro, since there is enough rainfall. In contrast, the stochastic programming policy hedges against the possibility of no rainfall in period 2 (scenario 1), and produces as much as possible from unit 1 in period 1. This is slightly inefficient relative to the perfect foresight policy, because this energy is released at an efficiency of $\eta = 0.8$ in scenario 2 of period 2.

---

When considering implementable policies that are easier to compute than the stochastic programming solution, one of the most straightforward decisions is to replace the uncertain parameters by their expected value, $\bar{\xi} = \mathbb{E}[\xi]$. This gives rise to the **expected (or mean) value problem**:

$$EV = \min_x z(x, \bar{\xi}).$$

Denote the optimal solution of the expected value problem as $x^\star(\bar{\xi})$. This is defined as the **expected value solution**. The expected value of using the EV solution measures the performance of $x^\star(\bar{\xi})$:

$$EEV = \mathbb{E}[z(x^\star(\bar{\xi})), \xi].$$

The **value of the stochastic solution** measures the performance advantage of the optimal solution of the stochastic program relative to the EV solution:

$$VSS = EEV - SP.$$

For maximization problems, the $VSS$ is defined the other way around, since $SP$ is better (i.e. greater) than $EEV$, $VSS = SP - WSS$.

The computation of $x^\star(\bar{\xi})$ is a relatively easy task, since $EV$ is a tractable linear program. And the computation of $EEV$ amounts to fixing the first-stage decision to $x^\star(\bar{\xi})$ and computing the optimal second-stage decisions for all scenarios $\omega \in \Omega$. This, too, is a relatively tractable task, for a reasonable number of scenarios $|\Omega|$.

---

**Example 8.13** *Value of the stochastic solution for the hydrothermal planning problem.* Consider the problem of example 8.12. The average rainfall in period 2 is $\bar{\xi} = 0.5 \cdot 0 + 0.5 \cdot 55 = 27.5$. Using the code of problem 8.7, we can compute the optimal hydro level of period 1 as 10 MWh. We can then fix the first-stage solution and compute the expected value of using the EV solution as:

$$EEV = \$148110.$$

This implies that the value of the stochastic solution for this example is zero:

$$VSS = SP - EEV = 148110 - 148110 = \$0.$$

In this specific example, the $EV$ solution coincides with that computed by the stochastic program. In other instances (e.g. capacity expansion planning, or the newsboy problem analyzed in problem 8.3), the value of the stochastic solution becomes more important.

---

In situations where the computation of expectations is computationally intractable (e.g. when the uncertain parameter $\xi$ is a continuous random variable), it is still possible to estimate $WS$ and $EEV$ through sample average approximation:

- For $i = 1, \ldots, K$
  - Sample $\xi_i$
  - Compute $x^\star(\bar{\xi})$
  - Compute $WS_i = z(x^\star(\xi_i), \xi_i)$ and $EEV_i = c^T x^\star(\bar{\xi}) + Q(x^\star(\bar{\xi}), \xi_i)$.
- Estimate $\bar{WS} = \frac{1}{K} \sum_{i=1}^{K} WS_i$ and $E\bar{E}V = \frac{1}{K} \sum_{i=1}^{K} EEV_i$.

Intuitively, the more samples $K$ that we use in this sample average approximation, the more accurately we can expect to estimate $WS$ and $EEV$. This intuition is formalized through the **central limit theorem**.

**Theorem 8.2 (Central limit theorem)**  *Suppose $X_1, X_2, \ldots$ is a sequence of independent, identically distributed random variables, with $\mathbb{E}[X_i] = \mu$ and $Var(X_i) = \sigma^2 < \infty$. Then, as $n$ approaches infinity, $\sqrt{n} \cdot \left( \frac{\sum_{i=1}^{n} X_i}{n} - \mu \right)$ converges in distribution to a normal random variable, $N\left(0, \sigma^2\right)$:*

$$\sqrt{n} \left( \frac{1}{n} \sum_{i=1}^{n} X_i - \mu \right) \xrightarrow{d} N\left(0, \sigma^2\right).$$

The central limit theorem is reassuring, because it demonstrates that as we increase the number of samples we approach the mean value that we wish to estimate, but in practice it can lead to slow convergence to the mean. For instance, rare but highly impactful outcomes may be sampled rarely. **Importance sampling** aims at remedying this drawback by placing greater importance on more impactful samples.

Importance sampling works as follows. Suppose that we wish to estimate $\mathbb{E}[C]$, where $C$ is distributed according to a density function $f$. Sample average approximation draws samples $C_i$ according to the distribution $f$ and estimates $\mathbb{E}[C]$ as $\sum_{i=1}^{N} \frac{1}{N} C_i$. Instead, importance sampling draws samples $C_i$ according to a distribution

$$g(x) = \frac{f(x) \cdot x}{\mathbb{E}[C]},$$

and estimates $\mathbb{E}[C]$ as

$$\frac{1}{N} \sum_{i=1}^{N} \frac{f(x_i) \cdot x_i}{g(x_i)}.$$

This is circular, since we are trying to estimate $\mathbb{E}[C]$ in the first place, but we can replace $\mathbb{E}[C]$ with a reasonable estimate in order to obtain an estimator with a lower variance.

## Problems

**8.1** *Two-stage deterministic hydrothermal planning.* Implement the model of example 8.1 in mathematical programming code, and confirm the results of the example.

**8.2** *Two-stage stochastic hydrothermal planning.* Implement the model of example 8.2 in mathematical programming code, and confirm the results of the example.

**8.3** *The newsboy problem.* Consider an instance of the newsboy problem where the newsboy can buy newspapers at $1 per copy, and sell them the next day at a price of $3 per copy. The demand is uniformly distributed in increments of 10 copies for 10 up to 100 newspapers. Implement a model of the newsboy problem in mathematical programming code. What is the optimal order quantity? What is the value of the stochastic solution? What is the expected value of perfect information?

**8.4** *Multi-stage stochastic hydrothermal planning.* Implement the model of example 8.8 in mathematical programming code, and confirm the results of the example.

**8.5** *Unfolding a lattice into a scenario tree.* Unfold the lattice of figure 8.7 into a scenario tree.

**8.6** *Computing EVPI.* Implement mathematical programming code that can be used for computing the expected value of perfect information in example 8.12.

**8.7** *Computing VSS.* Implement mathematical programming code that can be used for computing the expected value solution and the value of the stochastic solution in example 8.13.

**8.8** *The Markov property in three-stage trees.* Is it possible for a three-stage scenario tree to not satisfy the Markov property? Prove that it is the case using an example, or argue that it cannot be the case.

# Bibliography

**Section 8.1**   The seminal publication of Pereira and Pinto (1991) describes the hydrothermal planning problem as a multi-stage stochastic linear program and proposes the stochastic dual dynamic programming algorithm for solving the problem. An algorithmic strategy for treating fuel cost uncertainty is described in Gjelsvik, Belsnes, and Haugstad (1999). Alternative ways of modeling scenario trees are discussed in Shapiro, Dentcheva, and Ruszczynski (2009).

**Section 8.2**   Papers in which monthly time steps are used with a horizon of multiple years include Shapiro et al. (2013), De Matos, Philpott, and Finardi (2015), and Machado et al. (2021), thus leading to models in the order of 120 stages. Lattices are used in multi-stage stochastic programs by Löhndorf and Shapiro (2019) and Löhndorf and Wozabal (2021). The literature reports 20 to 100 random realizations per stage in lattice models of uncertainty; see Shapiro et al. (2013), De Matos, Philpott, & Finardi (2015), Machado et al. (2021), De Matos et al. (2010), and Flach, Barroso, and Pereira (2010). The state space dimension ranges in the few tens (4 to 44 in the previously cited papers, although in practice the dimension of the state space can be much greater). The connections between multi-stage stochastic programming and Markov decision processes, as well as SDDP and batch learning algorithms in Q learning, are discussed in (Ávila, Papavasiliou, and Löhndorf in press), and broader connections are discussed in Powell (2007). The equivalence between the centralized hydrothermal planning problem and a decentralized market equilibrium is described in Philpott, Ferris, and Wets (2016).

**Section 8.3**   The material in this section is based on Birge and Louveaux (2010).

# 9 Risk management

Chapters 4–7 have covered the operation of short-term (day-ahead and real-time) markets. However, the majority of electric power is actually traded before the day ahead or real time. Forward markets that are operated days, months, or years in advance are less centralized, and are the subject of this chapter.

In forward markets, mostly financial transactions take place. Therefore, forward markets are more weakly linked to the physical aspects of system operation. The most commonly used financial instruments in electricity markets are forward contracts, options, and their variations and combinations. Section 9.1 introduces forward contracts. Financial transmission rights are introduced in section 9.2. Section 9.3 introduces call options, and how they can be combined with forward contracts in order to create useful hedging instruments for market participants. In section 9.4 we present models of risk aversion and use them to describe how prices back-propagate from real-time markets to forward markets.

## 9.1 Forward contracts

**Forward contracts** are financial instruments that can be used for trading a commodity at a price that can be fixed in advance of delivery time. This makes it possible for parties to hedge against real-time price uncertainty. Forward contracts are used extensively in electricity markets in order to trade bulk quantities of power months or years in advance of delivery and in order to secure investment in generation capacity.

A forward contract is sold from the *seller* of the contract to the *buyer* of the contract and is characterized by (i) its selling price $f_t$, (ii) the quantity $x$ of the commodity being traded, and (iii) the delivery time $T$ of the commodity, also known as the **expiration date** of the forward contract.

The contract entitles the buyer of the contract to a payment equal to the real-time price of the commodity at delivery time for each unit of the commodity that has been bought through the forward contract. This implies that a forward contract is linked to the real-time market of a commodity. Since it is a purely financial obligation, the payoff of which is *derived* from the price of a commodity, it is referred to as a financial derivative.

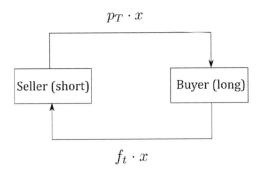

**Figure 9.1** Payments in a forward contract.

**Definition 9.1**    Forward Contract

*Seller.* The seller of a forward contract with expiration date $T$ sells the contract at time $t < T$ for a price $f_t$. The seller is said to have a **short position**.

*Buyer.* The buyer of a forward contract with expiration date $T$ buys the contract at time $t < T$ for a price $f_t$. The buyer is said to have a **long position**.

*Obligations and payoffs.* At time $t < T$ the buyer pays the seller $f_t \cdot x$. At time $t = T$ the seller of the contract pays the buyer $p_T \cdot x$, where $p_T$ is the spot price of the commodity.

The payments associated with a forward contract are shown in figure 9.1.

### 9.1.1    The virtues of forward contracts

A forward contract enables a seller and a buyer to trade a commodity at a price $f_t$ that can be agreed upon in advance. To see how this can be achieved, note that the producer of a commodity can sell (short) a forward contract at time $t < T$ for quantity $x$ and sell $x$ units of the commodity in the spot market in real time, while the consumer of a commodity can buy (long) the forward contract from the producer at time $t < T$ for quantity $x$, and buy $x$ units of the commodity from the spot market in real time.

The cash flows to the producer are:

- $+f_t \cdot x$ (from selling the forward contract)
- $+p_T \cdot x$ (from producing in the real-time market)
- $-p_T \cdot x$ (from settling the forward contract)

The cash flows to the consumer are:

- $-f_t \cdot x$ (from buying the forward contract)
- $-p_T \cdot x$ (from consuming in the real-time market)
- $+p_T \cdot x$ (from settling the forward contract)

Effectively, the producer is paid $f_t \cdot x$ and the consumer pays $f_t \cdot x$. Using a forward contract, the buyer and the seller effectively override the spot market and trade the commodity at a price $f_t$ that can be agreed upon bilaterally in advance of delivery time $T$.

The payoff of a producer that sells $x$ units of a commodity for a forward price $f_t$ at time $t$ is

$$R = f_t \cdot x + p_T \cdot (q - x),$$

where $q$ is the production in real time. In real time, the portion of revenues that relates to past decisions, $f_t \cdot x - p_T \cdot x$, is viewed as *sunk* from the point of view of the producer, in the sense that it has already been decided on and nothing can be done to change it. The only factor that can influence the revenues of the producer in real time is its real-time production, $q$. Note from the above equation that the seller receives the real-time price for its entire output at $T$. This implies that the incentives of the producer are not distorted, i.e. they are identical to the incentives that the producer would have had, had it not engaged in a forward contract. The producer always has the option of producing exactly the forward quantity, $q = x$, in which case the producer is paid exactly the forward price $f_t$ for each unit of output. In short, the forward contract hedges the producer from real-time price uncertainty while at the same time offering the producer the right incentives in real time. The argument and conclusion is identical for consumers.

The broad use of forward contracts for hedging risk has resulted in the definition of standardized forward contracts with rigid terms that are exchanged in a clearing house and for which the clearing house carries the **default risk**, which is the risk of the seller of the contract not paying its obligation at the expiration date. These standardized contracts are called **futures contracts**. Futures contracts are advantageous compared to forward contracts since the default risk is reduced (the default of the clearing house is much less likely than the default in bilateral agreements), liquidity is enhanced and there is no concern of credit-worthiness from the parties that engage in the trade of the contract since they are dealing with the clearing house. Nevertheless, forward contracts are more flexible than futures since parties can negotiate the exact terms of trade.

It is possible to integrate forward contracts for electricity with power system and power market operations. Entities that wish to override the market can trade a forward contract bilaterally. In real time the supplier submits a supply bid at the market floor and the consumer submits a bid at the market ceiling (in order to ensure that they are in the money) for the quantity they wish to trade.[1] The two parties then settle their forward contract outside of the market. The system operator charges agents the real-time price for their measured real-time energy production or consumption.

---

[1] Alternatively, and if the market rules allow it, market participants **self-schedule** their production and consumption, i.e. they inform the system operator about the quantity that they plan to produce or consume. In terms of the market model, these self-schedules are interpreted as fixed parameters, or equivalently as buy bids with infinite valuation and sell bids with minus infinite marginal cost. These are also referred to as price-taking offers, since, no matter what the price in the market turns out to be, these trades are planned to take place.

Futures contracts can be traded with the system operator. This in no way influences the physical scheduling of the system, although the forward positions provide information to the system operator about future supply and demand conditions in the system.

## 9.1.2  The price of forward contracts

A risk-neutral market agent values a forward contract according to the expectation of the commodity price at the expiration date $p_T$:

$$f_t = \mathbb{E}\left[p_T | \xi_{[t]}\right], \tag{9.1}$$

where $\xi_{[t]}$ describes the history of uncertainty up to time $t$ and $\xi_t$ describes the realization of uncertainty at time $t$, as in chapter 8.

---

**Example 9.1** *Forward contract pricing.* Consider a system with a linear demand function given by the following expression:

$$D(p) = 1620 - 4 \cdot p.$$

There are two generators in the system. The first generator has a capacity of 1880 MW and a marginal cost of 11.8 \$/MWh. The second generator has a capacity of 295 MW and a marginal cost of 65.1 \$/MWh. The first generator is unreliable, and its failure and restoration can be described by the Markov chain of figure 9.2. Here, 0.1 corresponds to the probability of generator failure when a generator is operational, and 0.5 corresponds to the probability of restoration when a generator has failed. Suppose that generator 1 is operational in period 0. Then the price of forward contracts that expire in period 2 can be computed moving backwards. Whenever the unreliable unit is off, a competitive market clears at 295 MW, at a price of 331.25 \$/MWh. Whenever the unreliable unit is on, the market clears at 1572.8 MW, at a price of 11.8 \$/MWh. The backward computation of forward prices is demonstrated in figure 9.3. The price in period 1 can be computed from equation (9.1) as

$$f_1 = \begin{cases} 0.9 \cdot 11.8 + 0.1 \cdot 331.25 = 43.745 \text{ \$/MWh,} & \xi_1 = \text{On,} \\ 0.5 \cdot 11.8 + 0.5 \cdot 331.25 = 171.525 \text{ \$/MWh,} & \xi_1 = \text{Off.} \end{cases}$$

**Figure 9.2** A Markov chain representing the operation of the unreliable unit.

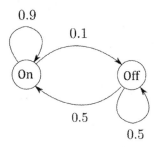

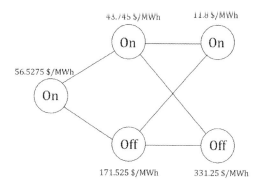

43.745 \$/MWh         11.8 \$/MWh

56.5275 \$/MWh

171.525 \$/MWh        331.25 \$/MWh

**Figure 9.3** Forward prices in example 9.1.

Similarly, the price in period 0 can be computed as

$$f_0 = 0.9 \cdot 43.745 + 0.1 \cdot 171.525 = 56.5275 \text{ \$/MWh}.$$

### 9.1.3  Contracts for differences and power purchase agreements

An alternative derivative that serves the same function as a forward contract is a **contract for differences (CfD)**. Like forward contracts, CfDs are characterized by their expiration date $T$, the quantity $x$, and the agreed price $f_t$.

**Definition 9.2**   Contract for differences

*Seller.* A seller sells a CfD with expiration date $T$ at time $t < T$ for $x$ units of a commodity.

*Buyer.* A buyer buys a CfD with expiration date $T$ at time $t < T$ for $x$ units of a commodity.

*Obligations and payoffs.* At time $T$ the seller pays the buyer $(p_T - f_t) \cdot x$, where $p_T$ is the spot price of the commodity at time $T$.

The CfD achieves the exact same purpose as a forward contract. A load can procure a CfD for $x$ units at $t < T$ and consume $x$ units in real time $T$. Then it pays a net of $f_t \cdot x$. Similarly, provided a supplier produces $x$ units of the commodity in real time, it is paid $f_t \cdot x$. The desirable properties of forward contracts, described in section 9.1.1, apply identically to CfDs.

**Power purchase agreements (PPAs)** are another instrument for ensuring trade at fixed prices. PPAs essentially amount to bilateral agreements for trading electricity at a fixed price. Although the definition is general, in practice PPAs are commonly used for financing renewable energy projects. The counter-parties are often owners of existing power production assets or developers of future power generation projects (who sell the PPAs) and large consumers of electricity, such as corporations with stewardship goals in the shift towards consuming renewable energy (who buy PPAs).

## 9.2    Financial transmission rights

The previous section demonstrates how forward contracts can be used for enabling bilateral trade at a fixed price without removing the incentive from market participants to operate efficiently in real time. Although forward contracts serve their intended purpose in a market with a common price for all locations, they fail to hedge perfectly in a system with locational marginal pricing.

---

**Example 9.2** *Forward contracts alone cannot hedge locational price differences.* Consider a system with locational marginal pricing. A generator in node A wishes to trade 400 MW with a load in node B at 40 $/MWh. This can be achieved if the generator sells a forward contract for $x = 400$ MW at price $f_t = 40$ $/MWh to the load. Suppose that there is no congestion and the real-time price turns out to be $p_A = p_B = 50$ $/MWh. The cash flows to the producer are:

- $+40 \cdot 400 = +\$16000$ (from selling the forward contract)
- $+50 \cdot 400 = +\$20000$ (from producing in the real-time market)
- $-50 \cdot 400 = -\$20000$ (from settling the forward contract)

The cash flows to the load are:

- $-40 \cdot 400 = -\$16000$ (from buying the forward contract)
- $-50 \cdot 400 = -\$20000$ (from consuming in the real-time market)
- $+50 \cdot 400 = +\$20000$ (from settling the forward contract)

The net effect of these transactions is that the generator is paid $16000 and the load pays $16000. The parties have achieved trade at 40 $/MWh, as originally intended. Suppose, instead, that there is congestion in the system with real-time prices $p_A = 36$ $/MWh and $p_B = 45$ $/MWh at the generator and load locations respectively. Suppose that the forward contract is for power in location A. The cash flows to the producer are:

- $+40 \cdot 400 = +\$16000$ (from selling the forward contract)
- $+36 \cdot 400 = +\$14400$ (from producing in the real-time market)
- $-36 \cdot 400 = -\$14400$ (from settling the forward contract)

The cash flows to the load are:

- $-40 \cdot 400 = -\$16000$ (from buying the forward contract)
- $-45 \cdot 400 = -\$18000$ (from consuming in the real-time market)
- $+36 \cdot 400 = +\$14400$ (from settling the forward contract)

Although the generator is paid 40 $/MWh, as intended, the load pays $19600, hence 49 $/MWh. Effectively, the load pays for the price difference $p_B - p_A = 9$ $/MWh between the generator and load location. It can be seen that a forward contract cannot hedge against locational price differences. The function of a financial transmission right is to hedge locational price differences when congestion arises.

---

The root of the problem in example 9.2 is the fact that forward contracts cannot hedge against locational price differences. In order to develop financial instruments that hedge against locational price differences, it is necessary to define rights for the usage of lines. How these rights are defined is a nontrivial issue.

The original proposal in the power industry, reasoning by analogy to other industries, was to define rights for transferring power along *paths*. These rights are referred to as **contract paths**. Contract paths are problematic because the amount of rights that can be issued cannot be defined independently of the overall state of the system, unlike in other industries. This is due to Kirchhoff's laws. The point is demonstrated by the following example.

---

**Example 9.3** *The weakness of contract paths.* Consider the network of figure 9.4. A generator in node 1 would like to trade with a load in location 3. Line 1–3 can only carry 600 MW of power, while line 2–3 can only carry 150 MW. The electrical characteristics of the lines in the network are identical. The idea of a contract path is to find a path along the electrical network that can presumably carry the power from the generator to the load location, and to buy rights on that path. In the case of the example, suppose that path 1–3 is considered. If 600 MW of rights are traded in the market, which is the physical capacity of line 1–3, then this can create a problem for line 2–3: a trade of 600 MW from node 1 to node 3 would result in a flow of 300 MW over line 2–3, which would overload the line. One might argue that the rights along the path should be equal to 300 MW, because in that case it is guaranteed that line 2–3 would not be overloaded. But if there are loads in location 2 trading with producers in location 3 then line 2–3 is relieved, and that would allow for more rights on path 1–3. The point of the example is that rights on path 1–3 are not a fixed number, but rather depend on the state of the system (i.e. all other planned trades between locations). This means that their definition is fundamentally problematic. As in the case of zonal pricing, either economic efficiency or system security is compromised.

---

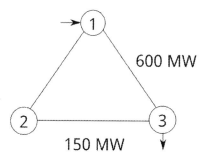

**Figure 9.4** The network of example 9.3. A generator in location 1 would like to trade with a load in location 3. Line 1–3 can carry 600 MW of power and line 2–3 can carry 150 MW. How many rights along path 1–3 should be issued?

The idea of **financial transmission rights** is to overcome the weakness of contract paths described above by defining long-term rights *between buses* of the network, instead of over paths.

**Definition 9.3**   Financial transmission rights (FTRs) are financial instruments that are defined as follows.

*Seller.* At time $t < T$ the seller sells a financial transmission right for shipping power from location A to location B for $x$ MW with expiration date $T$.

*Buyer.* At time $t < T$ the buyer of an FTR with expiration date $T$ buys the contract.

*Obligations and payoffs.* At time $T$ the seller pays the buyer of the FTR $(p_B - p_A) \cdot x$, where $p_A$, $p_B$ are the locational marginal prices of electricity in locations A and B, respectively.

---

**Example 9.4** *Hedging with FTRs.* Revisiting example 9.2, suppose that the load in location B buys a forward contract from the generator in location A as well as a financial transmission right for shipping power from A to B. The cash flows to the producer do not change, i.e. the producer sells power at 40 $/MWh. The cash flows to the load are:

- $-40 \cdot 400 = -\$16000$ (from buying the forward contract)
- $-45 \cdot 400 = -\$18000$ (from consuming in the real-time market)
- $+36 \cdot 400 = +\$14400$ (from settling the forward contract)
- $+9 \cdot 400 = +\$3600$ (from settling the FTR)

The net result of these transactions is that the load pays $16000. Hence, the FTR bundled with a forward contract enables the load to trade power bilaterally with the generator at a fixed price, independently of the market.

---

### 9.2.1   FTR auctions

In contrast to forward contracts, which can be traded among market agents bilaterally, the default seller of FTRs is the system operator. The reason for this is that FTRs need to be allocated in a way that ensures transmission constraints are respected, and the only entity that has a full picture of the physical constraints of the transmission network is the system operator. This property is referred to as **simultaneous feasibility**. FTRs are allocated through an FTR auction. The allocation of FTRs in an FTR auction must be such that the resulting flows satisfy the transmission constraints of the network.

FTRs are typically auctioned by the system operator to market participants, and can then be traded among market participants in secondary markets. Since the system operator is the original seller of the FTRs, the system operator is also responsible for paying off the FTRs in real time. An important property of FTRs is that, provided that the FTRs that are sold originally respect the transmission constraints of the network, the system operator collects adequate revenue in an LMP auction to cover the FTR payments. This property is referred to as **revenue adequacy** of FTRs.

**Proposition 9.4** *Financial transmission rights are revenue adequate.*

*Proof* This proposition is essentially a restatement of proposition 5.9.

$\square$

### 9.2.2 The virtues of FTRs

FTRs can be combined with forward contracts in order to enable two parties to trade power at a fixed price that the parties agree to in advance, even if price varies by location. This can be achieved, for instance, by having a producer sell a forward contract at its location, and having a load buy the forward contract and an FTR for shipping power from the location of the seller to its own location. Other ways of sharing the cost of transmission between counterparties can also be foreseen, but the coutnerparties have to collectively buy the necessary rights for getting the energy from the location of the seller to the location of the buyer.

FTRs maintain the incentives of market participants to produce/consume efficiently in real time. The argument is identical to the argument used for forward contracts.

Due to the fact that FTRs are purely financial instruments, and do not interfere with the physical operation of the network, they can be integrated easily with power system operations. By contrast, **physical transmission rights** (PTRs) are physical rights for using a transmission line that grant access to transmission lines *exclusively* to the holders of the rights, and provide no financial payoff. PTRs interfere with the physical operation of the network since the rights to access the lines cannot be transferred, unless the PTRs themselves are traded.

FTRs, like forward contracts, provide information to the system operator in advance of real-time operations. This information can support the efficient scheduling of units.

## 9.3 Callable forward contracts and reliability options

Callable forward contracts are an extension of forward contracts, introduced in section 9.1, that can be particularly useful for integrating demand response and hedging investment risk. In order to introduce callable forward contracts, it is first necessary to define call options.

**Definition 9.5** **Call options** are financial instruments that are characterized by a **strike price**, an expiration date, and an underlying commodity. They function as follows:

*Seller.* The seller of a call option with expiration date $T$ and strike price $k$ sells the option at time $t < T$ for an amount $x$ of an underlying commodity. The seller takes a *short position* in the call option.

*Buyer.* The buyer of a call option with expiration date $T$ at strike price $k$ buys the contract at time $t < T$ for an amount $x$ of an underlying commodity. The buyer takes a *long position* in the call option.

*Obligations and payoffs.* At time $t < T$ the buyer pays the seller the price of the call option. At time $T$ the seller of the option pays the buyer $\max(p_T - k, 0) \cdot x$, where $p_T$ is the spot price of the underlying commodity.

A call option entitles its buyer with the right, but not the obligation, to buy the underlying commodity from the seller at a certain strike price $k$ at expiration time. To see this, note that if the price of the commodity is less than $k$ then the call option has no value at expiration. If the spot price of a commodity is above $k$ then the buyer of the option receives $(p_T - k) \cdot x$ from the seller and can buy the commodity in the spot market, with a net expense of $k$ per unit of the traded commodity.

Call options can be used as instruments for hedging risk for buyers who do not want to be exposed to high real-time prices of commodities, as well as investors who build generation capacity. Call options can specifically be bundled with capacity markets in order to allow generators to trade the payoff of the market during periods of stress with a forward payment. Call options that serve this purpose are referred to as **reliability options**. Reliability options are discussed in further detail in section 11.3.2. As in the case of forward contracts and other financial instruments, there exists a secondary market for trading call options.

The payoff of a call option as a function of the price of the underlying commodity is shown in figure 9.5. Call options can be bundled with forward contracts in order to define a callable forward contract.

**Definition 9.6    Callable forward contracts** are defined as follows:

*Seller.* The seller of a callable forward with expiration date $T$ and strike price $k$ sells the contract at time $t < T$ for an amount $x$ of an underlying commodity.

*Buyer.* The buyer of a callable forward with expiration date $T$ and strike price $k$ buys the contract at time $t < T$ for an amount $x$ of an underlying commodity.

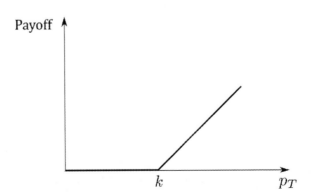

**Figure 9.5** The payoff of a call option.

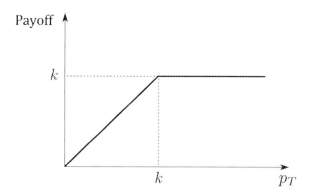

**Figure 9.6** The payoff of a callable forward.

*Obligations and payoffs.* At time $t < T$ the buyer pays the seller the price of the callable forward. At time $T$ the seller of the contract pays the buyer $\min(p_T, k) \cdot x$, where $p_T$ is the spot price of the underlying commodity.

The payoff of a callable forward as a function of the price of the underlying commodity is shown in figure 9.6. A callable forward is the bundle of a short position on a forward contract and a long position on a call option. This implies that at expiration the seller of the contract pays the buyer $p_T$ in order to settle the short position in the forward contract and receives a payment of $\max(p_T - k, 0)$ in order to settle the long position in the call option, i.e. the buyer of the callable forward receives a payment of $\min(p_T, k)$ from the seller.

The callable forward can be used in order to curtail the provision of a commodity to the buyer of the contract when the spot price of the commodity exceeds a strike price $k$. To see this, note that at expiration the buyer of the contract receives $p_T$ from the seller if $p_T \leq k$ and can use this payment to procure the commodity from the spot market. If $p_T > k$ then the buyer of the contract receives $k$.

### 9.3.1 The price of callable forwards

The price of a callable forward at any time $t$ prior to expiration depends on the predicted spot price of the commodity, as well as the strike price $k$. The conditional distribution of the spot price at expiration time is given by

$$Q_t(p) = \mathbb{P}\left[p_T \leq p | \xi_{[t]}\right],$$

where $\xi_{[t]}$ denotes the information in time $t$. We also define an associated density for this conditional distribution, which we assume exists:

$$q_t(p) = \frac{dQ_t(p)}{dp}.$$

Assuming that the price of the callable forward is the expected value of the payoff in expiration time, we get the following price for the forward and callable forward at time $t$:

$$f_t = \mathbb{E}[f_T|\theta_t] = \int_0^\infty pq_t(p)dp, \tag{9.2}$$

$$j_t(k) = \mathbb{E}[j_T(k)|\theta_t] = \int_0^\infty \min(p,k)q_t(p)dp. \tag{9.3}$$

Knowledge of the conditional density of the spot price $q_t(p)$ permits us to infer the price of the callable forward $j_t(k)$. Conversely, we can infer the density from the contract prices. To see this, we first integrate equation (9.3) by parts to obtain:

$$j_t(k) = k - \int_0^k Q_t(p)dp$$
$$= \int_0^k (1 - Q_t(p))dp. \tag{9.4}$$

We then differentiate with respect to $k$ to get

$$\frac{dj_t(k)}{dk} = 1 - Q_t(k), \tag{9.5}$$

and differentiating again with respect to $k$ we get

$$q_t(k) = -\frac{d^2 j_t(k)}{dk^2}. \tag{9.6}$$

We can then conclude that the price of the callable forward obeys the following properties:

- $j_t(k)$ is nondecreasing and concave in $k$. To see this, we use equations (9.5) and (9.6). This implies that the price of callable forwards increases with the strike price, since a higher strike price increases the payoff of the holder.
- $j_t(k) \leq k$ for all $k$. To see this, we use equation (9.4). This implies that the price of the callable forward cannot be greater than the strike price, since the callable forward cannot pay more than the strike price at delivery time.
- $\lim_{k\to\infty} j_t(k) = f_t$. This follows from equations (9.2) and (9.3). The result is intuitive since, as the strike price increases arbitrarily, the likelihood of the call option being exercised converges to zero, which makes the forward and callable forward indistinguishable.

---

**Example 9.5** *Price of a callable forward.* Consider a market with the following spot price (see figure 11.7 in example 11.9):

- 1000 $/MWh for hours 1–20
- 880.04 $/MWh for hour 21
- 160 $/MWh for hours 22–328
- 120.06 $/MWh for hour 329
- 80 $/MWh for hours 330–1752
- 25.21 $/MWh for hour 1753
- 25 $/MWh for hours 1754–7576

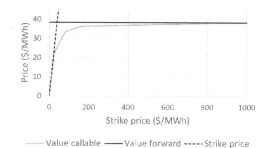

**Figure 9.7** Price of the callable forward contract as a function of strike price $k$ in example 9.5.

- 10.81 $/MWh for hour 7577
- 6.5 $/MWh for hours 7578–8760

The price of a forward contract for this market is

$$f_t = (20 \cdot 1000 + 1 \cdot 880.04 + 307 \cdot 160 + 1 \cdot 120.06 + 1423 \cdot 80$$
$$+1 \cdot 25.21 + 5823 \cdot 25 + 1 \cdot 10.81 + 1183 \cdot 6.5)/8760 = 38.50.$$

Let us consider a given strike price, e.g. 300 $/MWh. The payoff of the callable forward for this specific strike price is:

$$j_t(300) = (20 \cdot 300 + 1 \cdot 300 + 307 \cdot 160 + 1 \cdot 120.06 + 1423 \cdot 80$$
$$+1 \cdot 25.21 + 5823 \cdot 25 + 1 \cdot 10.81 + 1183 \cdot 6.5)/8760 = 36.84.$$

The price of the call option with a strike price of 300 $/MWh is essentially the difference between the forward and the callable forward, i.e. $f_t - j_t(k) = 1.66$ $/MWh. We can apply the same calculations repeatedly for different values of strike price $k$. The resulting value of the callable forward is presented in figure 9.7. We verify the properties that are derived earlier: the price of the callable forward never exceeds the strike price, it tends to the forward price, and it is nondecreasing and concave in $k$.

## 9.3.2 The virtues of callable forward contracts

Callable forward contracts can be used in order to secure a compensation $k$ in case the real-time price of electricity exceeds $k$. Without such contracts, loads with valuation $v$ that bid truthfully do not receive any power whenever $p_T \geq v$. By procuring a callable forward with a strike price $k$ equal to their valuation, $k = v$, loads ensure that even if their bid is out of the money they receive a compensation equal to their valuation for power. Effectively, callable forwards with $k = v$ ensure that loads *always* receive the full value of power supply, regardless of the real-time price of electricity.

The system operator also benefits from callable forward contracts. Loads that procure these contracts engage in long-term commitments for demand response.

By selling callable forward contracts, the system operator receives information about the shape of the demand function in advance of operations. This can be beneficial in terms of planning the capacity expansion of the system optimally, especially in systems with large amounts of renewable energy sources.

Efficiency dictates that when consumers are presented with a menu of callable forward contracts, they select a callable forward contract that allocates the commodity if the spot price of the commodity is lower than their valuation and does not allocate the commodity otherwise. For consumers with valuation $v$, this translates to self-selecting a callable forward with strike price $k = v$.

Suppose that consumers are risk neutral and callable forwards are priced according to their expected payoff. Denote $B_t(k)$ as the benefit of a consumer who selects a callable forward with strike price $k$ in period $t$. Then the payoff of the consumer is $v$ with probability $Q_t(k)$ (if $p \leq k$), and $k$ otherwise. This gives an expected payoff of

$$\mathbb{E}[B_t(k)|\theta_t] = Q_t(k) \cdot v + (1 - Q_t(k))k - j_t(k)$$
$$= k + Q_t(k)(v - k) - j_t(k). \tag{9.7}$$

We therefore get

$$\frac{d\mathbb{E}[B_t(k)|\theta_t]}{dk} = 1 - \frac{dj_t(k)}{dk} - Q_t(k) + (v - k) \cdot q_t(k)$$
$$= (v - k) \cdot q_t(k), \tag{9.8}$$

which follows from equation (9.5).

We assume that $q_t(k) > 0$ for all $k > 0$. Then the derivative of the payoff with respect to $k$ is zero at $v = k$, and we have $\frac{d\mathbb{E}[B_t(k)|\theta_t]}{dk} > 0$ for $k < v$ and $\frac{d\mathbb{E}[B_t(k)|\theta_t]}{dk} < 0$ for $k > v$. We therefore conclude that $k = v$ is the unique maximizer of the expected consumer payoff. Moreover, we note from equation (9.7) that the expected payoff for $k = v$ is $v - j_t(v)$. From equation (9.4) and the fact that $q_t(k) > 0$ we conclude that the payoff of self-selecting the callable forward has a strictly higher payoff than not buying a callable forward, which yields a value of 0. We have therefore shown that consumers self-select to reveal the true information about their valuation for power, through their choice of strike price.

## 9.4    Modeling risk aversion

Risk measures are used to represent the way risk-averse agents rank lotteries, i.e. assets that give them random payoffs. A **risk measure** $\mathcal{R}$ is formally defined as a mapping from a real-valued random variable $\xi \colon \Omega \to \mathbb{R}$ to a real number. Here, $\Omega$ is the sample space of a probability space. We interpret $\xi$ as a lottery. Intuitively, the risk measure assigns a risk-adjusted score to the lottery. The ensuing discussion considers the lottery as a payment *from* the owner of the lottery, although the same analysis applies in the case where the lottery corresponds to a payment *to* the owner of the lottery.

**Example 9.6** *Forward contract as a lottery.* Consider a forward contract for 1 MW of electricity at a future delivery date during the winter. Suppose that there are two possible states of the world. In the first state of the world, electricity prices $\xi$ in a system dominated by electric heating are high, whereas in the second state of the world they are low. We can specifically define $\Omega = \{\text{Cold, Hot}\}$. The forward contract obliges its seller to pay back the realized price of electricity at the delivery date. Denoting the payoff of the forward contract as $\xi$, we have $\xi(\text{Cold}) = 100$ \$/MWh, and $\xi(\text{Hot}) = 50$ \$/MWh. For the sake of subsequent examples, we further assume that each of the weather outcomes is equally likely.

**Example 9.7** *Call option as a lottery.* Returning to the previous example, suppose that an agent has sold a call option on electricity with a strike price of $k = 70$ \$/MWh. In this case, $\xi(\text{Cold}) = \max(100-70, 0) = 30$ \$/MWh, and $\xi(\text{Hot}) = \max(50-70, 0) = 0$ \$/MWh.

**Example 9.8** *Expectation as a risk measure.* One of the most widely used risk measures is expectation. Returning to the previous examples, let us assume that $\mathcal{R}(\xi) = \mathbb{E}[\xi]$. Then, the risk measure applied to the payoff of the forward is $\mathcal{R}(\xi) = 75$ \$/MWh, while the risk measure applied to the payoff of the call option is $\mathcal{R}(\xi) = 15$ \$/MWh.

**Example 9.9** *Worst-case payoff as a risk measure.* Let us consider the worst-case outcome as a risk measure: $\mathcal{R}(\xi) = \max_{\omega \in \Omega} \xi(\omega)$. We then have $\mathcal{R}(\xi) = 100$ \$/MWh in the case of settling a forward contract, and $\mathcal{R}(\xi) = 30$ \$/MWh in the case of settling a call option.

**Definition 9.7** $\mathcal{R}(\cdot)$ is a **coherent risk measure** (CRM) if the following hold:

1. Subadditivity: $\mathcal{R}(\xi + \zeta) \leq \mathcal{R}(\xi) + \mathcal{R}(\zeta)$ for any random variables $\xi$ and $\zeta$. The intuitive interpretation of this condition is that pooling risk is good.
2. Positive homogeneity of degree one: $\mathcal{R}(\lambda \cdot \xi) = \lambda \cdot \mathcal{R}(\xi)$ for all $\lambda \geq 0$. This property implies that discounting costs discounts risk.
3. Monotonicity: $\mathcal{R}(\xi) \leq \mathcal{R}(\zeta)$ whenever $\xi \preceq \zeta$, where $\preceq$ indicates **first-order stochastic dominance**, i.e. $\mathbb{P}[\xi \leq t] \geq \mathbb{P}[\zeta \leq t], \forall t \in \mathbb{R}$. This corresponds to the fact that lower costs imply lower risk.
4. Translation invariance: $\mathcal{R}(\xi + t) = \mathcal{R}(\xi) + t$ for any $t \in \mathbb{R}$. This implies that fixed costs add a fixed amount of risk.

Stochastic dominance, which is used in definition 9.7, is a relation between random variables. It is interpreted intuitively as a generalization of inequalities to the case of distributions: a random variable $\xi$ stochastically dominates another one $\zeta$ if the probability of $\xi$ being less than $t$ is greater than that of $\zeta$ being less than $t$ for any $t \in \mathbb{R}$.

---

**Example 9.10** *Stochastic dominance of electricity prices.* Consider the electricity price distribution of example 9.6, and let us compare this price $\zeta$ to that of another market, in which the price is equal to 50 \$/MWh with a probability of 0.25 and equal to 120 \$/MWh with a probability of 0.75. Then, $\xi \preceq \zeta$. For instance, for $t = \$60$, we have that $\mathbb{P}[\xi \leq 60] = 0.5 > \mathbb{P}[\zeta \leq 60] = 0.25$.

---

Expectation and worst-case payoff are examples of coherent risk measures. Expectation is treated in the following example, while worst-case payoff is treated in problem 9.2. Two interesting examples of risk measures that are not coherent include value at risk and the Markowitz risk measure. Each of these is treated in further detail in the following.

---

**Example 9.11** *Expectation is a coherent risk measure.* In order to establish that expectation is a coherent risk measure, we need to prove subadditivity, positive homogeneity of degree one, monotonicity, and translation invariance. For subadditivity, we observe that $\mathbb{E}[\xi + \zeta] = \mathbb{E}[\xi] + \mathbb{E}[\zeta]$. For positive homogeneity of degree one, we have that $\mathbb{E}[\lambda \cdot \xi] = \lambda \cdot \mathbb{E}[\xi]$. For monotonicity, let us denote by $F$ the cumulative distribution function of $\xi$, by $G$ the cumulative distribution function of $\zeta$, and let us assume that both are continuous and strictly increasing in $\mathbb{R}$, in order to simplify the exposition. We first note that, for every upper bound on $x$ of $\xi$, the following upper bound $y$ of $\zeta$ corresponds to the same probability $F(x)$ of materializing:

$$y(x) = G^{-1}(F(x)).$$

Since $G(x) \leq F(x)$, and since $G^{-1}$ is monotonically increasing, we have that

$$y(x) = G^{-1}(F(x)) \geq x.$$

Thus,

$$\mathbb{E}[\zeta] = \int_{\mathbb{R}} y dG(y) = \int_{\mathbb{R}} y(x) dG(y(x)) = \int_{\mathbb{R}} y(x) dF(x) \geq \int_{\mathbb{R}} x dF(x) = \mathbb{E}[\xi].$$

For translation invariance, we have that $\mathbb{E}[\xi + t] = \mathbb{E}[\xi] + t$.

---

It can be shown that subadditivity and positive homogeneity imply that $\mathcal{R}$ is a convex function (see problem 9.3). Convexity corresponds intuitively to the fact that the marginal cost of risk is increasing. $\mathcal{R}$ is a **convex risk measure** if it satisfies conditions 1–3 of definition 9.7.

## 9.4.1  Value at risk and conditional value at risk

**Value at risk (VaR)** is a risk measure that is commonly used in finance, where $VaR_\alpha$ is the greatest loss in portfolio value that can occur with a given probability $\alpha$:

$$VaR_\alpha(\xi) = \min\{t | \mathbb{P}[\xi \leq t] \geq \alpha\}.$$

It is essentially the $\alpha$-quantile of the distribution of the lottery $\xi$.

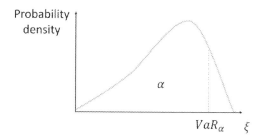

**Figure 9.8** A graphical illustration of the value at risk.

The risk measure is illustrated graphically in figure 9.8. In order to compute the value at risk for a certain level $\alpha$, we scan the horizontal axis from left to right. We stop at the point where the probability mass accumulated to the left is at least $\alpha$. These are the worst-case losses that an investor faces with a probability of $\alpha$.

---

**Example 9.12** *Value at risk.* Consider the forward contract of example 9.6. The value at risk for $\alpha = 0.1$, $VaR_{0.1}(\xi)$, is 50 \$/MWh. In intuitive terms, it states that if the investor observes the payoff of the forward contract that it is obliged to settle for 1000 realizations of the market and picks the best 100 of them, then the cost it is required to pay off can be as high as 50 \$/MWh. The value at risk for $\alpha = 0.9$, $VaR_{0.9}(\xi)$, is 100 \$/MWh. This states that if the investor observes the payoff of the forward contract that it is obliged to settle for 1000 realizations of the market and picks the best 900 of them, then the cost it is required to pay off can be as high as 100 \$/MWh.

---

Value at risk suffers from the fact that it can be highly sensitive to data, and therefore unstable. To see that the value at risk is not stable, consider a set of nearly identical cumulative density functions the positive tail of which is flat. One would desire that these distributions correspond to nearly identical risk. Whereas the area under the cumulative distribution function that lies to the right of the value at risk will not vary strongly, these density functions can be constructed such that the value at risk itself can vary substantially. As Rockafellar argues, *"This degree of instability is distressing for a measure of risk on which enormous sums of money might be riding."*

An additional feature of VaR that is noteworthy is that VaR is not a coherent risk measure. In particular, VaR does not obey subadditivity, although the reader is asked to verify that it obeys the other properties of coherent risk measures (see problem 9.4).

---

**Example 9.13** *Value at risk is not subadditive.* Consider two possible states of the world, $\Omega = \{1, 2\}$, and two random variables, $\xi$ and $\zeta$. We have that each state of the world occurs with the same probability of 0.5. Furthermore, we have that $\xi(1) = 10$, $\xi(2) = 100$, $\zeta(1) = 100$, $\zeta(2) = 10$. Then, $VaR_{0.1}(\xi) = 10$, $VaR_{0.1}(\zeta) = 10$, and $VaR_{0.1}(\xi + \zeta) = 110$, which is greater than $VaR_{0.1}(\xi) + VaR_{0.1}(\zeta)$.

---

Having defined value at risk, it is now possible to define conditional value at risk. This risk measure overcomes a number of the aforementioned drawbacks of value at risk, in the sense that it is more stable with respect to data and also turns out to be a coherent risk measure (see problem 9.5).

**Conditional value at risk (CVaR)** is the expectation of losses, conditional on losses being greater than $VaR$:

$$CVaR_\alpha(\xi) = \mathbb{E}_{P_\alpha}[\xi],$$

where $P_\alpha$ is defined as

$$P_\alpha(t) = \begin{cases} 0, & \text{if } t < VaR_\alpha(\xi), \\ \frac{\mathbb{P}[\xi \le t] - \alpha}{1 - \alpha}, & \text{if } t \ge VaR_\alpha(\xi). \end{cases} \tag{9.9}$$

Another important advantage of CVaR relative to VaR is that it can be expressed as a linear program. The following result is due to Rockafellar and Uryasev (2002):

**Proposition 9.8**

$$CVaR_\alpha(\xi) = \min_t \left\{ t + \frac{1}{1 - \alpha} \mathbb{E}_P[(\xi - t)^+] \right\},$$

*with the optimal solution of t providing the $VaR_\alpha$.*

---

**Example 9.14** *Computation of CVaR by hand and as a linear program.* Consider the distribution of costs presented in table 9.1. We then have

$$CVaR_{0.96} = \$1000$$
$$CVaR_{0.86} = \frac{4}{14} \cdot (1000) + \frac{10}{14} \cdot 0 = \$285.7.$$

The intuition for $CVaR_{0.96}$ is that the 4% least favorable outcomes correspond to the unique realization where cost is equal to \$1000. The conditional probability of this event occurring is 100%. Similarly, in order to compute $CVaR_{0.86}$, if the 14% least favorable outcomes occur, this corresponds to the outcomes with costs of \$1000 and \$0. The conditional distribution then assigns a probability of (4/14) to the outcome with a cost of \$1000, and a probability of (10/14) to the outcome with a cost of \$0. The linear programming formulation for computing $CVaR_{0.86}$ can be expressed as follows:

**Table 9.1 Scenarios in example 9.14.**

| Cost (\$) | Probability (%) |
|---|---|
| 1000 | 4 |
| 0 | 10 |
| −1000 | 12 |
| −2000 | 14 |
| −3000 | 60 |

$$\min_{t,y} t + \frac{1}{1-0.86}(0.04 \cdot y_1 + 0.1 \cdot y_2 + 0.12 \cdot y_3 + 0.14 \cdot y_4 + 0.6 \cdot y_5)$$

$$y_1 \geq 1000 - t, y_2 \geq 0 - t, y_3 \geq -1000 - t, y_4 \geq -2000 - t, y_5 \geq -3000 - t$$

$$y \geq 0.$$

The objective function value of this model equals \$285.7, which is $CVaR_{0.86}$. The optimal value of $t$, which is equal to 0, is the VaR.

---

The representation of CVaR (and other risk measures) as a linear program opens a wide range of modeling possibilities. In particular, CVaR can lead to computationally tractable models of optimization under risk aversion and equilibria under risk aversion. Recently, this has been exploited in the development of various applications in electricity markets, including stochastic equilibrium models for capacity expansion and the modeling of market design options in real-time markets.

## 9.4.2 Worst-case characterization of coherent risk measures

$\mathcal{R}$ is a coherent risk measure if and only if there exists a class of probability measures $\mathcal{M}$ such that $\mathcal{R}(\xi)$ equals the highest expectation of $\xi$ with respect to members of this class:

$$\mathcal{R}(\xi) = \max_{q \in \mathcal{M}} \mathbb{E}_q[\xi] = \max_{q \in \mathcal{M}} \sum_{\omega \in \Omega} q_\omega \cdot \xi(\omega).$$

The vector $\bar{q}$ that maximizes the above expression is the **risk-adjusted probability measure**.

---

**Example 9.15** *Worst-case characterization of CVaR and expectation.* The probability distribution $P_\alpha$ that is defined in equation (9.9) is the vector $\bar{q}$ that maximizes the expectation under the following family of probability distributions:

$$\mathcal{M} = \left\{ q : q_\omega \leq \frac{P_\omega}{\alpha}, \sum_{\omega \in \Omega} q_\omega = 1, q \geq 0 \right\}.$$

Concretely, we allow ourselves to redistribute the mass of all events by increasing the original probabilities $P_\omega$ by a factor of up to $1/\alpha$. If our goal is to maximize the damage implied by the expectation of cost $\xi(\omega)$, we can maximize this expectation by "pushing" as much probability mass as possible to the largest realizations of $\xi(\omega)$. This gives rise to the probability measure of equation (9.9), and constitutes the risk-adjusted probability measure $\bar{q}$. It is a pessimistic, or risk-averse, outlook on the realization of the random variable $\xi(\omega)$. Note that the special case where $\alpha = 1$ corresponds to the case where we have no real freedom in moving probability mass around: we can only increase the original probability measure $P$ by a factor of 1 for each realization, and the total probability mass has to equal 1, so our only choice is to stay with the original probability measure $P$. This corresponds to a neutral attitude towards risk, with $CVaR_1$ being identical to the expectation under risk measure $P$, i.e. $\mathbb{E}_P$.

---

In order to appreciate the strength of the above result, we use it to immediately arrive to the following conclusions: expectation and CVaR are coherent risk measures. Setting $\mathcal{M} = \{P\}$ proves that $\mathbb{E}_P[\cdot]$ is a coherent risk measure. Setting $\mathcal{M} = \{q : q_\omega \leq \frac{P_\omega}{\alpha}, \sum_{\omega \in \Omega} q_\omega = 1, q \geq 0\}$ proves that $CVaR_\alpha(\cdot)$ is a coherent risk measure.

The vector $\bar{q}$ is a subgradient of the function $\mathcal{R}(\xi)$. Moreover, the subgradient of the risk measure with respect to a parameter $a$ can be obtained using the chain rule:

$$\frac{\partial \mathcal{R}(\xi)}{\partial a} = \sum_{\omega \in \Omega} \frac{\partial \mathcal{R}}{\partial \xi} \frac{\partial \xi}{\partial a} = \mathbb{E}_{\bar{q}} \left[ \frac{\partial \xi}{\partial a} \right].$$

---

**Example 9.16** *Subgradient of risk-adjusted payoff.* Let us consider an agent that sells $a$ MW of electricity in a day-ahead forward market. The cost of settling the forward contract in real time can be described as

$$\xi(\omega) = \lambda^{RT}(\omega) \cdot a.$$

Selling the contract in the forward day-ahead market means that the agent is obliged to pay the real-time price, $\lambda^{RT}(\omega)$, for every MW that it has sold forward. Let us assume that real-time prices are distributed uniformly between the values of $\{10, 20, \ldots, 100\}$ \$/MWh. And suppose that the agent has a risk attitude of $CVaR_{0.8}$. This means that $P_{0.8}$ in equation (9.9) is the probability mass that increases the worst possible outcomes by a factor of $1.25$, i.e. $\bar{q} = \{0, 0, 0.125, \ldots, 0.125\}$. Moreover, the subgradient of the payoff $\xi$ with respect to the forward position $a$ is simply $\lambda^{RT}(\omega)$: $\frac{\partial \xi}{\partial a} = \lambda^{RT}$. In intuitive terms, selling one more MW in the forward day-ahead market costs the agent $\lambda^{RT}(\omega)$ when outcome $\omega$ materializes. Then, the risk-adjusted cost of selling one more MW in the forward market is equal to

$$\frac{\partial \mathcal{R}(\xi)}{\partial \alpha} = \mathbb{E}_{\bar{q}} \left[ \frac{\partial \xi}{\partial \alpha} \right] = \mathbb{E}_{\bar{q}} \left[ \lambda^{RT} \right] = 0.125 \cdot (30 + 40 + \cdots + 100) = \$65.$$

Even though the expected real-time price is equal to \$55, the risk aversion of the agent means that it assigns a risk-adjusted cost for selling one MW in the forward market that is equal to \$65.

---

## 9.4.3 Back-propagation

Risk measures provide a quantitative framework for explaining the **back-propagation** of prices to forward markets. Back-propagation is the process by which the price of forward contracts forms as a function of the distribution of the real-time price of the underlying product. We encounter back-propagation in specific settings of risk-neutral agents previously in the textbook. For instance, in section 9.1 we argue that forward contract prices for electricity should equal the expected real-time price of electricity. And in section 9.3 we develop a similar argument for callable forward contracts. Although this statement is not justified quantitatively up to now, the intuitive argument relates to necessary conditions for a market equilibrium to hold. For instance, if the forward price of electricity is much higher than the expected real-time price of electricity, then arbitrageurs are likely to sell forward contracts and close their position in real time, thereby earning an expected profit. This process of selling forward contracts exerts a downward pressure on forward prices (and has the contrary effect on real-time prices), and tends to align forward contract prices to expected real-time prices. In the opposite case, when forward contract prices are significantly lower than expected real-time prices, then arbitrageurs are expected to buy forward contracts and close their position in real time, thereby earning again a positive profit in expectation. This exerts an upward pressure on forward prices (and the contrary on real-time prices), until forward prices align to expected real-time prices.

In order to represent back-propagation quantitatively, we return to example 9.16.

---

**Example 9.17** *Back-propagation of real-time prices to forward markets.* Let us consider the problem of an agent who is deciding how much electricity to sell in the day-ahead forward market. And suppose that the agent does not have any trading restrictions: it can sell any amount of MWs that it wants in the day-ahead market.[2] The surplus maximization problem of the agent in the day-ahead market can then be expressed as follows:

$$\max_{a} \lambda^{DA} \cdot a - \mathcal{R}(\xi(a)),$$

where $\xi(a) = \lambda^{RT} \cdot a$, as explained in example 9.16. Note that we highlight the dependency of $\xi$ on $a$ with this notation. The first-order optimality conditions of this surplus maximization problem are expressed by setting the subgradient equal to zero:

$$\lambda^{DA} = \frac{\partial \mathcal{R}}{\partial \alpha} = \mathbb{E}_{\bar{q}}\left[\lambda^{RT}\right].$$

In the specific case of the agent of example 9.16, this means that the position of the agent is determined by the risk-adjusted expectation of the real-time price. Concretely, if the day-ahead price is greater than $65, then this cannot be an equilibrium, because the agent would take an arbitrarily short position in the day-ahead market (i.e. it would sell an arbitrarily large quantity) in order to close its position in real time at an unlimited profit (i.e. it would buy back cheaper in real time). In the opposite case, if the day-ahead price is less than $65, the agent would take an

---

[2] Agents often have trading restrictions. Apart from the fact that they cannot put arbitrary sums of money on the line, there are additional restrictions on **virtual trading**. Virtual trading refers to the practice whereby agents are allowed to buy or sell electricity in day-ahead and other forward markets, even if they do not own actual physical assets, i.e. even if they do not plan to actually produce or consume the electricity that they sell or buy forward, but simply enter a purely financial transaction. The goal of virtual trading is to enhance price discovery and put the "wisdom of the crowds" to work towards closing gaps between real-time and forward markets.

arbitrarily long position in the day-ahead market (i.e. it would buy an arbitrarily large quantity) in order to close its position in real time at an unlimited profit (i.e. it would sell back more expensive in real time).

---

The above discussion highlights that the real-time market should be the focal point of electricity market design, because real-time prices drive forward markets. Reasoning from the end of time backwards is crucial, and is a principle that is often violated in practice. A prevalent misunderstanding in electricity markets, for instance, is that day-ahead market design carries the same or more weight than real-time market design. The fact that the volumes traded in the day-ahead market are greater than those of real time is sometimes used in order to rationalize this perverse point of view. This is reflected, for instance, in how the evolution of market integration in Europe prioritized the coupling of day-ahead markets through the single day-ahead coupling (SDAC), before turning attention to the coupling of real-time markets through MARI and PICASSO, which are the cross-border European markets for the activation of frequency restoration reserves. The challenges that are emerging in the cross-border coupling of European real-time markets are now becoming apparent, and are requiring a reconsideration of various aspects of the day-ahead design, e.g. as it relates to the treatment of transmission constraints (see the discussion on nodal versus zonal pricing in chapter 5).

### 9.4.4    Other ways of representing risk aversion

A common way of representing risk aversion in economics uses convex **utility functions**. According to this model of risk aversion, given two lotteries that result in random costs $\xi$, $\zeta$, an agent with a convex utility function $U: \mathbb{R} \to \mathbb{R}$ prefers $\xi$ to $\zeta$ provided that $\mathbb{E}_P[U(\xi(\omega))] \leq \mathbb{E}_P[U(\zeta(\omega))]$. The risk measure is therefore given as

$$\mathcal{R}(\xi) = \mathbb{E}[U(\xi(\omega))].$$

---

**Example 9.18** *Risk aversion using a piecewise linear convex utility function.* Compare a lottery $\xi$ that requires its owner to pay $100 or −$100 with equal probability, versus a certain payment $\zeta$ of $0. Consider a decision-making agent with utility function $\max(x, 0.5x)$. The agent evaluates the risk of the lottery as $\mathcal{R}(\xi) = 0.5 \cdot (100 - 50) = $25$. Instead, the sure payoff achieves a risked cost of $R(\zeta) = $0$. Since $R(\xi) \geq R(\zeta)$, the sure payoff is preferred to the risky lottery. The choice is depicted graphically in figure 9.9.

---

Another notable risk measure is the **Markowitz risk measure**, which is defined as follows:

$$\mathcal{R}(\xi) = \mathbb{E}[\xi] + \beta \cdot var(\xi).$$

Here, $\beta$ is a fixed parameter, and $var(\xi)$ is the variance of the random variable $\xi$.

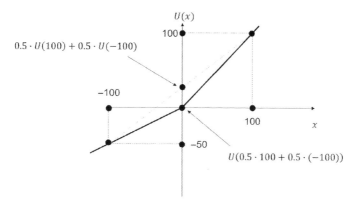

**Figure 9.9** Representation of risk aversion with a convex utility function: the sure payoff of $0 is preferable to a 50–50 lottery that pays off $100 and −$100 respectively.

**Example 9.19** *Markowitz risk measure.* Consider the lottery of example 9.18. The expected value and variance of the lottery are respectively computed as follows:

$$\mathbb{E}[\xi] = 0.5 \cdot 100 + 0.5 \cdot (-100) = 0,$$
$$var(\xi) = 0.5 \cdot (100 - 0)^2 + 0.5 \cdot (-100 - 0)^2 = 10000.$$

Given $\beta = 0.05$, the Markowitz risk measure of the lottery is computed as follows:

$$\mathcal{R}(\xi) = 0 + 0.05 \cdot 10000 = 500.$$

In problem 9.7 we show that the Markowitz risk measure is not a coherent risk measure.

## Problems

**9.1** *Virtues of CfDs.* Why does a CfD exist, if it achieves the same purpose as a forward contract?

**9.2** *Worst-case payoff is a CRM.* Prove that worst-case payoff is a coherent risk measure.

**9.3** *Subadditivity and positive homogeneity imply convexity.* Prove that subadditivity and positive homogeneity in the definition of a coherent risk measure imply that the risk measure is convex.

**9.4** *Properties of VaR.* Prove that VaR is positive homogeneous, monotonic, and translation invariant.

**9.5** *CVaR is a coherent risk measure.* Prove that CVaR is a coherent risk measure.

**9.6** *Analytical derivation and mathematical programming formulation of CVaR.* Compute $CVaR_{0.84}$ for the values of table 9.1 both analytically as well as using mathematical programming software. Confirm that you get identical results in both cases.

**9.7** *The Markowitz risk measure is not a coherent risk measure.* Prove that the Markowitz risk measure violates monotonicity.

**9.8** *Cash flows in forward contracts versus CfDs.* Consider a supplier who wishes to procure 3000 MWh of electricity 12 months from now, where the market 12 months from now will have a marginal cost of $20 + 0.1 \cdot Q$ \$/MWh and a demand of $Q$, which is normally distributed with a mean value of 1000 MW and a standard deviation of 200 MW (for every hour of the month, accounting also for the consumption of the supplier in question).

1. Describe the cash flows (now and at the expiration of the contract) that the supplier is exposed to if it engages in (i) a forward contract, or (ii) a contract for differences, and if demand 12 months from now turns out to be 1500 MW (for each hour of the month) and the supplier ends up consuming 3100 MWh during the month.
2. Do the total cash flows of the two financial instruments have any difference?
3. Which of the two financial instruments leads to lower cash reserve for the supplier over the course of the year?

# Bibliography

**Section 9.1**    The material in this section is based on Kaye, Outhred, and Bannister (1990). Contracts for differences and power purchase agreements are surveyed in UK Government (2013).

**Section 9.2**    The material in this section is based on Hogan (1992) and Lyons, Fraser, and Parmesano (2000). Example 9.3 is drawn from Hogan (1992).

**Section 9.3**    The material in this section is based on Gedra and Varaiya (1993).

**Section 9.4**    The development on risk measures is inspired by Ralph and Smeers (2011, 2015). Risk measures are also discussed in Birge and Louveaux (2010). Coherent risk measures are defined originally by Artzner et al. (1999). The excerpt from Rockafellar regarding value at risk in section 9.4.1 and proposition 9.8 are due to Rockafellar and Uryasev (2002). Conditional value at risk is used for the analysis of stochastic equilibrium in capacity expansion by Ehrenmann and Smeers (2011a). Stochastic equilibrium based on coherent risk measures is also used for the analysis of real-time market design in Papavasiliou, Smeers & de Maere-d'Aertrycke (2021). The result of the subgradient of a risk measure with respect to a parameter that is used in section 9.4.3 in order to quantitatively represent back-propagation is based on Shapiro et al. (2009). The result applies to the more general setting of convex risk measures (Follmer & Schied 2002). The argument that "the last should be first" has been repeatedly conveyed by Hogan over years, and multiple presentations, in order to emphasize the crucial role of getting the design of the real-time market right; see for instance slide 5 in Hogan (2016a). More complex interactions between real-time and forward markets can occur due to market power, as described in Allaz and Vila (1993). The virtues of forward trading are highlighted in Hogan (2016b), while Parsons et al. (2015) provides a critical review of designs that encompass virtual trading. Virtual trading is modeled using coherent risk measures in Papavasiliou, Smeers & de Maere-d'Aertrycke (2021). Problem 9.7 is based on Ralph and Smeers (2015).

# 10   Demand response

The paradigm of power system operations since their inception has been load-following, meaning that production adjusts itself to the instantaneous load in the system. This is related to the separation of commercial and residential consumers from electricity prices. Electricity prices at the retail level (which typically corresponds to commercial and residential consumption) are often fixed. This implies that, even if consumers would be able to quantify their valuation for power, which might vary throughout the day, this information would be lost during power system operations because consumers face a fixed retail price.

**Demand response** refers to the active participation of consumers in the efficient consumption of electricity as well as the provision of ancillary services. Demand response can be delivered typically through efficiency (an overall reduction in consumption), peak load shaving (reducing the amount of peak power consumption), or load shifting (moving demand to hours when the system is less stressed). Responsive demand serves as an additional source of flexibility in the system, and is therefore perceived as especially valuable for supporting the integration of renewable energy resources.

Traditionally, consumers face fixed retail prices that are disconnected from the real-time conditions of the system. This results in unresponsive demand in the system, which complicates both the operation of electric power systems and electricity markets.

In theory, real-time pricing is the most straightforward mechanism for integrating demand response in electricity markets. According to real-time pricing, consumers pay the price for electricity, as this is determined in the real-time electricity market. Real-time pricing requires active monitoring of electricity prices and reactiveness on the side of loads. The process of reacting to real-time electricity prices can be quite complex and risky for small consumers. A relatively small fraction of consumers, typically large industrial customers, are exposed to the wholesale electricity market price.

**Time of use pricing**, which is introduced in section 10.1, is an alternative pricing method that fixes the price of electricity in advance, with the price varying according to the time of day. The motivation of time of use pricing is to level out peak demand and minimize the operating and investment cost of operating the system. Customers who contribute to the system peak consumption essentially pay an additional charge in order to cover the capital costs of the technology that is used for satisfying their demand.

There exists a variety of alternative mechanisms for inducing demand response. **Critical peak pricing** refers to time of use pricing with a maximum number of critical peak events, during which consumers pay a very high price for power. Real-time demand reduction programs are programs in which consumers pay the real-time price for any consumption above an administratively set baseline. **Interruptible service** programs allow the utility to interrupt service to consumers with very short notice, in exchange for a reduction in their electricity rates or a periodic payment.

Interruptible service programs, which represent an intermediate mechanism between real-time pricing and time of use pricing, are driven by the principle that consumers view electricity as a service, the features of which can be differentiated, rather than a commodity, the real-time price of which fluctuates continuously. The power system economics literature often uses reliability of supply as the feature that differentiates service. This is the motivation of priority service pricing, which is developed in section 10.2.

## 10.1  Time of use pricing

Time of use pricing is a specific type of peak-load pricing. The idea of the time of use pricing model that is developed in this section is to determine the optimal amount of capacity that is needed to satisfy demand, where different parts of the day correspond to different demand functions. Once the optimal level of capacity is determined, equilibrium prices are found that ensure that welfare is maximized given that the capacity in the system is equal to its optimal level. The retail pricing of electricity is broken into two components. (i) The energy component is a charge that is proportional to the amount of power consumption, and differs depending on the time of day. (ii) The capacity component is applied to consumers who contribute to the need of installing additional capacity in the system.

The idea is best illustrated in the following simple context. Consider two marginal benefit functions $MB_1(d)$ and $MB_2(d)$, corresponding to two different parts of the day, peak and off-peak, respectively. Peak and off-peak periods last for a fraction $\tau_1$, $\tau_2$ of the day, respectively. Suppose that there is a marginal investment cost $MI(x)$ for expanding capacity in the system, and a marginal cost $MC(p)$ for supplying power. Suppose that $MI(x) > 0$ for all $x$. Also suppose that $MB_1(0) > MC(0) + MI(0)/\tau_1$. The cost and benefit functions are "well behaved": $MB_i$ is decreasing, $MI$ and $MC$ are increasing.

The welfare maximization problem that maximizes consumer benefit minus production cost (where production cost consists of investment and operating cost) can be described as follows:

$$\max \tau_1 \cdot \int_0^{p_1} MB_1(q)dq + \tau_2 \cdot \int_0^{p_2} MB_2(q)dq$$

$$- \int_0^x MI(q)dq - \tau_1 \int_0^{p_1} MC(q)dq - \tau_2 \int_0^{p_2} MC(q)dq$$

$$(\rho_1 \cdot \tau_1): \quad p_1 \leq x$$

$$(\rho_2 \cdot \tau_2): \quad p_2 \leq x$$

$$p_1, p_2, x \geq 0,$$

where $x$ denotes the amount of constructed capacity, $p_1$ is the power production in peak hours, and $p_2$ is the production in off-peak hours. In the optimal solution, either $p_1 = x$, $p_2 = x$, or both, since otherwise it would be possible to decrease investment cost without altering the production in peak and off-peak hours.

The KKT conditions of the welfare maximization problem are given by (note that the dual multipliers have been scaled by $\tau_i$):

$$0 \leq \rho_1 \perp x - p_1 \geq 0,$$

$$0 \leq \rho_2 \perp x - p_2 \geq 0,$$

$$0 \leq p_1 \perp -MB_1(p_1) + MC(p_1) + \rho_1 \geq 0,$$

$$0 \leq p_2 \perp -MB_2(p_2) + MC(p_2) + \rho_2 \geq 0,$$

$$0 \leq x \perp MI(x) - \rho_1\tau_1 - \rho_2\tau_2 \geq 0.$$

**Proposition 10.1** *Suppose that electricity is priced at the marginal variable cost $MC(p_i)$ for each period i. This will result in suboptimal investment if the system is built so as to make sure that no demand can be left unserved.*

*Proof* Since $MB_1(0) > MC(0) + \frac{MI(0)}{\tau_1}$, the optimal solution must have $x > 0$.

If power is priced at marginal cost and capacity is built in order to ensure that the entire demand is served, then $MC(p_1) = MB_1(p_1)$, $MC(p_2) = MB_2(p_2)$, and $x = \max(p_1, p_2)$. Suppose that this is the optimal solution. The KKT conditions can only be satisfied if $\rho_1 = \rho_2 = 0$.

Suppose that the $x$ obtained from marginal fuel cost is optimal. From complementarity it follows that

$$MI(x) = \rho_1 \cdot \tau_1 + \rho_2 \cdot \tau_2.$$

And since $MI(x) > 0$ for all $x \geq 0$, it must be the case that $\rho_i > 0$ for $i = 1$, or $i = 2$, or both. This leads to a contradiction. It must therefore be the case that the capacity resulting from marginal fuel cost is suboptimal.     □

The multiplier $\rho_i$ can be interpreted as a charge above the marginal cost of the marginal technology, $MC(p_i)$. For constant marginal investment cost, $MI(x) = MI$, it can be seen that these additional charges are exactly equal to the capital investment costs. Therefore, consumers who contribute to peak load are charged for the additional capacity that they require and the charges are exactly sufficient to cover the capital cost of this additional technology.

**Example 10.1** *Time of use pricing where peak demand bears capital costs.* Consider a market with the following characteristics: $MI(x) = 5$ $/MWh, $MC(p) = 80$ $/MWh, $MB_1(d) = \max(1000-d, 0)$ $/MWh, $MB_2(d) = \max(500-d, 0)$ $/MWh. Peak periods correspond to $\tau_1 = 20\%$ of the day, off-peak periods correspond to $\tau_2 = 80\%$ of the day. The optimal investment for this market is 895 MW. Find the price of power at each period of the day in order to ensure an optimal level of consumption for each period of the day given that capacity is constructed up to its optimal level.

**Solution**
It has been argued that the production in peak or off-peak hours (or both) has to equal the installed capacity. Since the optimal investment is 895 MW, then either $p_1 = 895$ MW, $p_2 = 895$ MW, or both. It can be checked that $MB_1(895) = 105$ $/MWh and $MB_2(895) = 0$ $/MWh. Obviously $p_2 < x$ because the marginal benefit at 895 MW in off-peak periods is zero, whereas $MC(895) = 80$ $/MWh. Therefore, $p_1 = 895$ MW, and the price in peak periods is 105 $/MWh. From the KKT conditions it follows that

$$MB_2(p_2) = MC(p_2),$$

and therefore the price in off-peak periods is 80 $/MWh. Note that consumers in off-peak periods are paying the marginal cost for off-peak hours, $MC(p_2)$. Instead, peak consumers are paying the marginal cost of peak hours $MC(p_1)$ *plus* a capacity charge $MI/\tau_1 = 25$ $/MWh, which is required for covering the capital cost of the investment needed in order to cover their peak demand. Figure 10.1 compares the time of use demand to the demand that would occur under uniform retail pricing over all hours of the day, where the retail rate is set as the weighted average of the price of the peak and off-peak hours: $0.2 \cdot 105 + 0.8 \cdot 80 = 85$ $/MWh. Note that time of use pricing depresses consumption in peak hours, and increases it in off-peak hours, relative to fixed retail pricing.

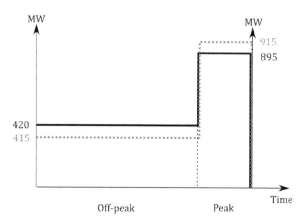

**Figure 10.1** Demand under fixed retail pricing (gray dashed curve) and time of use pricing (black solid curve) for example 10.1.

**Example 10.2** *Time of use pricing where peak and off-peak demand bear capital costs.*
Consider the market of example 10.1, with a marginal benefit function for off-peak
hours given by $MB_2(p_2) = 980 - p$ $/MWh. The remaining features of the market
remain identical to the previous example. A price of 80 $/MWh in the off-peak hours
is not possible in this case because it would violate the amount of installed capacity.
Indeed, the optimal investment amounts to 899 MW. Consumption in the peak and
off-peak hours is equal, $p_1 = p_2 = 899$ MW. The peak and off-peak consumers
share the capital cost of the installed capacity. Peak consumers are charged $\rho_1/\tau_1 =$
21 $/MWh above the marginal cost of 80 $/MWh. Off-peak consumers are charged
$\rho_2/\tau_2 = 1$ $/MWh above the marginal cost.

## 10.2 Priority service pricing

The idea of priority service contracts is to sell power as a service of differentiated
quality to consumers, instead of selling it as a commodity the price of which
varies over time. This paradigm has been applied successfully to other industries.
Services differ from commodities by the fact that they can be differentiated (and
priced according to their specific features). Electricity produced in a given hour in
a given location is a commodity, since the consumer is indifferent if this electricity
is produced by a nuclear plant, coal plant, or gas plant. Instead, electricity that
is supplied with 100% reliability is different from electricity supplied with 50%
reliability. One can price electricity as a commodity, the price of which varies over
time according to the conditions of the system, or one can price electricity as a
service, with higher reliability corresponding to a higher price. The former approach
is the one adopted in wholesale electricity markets, and it is how large industrial
consumers who participate actively in electricity markets buy power. The latter
approach is the one that is explored in this section.

More specifically, priority service pricing is based on the idea of selling power
at a reliability level $r$, and pricing the reliability as $p(r)$. The pricing of reliability
determines the reliability level that consumers choose. The key in the design of the
menu is to ensure that customers select reliability in such a way that the system
can deliver the promised reliability. For example, if extremely reliable service is very
cheap, then it is likely that most consumers will choose extremely reliable service.
However, if the system supply function consists exclusively of unreliable renewable
generators, then the promised reliability cannot be delivered and the priority service
menu would have to be changed.

The first step, therefore, in designing reliability service contracts is to determine
the reliability that the system can provide. Two elements are required in order to
achieve this. The first is the demand function of the system, denoted $D(v)$, that maps
valuation $v$ to the demand (in MW) of consumers who value power at $v$ or above.
The second is the reliability of supply, denoted $F(L)$, which indicates the probability

of the system being able to serve $L$ MW of demand.[1] These two functions can be used to define the mapping $F(D(v))$. This mapping determines the probability that consumers who value power at $v$ or more are served. This is an increasing function of $v$, since the higher the valuation the lower the total demand of consumers who value power at $v$ or above, and therefore the higher the probability that these customers are served.

---

**Example 10.3** *Computation of $F(D(v))$.* Consider the system of example 9.1. The system is characterized by the following demand function:

$$D(v) = 1620 - 4 \cdot v.$$

Recall that the system consists of two technologies. One technology is reliable and has a capacity of 295 MW. The other technology is unreliable, and has a capacity of 1880 MW. The reliability model of this technology is described by the Markov chain of figure 9.2. To compute the stationary distribution of the Markov chain that describes the availability of the unit, the following linear system is solved:

$$0.9 \cdot \pi_{\text{on}} + 0.5 \cdot \pi_{\text{off}} = \pi_{\text{on}},$$

$$\pi_{\text{on}} + \pi_{\text{off}} = 1,$$

which results in $\pi_{\text{off}} = 0.167$, $\pi_{\text{on}} = 0.833$. The function $F(L)$ is given as follows:

$$F(L) = \begin{cases} 1, & L \le 295 \text{ MW}, \\ 0.833, & 295 \text{ MW} < L \le 2175 \text{ MW}, \\ 0, & L > 2175 \text{ MW}. \end{cases}$$

The service reliability of the system as a function of valuation can then be expressed as follows:

$$F(D(v)) = \begin{cases} 0.833, & 0 \text{ \$/MWh} \le v \le 331.25 \text{ \$/MWh}, \\ 1, & 331.25 \text{ \$/MWh} < v \le 405 \text{ \$/MWh}. \end{cases}$$

---

The goal of priority service pricing is to sell power as a service with different quality, with higher quality corresponding to higher price. "Quality," in the case of priority service contracts, is the reliability of service. Power of higher reliability is priced higher. Note that the pricing of reliability determines the level of reliability chosen by consumers. Priority service pricing aims at designing a price menu $p(r)$ that ensures that customers choose reliability levels that are consistent with what the system has to offer.

The question, then, becomes: what is the best that the system can offer? If the designer of the price menu knew, in advance, the valuation of individual customers, then it could dispatch customers in order of increasing valuation since this maximizes the benefit of power consumption: loads with higher valuation consume first. The function $F(D(v))$, defined earlier, achieves exactly this goal. If a consumer with

[1] In particular, $F(L) = \mathbb{P}[C \ge L]$, where $C$ is the available capacity of the system.

valuation $v$ selects reliability level $F(D(v))$ then, by serving customers in order of decreasing reliability choices, the designer of the menu is able to deliver the promised level of reliability, and to maximize the value of the produced power by ensuring that consumers with higher valuation for power get served first.

The next question is, how can the pricing menu $p(r)$ be designed so that a consumer with valuation $v$ selects a contract with reliability $F(D(v))$? This requires examining the optimization problem faced by a customer who knows its private valuation $v$ and is considering different options of reliability levels $r$:

$$\max_{0 \leq r \leq 1} r \cdot v - p(r).$$

The conditions for an interior solution ($0 < r < 1$) that is preferred to not procuring a contract at all are given as follows:

$$v - p'(r) = 0, \tag{10.1}$$

$$r \cdot v - p(r) \geq 0. \tag{10.2}$$

In this case, a load with valuation $v$ will have benefits from procuring a reliability contract (this follows from equation (10.2)) and will choose a reliability level $r$ (this follows from condition (10.1)). By replacing $r$ with $F(D(v))$ in equation (10.1), a differential equation is obtained that induces customers with valuation $v$ to choose reliability $r(v)$.

The differential equation (10.1) can be integrated over $v$ in order to arrive at a price menu $\hat{p}(v)$ that maps the valuation of customers (\$/MWh) to the price they pay (\$/MW) per unit of power consumption per unit of time. Integrating equation (10.1), the following price menu is obtained:

$$\hat{p}(v) = p_0 + \int_{v_0}^{v} y \cdot dr(D(y)), \tag{10.3}$$

where $v_0$ and $p_0$ are free parameters. The price $p_0$ is charged independently of $v$; therefore it represents a fixed charge. This fixed charge determines $v_0$, the valuation of the customer with the lowest willingness to pay who still chooses to buy a priority service contract. This valuation $v_0$ is calculated at the point for which equation (10.2) holds as an equality:

$$v_0 \cdot r(v_0) - p_0 = 0. \tag{10.4}$$

Customers with $v < v_0$ do not procure reliability contracts. Having expressed the price menu $\hat{p}(v)$ as a function of $v$, it is now possible to express it as a function of reliability $r$, by using the following mapping:

$$\{F(D(v)), \hat{p}(v), v \in [v_0, V]\},$$

where $V$ is the maximum valuation. Reliability for $v < v_0$ is equal to zero, since consumers do not subscribe to a contract.

Note that the design of the menu relies on aggregate information that is available to the menu designer: the reliability of the system $F(L)$, and the demand function of the system $D(v)$.

It is important to note that $F(D(v))$ is an increasing function of $v$. Therefore, customers with higher valuation will select higher reliability. This implies that, in order to maximize the value of the available power, whenever there is a shortage the designer of the menu should curtail customers in order of decreasing reliability selections. It is remarkable to note that although the designer of the menu is not aware of the valuations of individual customers when it designs the menu, it is able to dispatch these customers efficiently *after* they have selected their preferred level of reliability. Therefore, the menu is capable of extracting a great amount of information from consumers.

---

**Example 10.4** *Computation of an optimal priority service menu.* Consider again the system of example 9.1, for which the function $r(D(v))$ is computed in example 10.3. Suppose that the menu designer chooses to exclude customers with a valuation below $v_0 = 10$ \$/MWh from service, in order to be able to charge a higher fixed cost to all other customers. Equation (10.4) yields the following fixed price:

$$p_0 = 10 \cdot 0.833 = 8.33 \text{ \$/MWh.}$$

Equation (10.3) yields

$$\hat{p}(v) = p_0 + \int_{v_0}^{v} u \cdot dr(u)$$

$$= \begin{cases} 8.33 \text{ \$/MWh,} & 10 \text{ \$/MWh} \le v \le 331.25 \text{ \$/MWh,} \\ 8.33 + 331.25 \cdot 0.167 \text{ \$/MWh,} & 331.25 \text{ \$/MWh} < v \le 405 \text{ \$/MWh,} \end{cases}$$

$$= \begin{cases} 8.33 \text{ \$/MWh,} & 10 \text{ \$/MWh} \le v \le 331.25 \text{ \$/MWh,} \\ 63.65 \text{ \$/MWh,} & 331.25 \text{ \$/MWh} < v \le 405 \text{ \$/MWh.} \end{cases}$$

This translates to the following service menu $p(r)$:

$$p(r) = \begin{cases} 8.33 \text{ \$/MWh,} & r = 0.833, \\ 63.65 \text{ \$/MWh,} & r = 1, \end{cases}$$

which corresponds to a menu with two options. Indeed, consider the choice of a load with valuation $v$. The load will solve the following optimization:

$$\max(0, 0.833 \cdot v - 8.33, v - 63.65).$$

This implies the following:

- In order for a load to choose $r = 0$, it must be the case that $0.833 \cdot v - 8.33 \le 0$ and $v - 63.65 \le 0$, i.e. $v \le 10$.
- In order for a load to choose $r = 0.833$, it must be the case that $0 \le 0.833 \cdot v - 8.33$ and $v - 63.65 \le 0.833 \cdot v - 8.33$, i.e. $10 \le v \le 331.25$.
- In order for a load to choose $r = 1$, it must be the case that $0 \le v - 63.65$ and $0.833 \cdot v - 8.33 \le v - 63.65$, i.e. $v \ge 331.25$.

If, instead, the menu designer would like all customers to engage in reliability contracts, i.e. $v_0 = 0$, then $p_0 = 0$ \$/MWh and

$$p(r) = \begin{cases} 0 \text{ \$/MWh,} & r = 0.833, \\ 55.32 \text{ \$/MWh,} & r = 1. \end{cases}$$

## Problems

**10.1** *Priority service pricing with cost considerations.* How does the analysis of section 10.2 change if the designer of the menu wants to supply power to customers *only if* it is cost efficient (i.e. only if $MC(D(v)) \leq v$)?

**10.2** *Time of use pricing where peak demand bears capital costs.* Implement a mathematical programming code that reproduces the results of example 10.1.

**10.3** *Time of use pricing where peak and off-peak demand bear capital costs.* Implement a mathematical programming code that reproduces the results of example 10.2.

**10.4** *Optimality conditions of consumers in priority service pricing.* Suppose that a consumer with valuation $v$ and a load equal to $D$ MW is presented with a priority service menu of finite options, $\{(p_i, r_i), i \in I\}$, where $I$ is the set of available options and $p_i$ and $r_i$ are the respective price and reliability of option $i$. We can describe the choice problem of this consumer as:

$$\max_s \sum_{i \in I} (r_i \cdot v \cdot s_i - p_i \cdot s_i)$$

$$(\gamma): \quad \sum_{i \in I} s_i \leq D$$

$$s \geq 0.$$

Derive the KKT conditions of the model. Prove that there exists an optimal solution in which either $s_i = D$ or $s_i = 0$ for all $i \in I$. What is the interpretation of this result?

**10.5** *Priority service pricing as a mathematical program with complementarity constraints.* The priority service pricing problem can be described as a problem where the designer of the contract decides on the price of the different options in the menu, aiming to ensure that options are served in the correct priority, while respecting the optimality conditions by which loads choose an option in the menu as well as the dispatch constraints of the system. This amounts to a mathematical program with complementarity constraints. Propose a model and reproduce the results of example 10.4, where three options are offered, with probabilities of service equal to 0%, 83.3%, and 100%, respectively. You can approximate the demand function of the system with 1620 vertical load slices that have a width of 1 MW each. Make sure to introduce an option with zero reliability that has a price of zero in order to model the fact that consumers may opt not to engage in a contract, and also note that you should fix the price of the option that has a reliability of 83%, which corresponds to fixing the constant price $p_0$ in the priority service menu.

## Bibliography

**Section 10.1**  Time of use pricing was originally proposed by Boiteux (1960). A survey of demand response options is provided by Borenstein et al. (2002).

**Section 10.2**  Priority service pricing was proposed in the seminal work of Chao and Wilson (1987), and a simplified exposition of the concept is provided by Oren (1987). The concept is generalized to both time and reliability differentiation in multilevel demand subscription by Chao et al. (1986). The essence of these investigations is how to design contracts that extract information from consumers about their valuation for power based on nonlinear pricing principles (Wilson 1993). Priority service is considered in the long-term model of Joskow and Tirole (2007), and underlies the proposal of Papalexopoulos, Beal, and Florek (2013) for mobilizing residential demand response at mass scale. Priority service pricing is cast as a leader-follower Stackelberg game in Mou, Papavasiliou, and Chevalier (2019) and the role of service charges is investigated by Gérard and Papavasiliou (2022). Multilevel demand subscription is compared to priority service by Gérard et al. (2022).

# 11     Generation capacity expansion

This chapter focuses on long-term models. Section 11.1 introduces the capacity expansion planning problem as a centralized planning problem. Section 11.2 presents the market equilibrium model of agents who invest in generation capacity and relates this model to the centralized expansion planning model. Market design issues that relate to the provision of sufficient long-run capacity are presented in section 11.3.

## 11.1   Generation capacity expansion planning

The capacity expansion planning problem is the problem of constructing power generation capacity in order to ensure that the future demand of the system can be satisfied at minimum investment and operating cost.

In its simplest form, the problem can be cast as a two-stage optimization problem where the first stage consists of capacity expansion decisions and the second stage consists of operating decisions given the decided capacity expansion.

It is useful to distinguish load from demand in the ensuing discussion. **Load** is defined as the amount of power that would be consumed if energy were supplied at zero price. This is to be distinguished from **demand**, which is the consumption of power that occurs at a given price. Demand is equal to the supply in the system and cannot be greater than load.

---

**Example 11.1** *Load versus demand.* Suppose that a system consists of a single generator with a capacity of 100 MW and a demand function $D(v) = 110 - 5 \cdot v$, where $v$ is the valuation. Load is then equal to 110 MW, although demand in the system cannot exceed 100 MW.

---

The basic ingredients of a capacity expansion planning model are the investment cost and marginal cost of generators and the profile of loads. Marginal cost is measured in $/MWh. Investment cost is typically converted to an hourly running cost for supporting an investment in 1 MW of capacity (see section 3.1.1). Investment cost is therefore also measured in $/MWh. In our simplified analysis, in order for a technology to be worth considering, its fuel cost should be lower than the fuel cost of any competing technology that has a higher capital cost. Stated otherwise,

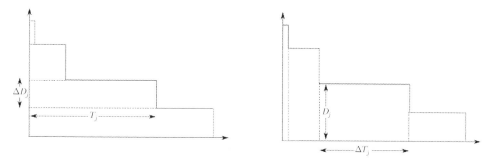

**Figure 11.1** (Left) A horizontal partition of the load duration curve into load slices. (Right) A vertical partition of the load duration curve into time slices.

technologies can be arranged in order of increasing capital cost and decreasing fuel cost: if $I_1 \leq I_2 \leq \cdots \leq I_n$, then $C_1 \geq C_2 \geq \cdots \geq C_n$.

Load is represented using a load duration curve (see section 1.2 for a detailed discussion of load duration curves). Load duration curves represent the number of hours in the year that load is greater than or equal to a given level. Load duration curves are obtained by sorting the time series of system load in decreasing order. A step-wise approximation of the load duration curve is used as input for capacity expansion planning models.

The load duration curve can be partitioned in two ways, as shown in figure 11.1. A horizontal partition splits the curve into $m$ load blocks, each of which lasts for a duration $T_j$. Each load block adds an amount of power demand $\Delta D_j$ to the system. This is the representation that is used in section 1.2. The vertical partition splits the load duration curve into $m$ time slices, each of which lasts $\Delta T_j$. Each slice $j$ requires a *total* amount of power $D_j$. This partitioning is appropriate for analyzing the long-run investment equilibrium of the market, since each vertical slice corresponds to an interval of time with the same price in the energy market.

The capacity expansion planning model using a vertical partition is formulated as follows. Denote $x_i$ as the investment in technology $i$ (in MW), $p_{ij}$ as the amount of power allocated from technology $i$ to block $j$ (in MWh), and $d_j$ as the load of block $j$ that is actually served. The vertical partition of the load duration curve leads to the following two-stage deterministic capacity expansion planning model:

$$\max_{p,d,x} \sum_{j=1}^{m} \Delta T_j \cdot \left( V_j \cdot d_j - \sum_{i=1}^{n} MC_i \cdot p_{ij} \right) - \sum_{i=1}^{n} I_i \cdot x_i$$

$$(\Delta T_j \cdot \rho_j): \quad d_j - \sum_{i=1}^{n} p_{ij} = 0, j = 1, \ldots, m$$

$$(\Delta T_j \cdot \mu_{ij}): \quad p_{ij} \leq x_i, i = 1, \ldots, n, j = 1, \ldots, m$$

$$(\Delta T_j \cdot v_j): \quad d_j \leq D_j, j = 1, \ldots, m$$

$$p, d, x \geq 0,$$

where $\Delta T_j$ is the *fraction* of time that is occupied by vertical slice $j$ of the load duration curve, $D_j$ is the load of block $j$ (in MW), $I_i$ is the hourly investment cost of technology $i$ (in \$/MWh), and $C_i$ is the marginal cost of technology $i$ (in \$/MWh). The value of lost load (VOLL) $V_j$, which is defined in chapter 3, represents the average value of power to consumers. Recall that VOLL is an estimate of *average* consumer value for power since, when rationing occurs, it will be performed randomly since the system operator cannot distinguish consumers with high value for power from consumers with low value for power in a system without price-responsive consumers. Demand response can be incorporated into the model by replacing VOLL with a marginal benefit function.

Note that the dual multipliers of the corresponding constraints are scaled by $\Delta T_j$ for every $j = 1, \ldots, m$. The reason for doing this is that it allows us to avoid carrying the constant $\Delta T_j$ in subsequent expressions, when we analyze the KKT conditions of the problem. The scaling of the dual multiplier can be thought of as replacing, in advance, the corresponding dual multiplier $\tilde{v}_j$ of a given constraint by the product of a constant $\Delta T_j$ and a new dual variable $v_j$, where $\tilde{v}_j = \Delta T_j \cdot v_j$.

The trade-off of the capacity expansion planning problem is to choose technologies that are cheaper to construct and more expensive to operate, versus technologies that are more expensive to construct but cheaper to operate. This is best understood by considering a horizontal partition of the load duration curve. Base-load horizontal load slices with a long duration $T_j$ are better served by technologies with low fuel cost $MC_i$. Peak-load horizontal load slices with a short duration $T_j$ are better served by generators which are cheaper to construct, since their fuel cost does not influence their total cost too much. Following this logic, a graphical solution to the capacity expansion planning problem using *screening curves* is possible. Screening curves are discussed in detail in section 1.2.

The preceding model ignores a number of details, but can be modified in order to accommodate them. Transmission constraints and availability factors of different technologies can be modeled explicitly. The incorporation of multiple time stages in the capacity expansion planning model allows a representation of the long-term evolution of equipment costs, the long-term evolution of the load duration curve (due to changes in industry activity, energy savings, or rate policies), as well as the appearance of new technologies or the retirement of current equipment. For similar reasons (uncertainty in evolution of equipment costs, uncertainty in the shape of the load duration curve, etc.), it may be desirable to introduce uncertainty into the model. This leads to a multistage stochastic programming problem.

---

**Example 11.2** *Optimal expansion using a linear program.* Recall the system of section 1.2. The technologies that are available for serving demand are presented in table 11.1. Note that both marginal costs and fuel costs are expressed in \$/MWh. The load duration curve of the system is presented in figure 11.2, along with a step-wise approximation that is used as input to the expansion planning model. Referring to the left part of figure 11.1, the load duration curve presented in figure 11.2 can be represented equivalently in two ways:

- Horizontal representation

**Table 11.1 The set of options for serving demand in example 11.2.**

| Technology | Fuel cost ($/MWh) | Inv cost ($/MWh) |
|---|---|---|
| Coal | 25 | 16 |
| Gas | 80 | 5 |
| Nuclear | 6.5 | 32 |
| Oil | 160 | 2 |

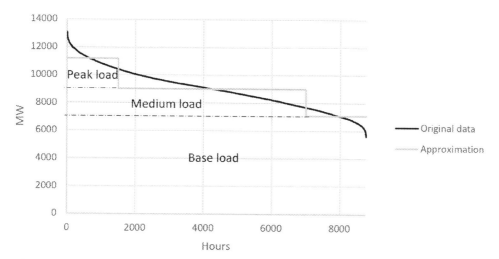

**Figure 11.2** The load duration curve used in example 11.2. The stepwise approximation indicates that load ranges are 0–7086 MW for the entire year, 7086–9004 MW for 7000 hours, and 9004–11169 MW for 1500 hours.

- – Base-load horizontal slice with a duration $T_1$ equal to 100% (8760 hours) and a load level $\Delta D_1$ equal to 7086 MW
- – Medium-load horizontal slice with a duration $T_2$ equal to 79.91% (7000 hours) and a load level $\Delta D_2$ equal to 1918 MW
- – Peak horizontal slice with a duration $T_3$ equal to 17.12% (1500 hours) and a load level $\Delta D_3$ equal to 2165 MW
- • Vertical representation
  - – Base-load vertical slice with a duration $\Delta T_1$ equal to 20.09% (1760 hours) and a load level $D_1$ equal to 7086 MW
  - – Medium-load vertical slice with a duration $\Delta T_2$ equal to 62.79% (5500 hours) and a load level $D_2$ equal to 9004 MW
  - – Peak vertical slice with a duration $\Delta T_3$ equal to 17.12% (1500 hours) and a load level $D_3$ equal to 11169 MW

The optimal solution of the problem is shown in table 11.2. Nuclear is used for satisfying base load, coal is used for satisfying medium load, and gas is used for satisfying peak load. Oil is not used.

**Table 11.2** *Optimal assignment of capacity for example 11.2.*

|  | Duration (hours) | Level (MW) | Technology |
|---|---|---|---|
| Base load | 8760 | 0–7086 | Nuclear |
| Medium load | 7000 | 7086–9004 | Coal |
| Peak load | 1500 | 9004–11169 | Gas |

The KKT conditions of the capacity expansion planning problem can be described as follows:

$$d_j - \sum_{i=1}^{n} p_{ij} = 0, j = 1, \ldots, m,$$

$$0 \le \Delta T_j \cdot \mu_{ij} \perp x_i - p_{ij} \ge 0, i = 1, \ldots, n, j = 1, \ldots, m,$$

$$0 \le \Delta T_j \cdot v_j \perp D_j - d_j \ge 0, j = 1, \ldots, m,$$

$$0 \le p_{ij} \perp \Delta T_j \cdot MC_i + \Delta T_j \cdot \mu_{ij} - \Delta T_j \cdot \rho_j \ge 0, i = 1, \ldots, n, j = 1, \ldots, m,$$

$$0 \le d_j \perp -\Delta T_j \cdot V_j + \Delta T_j \cdot v_j + \Delta T_j \cdot \rho_j \ge 0, j = 1, \ldots, m,$$

$$0 \le x_i \perp I_i - \sum_{j=1}^{m} \Delta T_j \cdot \mu_{ij} \ge 0, i = 1, \ldots, n.$$

**Proposition 11.1**  *The optimal consumption $d_j$ can be characterized as follows:*

- *If $0 < d_j < D_j$ then $V_j = \rho_j$*
- *If $d_j = 0$ then $V_j \le \rho_j$*
- *If $d_j = D_j$ then $V_j \ge \rho_j$*

*For generators for which $x_i > 0$, the optimal production can be characterized as follows:*

- *If $0 < p_{ij} < x_i$ then $MC_i = \rho_j$*
- *If $p_{ij} = 0$ then $MC_i \ge \rho_j$*
- *If $p_{ij} = x_i$ then $MC_i \le \rho_j$*

*The optimal investment is characterized as follows:*

- *If $x_i = 0$ then $I_i \ge \sum_{j=1}^{m} \Delta_j \cdot \mu_{ij}$*
- *If $x_i > 0$ then $I_i = \sum_{j=1}^{m} \Delta_j \cdot \mu_{ij}$*

*Proof*  The proof follows from the KKT conditions.  □

The proposition implies that demand with a valuation above the threshold $\rho_j$ consumes power, whereas demand with a valuation below this threshold does not consume power. Generators with a fuel cost below the threshold $\rho_j$ produce power at peak output, whereas generators with a fuel cost above this threshold do not produce power. The thresholds $\sum_{j=1}^{m} \Delta_j \cdot \mu_{ij}$ determine whether a technology $i$ is worth building or not.

**Proposition 11.2**  *Suppose that $I_i > 0$ for all technologies, then if a technology is constructed ($x_i > 0$) it is operated up to its full capacity ($p_{ij} = x_i$ for some $j$).*

*Proof*   Suppose that there exists a generator for which $p_{ij} < x_j$ for all $j = 1, \ldots, m$. Then from the KKT conditions it can be concluded that $\mu_{ij} = 0$ for all $j$ and $I_i = \sum_{j=1}^{m} T_j \cdot \mu_{ij} = 0$, which leads to a contradiction.                                      □

## 11.2   Investment in power generation capacity

In a market-based environment, generators are constructed based on the prediction of investors about future revenues from the energy market. In the following, it is shown that, in an ideal market with elastic demand, the investment result is identical to the optimal expansion plan resulting from centralized planning.

**Energy-only markets** rely exclusively on energy market price spikes in order to finance capital costs for investing in capacity. This financing is achieved through **scarcity rents**, which are defined as energy market revenues minus variable costs (startup and minimum load costs are ignored in this section). Scarcity rent can occur by producers who internalize their investment cost in their bids, or through demand curtailment. Scarcity rent in the case of demand curtailment is illustrated in figure 11.3. Note the correspondence between figure 11.3 and figure 4.3, and the equivalence between scarcity rent and producer surplus in the two figures.

Some reflection reveals that scarcity rents are inevitable in a market-based system. Note that if peak energy prices never exceed the marginal cost of the peak generator in the system, then this peak generator would not be able to cover its investment costs and would therefore be driven out of the market as a nonviable investment.

A long-run equilibrium model of the energy-only market consists of generators that decide on capacity investment and production quantities, and consumers who

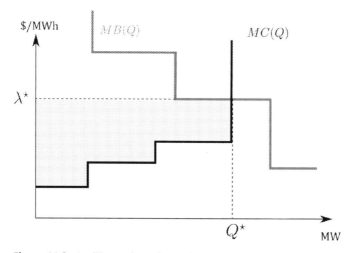

**Figure 11.3**  An illustration of scarcity rent. Here, $MC(\cdot)$ represents the marginal cost curve of the system, $MB(\cdot)$ is the marginal benefit, and $\lambda^\star$ and $Q^\star$ correspond to the market clearing price and quantity, respectively. The area shaded in gray is the scarcity rent earned by generators.

procure power from generators. Generators determine production and capacity expansion decisions jointly in order to maximize profit. Denote $\rho_j$ as the energy price that prevails in the time periods corresponding to the vertical slice $j$ of the load duration curve. The generator profit maximization problem of technology $i$ can be described as follows:

$$\max_{p,x} \sum_{j=1}^{m} (\rho_j - C_i) \cdot \Delta T_j \cdot p_{ij} - I_i \cdot x_i$$
$$(\mu_{ij} \cdot \Delta T_j): \quad p_{ij} - x_i \leq 0, j = 1, \ldots, m$$
$$p, x \geq 0.$$

The surplus maximization problem of consumers in load slice $j$ can be expressed as follows:

$$\max_{d} (V_j - \rho_j) \cdot d_j$$
$$(\nu_j \cdot \Delta T_j): \quad d_j - D_j \leq 0$$
$$d_j \geq 0.$$

The price adjustment process of the energy market for every load slice $j$ is expressed by the following condition:

$$d_j - \sum_{i=1}^{n} p_{ij} = 0.$$

**Definition 11.3**    An **energy-only market equilibrium** is a combination of investment and power allocation decisions and energy prices $(x^\star, p^\star, d^\star, \rho^\star)$ such that (i) given the prices, agents maximize individual surplus, and (ii) the energy market clears during all time periods.

**Proposition 11.4**    *Energy-only markets can support the optimal investment and result in an optimal short-term allocation of energy.*

*Proof*    See problem 11.1.    □

As we discuss in section 11.3, proposition 11.4 is an important starting point for analyzing regulatory interventions to electricity markets that concern investment. Crucially, the proposition can be generalized. Concretely, the equivalence of centralized planning and energy-only markets in the presence of demand uncertainty and transmission constraints can be established using the same reasoning. This means that properly designed markets can induce an optimal mix of investment at the right quantities in the right locations. This vision has been a major driver for the push to the deregulation of the electricity sector.

From the KKT conditions of the generator profit maximization problem it can be noted that if $p_{ij} > 0$ then

$$\mu_{ij} = \rho_j - MC_i.$$

This corresponds exactly to the scarcity rent that is defined earlier. The following investment criterion for generators can be obtained from the KKT condition of the generator profit maximization problem:

$$0 \leq x_i \perp I_i - \sum_{j=1}^{m} \mu_{ij} \cdot \Delta T_j \geq 0.$$

This condition can then be interpreted as follows: if investment cost for technology $i$ cannot be covered by scarcity rents, then one should not invest in technology $i$. If one invests in a certain technology, then competition drives energy prices down to the level where investment costs are exactly covered by scarcity rents.

An advantage of the energy-only market is that it leads to the optimal investment and operation of the system under perfect competition. However, the energy-only market design can suffer in practice due to the low elasticity of demand and also due to the fact that demand curtailment cannot be enforced in real time. Inelastic demand results in substantial price volatility, which can result in reluctance among investors and a higher risk premium for investment in electricity capacity. This handicap can be mitigated by alternative market designs that rely on operating reserve demand curves. The inability to curtail demand implies that there are moments when the system does not have enough capacity to cover the entire load. At those moments the system operator needs to decide what the energy market price should be. This decision, which is regulatory, strongly influences investment incentives. These issues are discussed in further detail in section 11.3.

---

**Example 11.3** *Long-term equilibrium prices.* We return to example 11.2. The long-term equilibrium clearing prices are as follows:

- Base load time periods: 12.56 $/MWh
- Medium load time periods: 27.52 $/MWh
- Peak load time periods: 109.20 $/MWh

Table 11.3 presents the production of each technology. We notice that there is no period during which the market clearing price is equal to the marginal cost of any given unit; therefore all technologies that are operating in a given period are collecting rents that contribute to the repayment of their investment cost. We further observe that all units are given short-term incentives that are compatible with the optimal

**Table 11.3 Production of each technology during each time slice of example 11.3.**

|         | Base load (MWh) | Medium load (MWh) | Peak load (MWh) |
|---------|-----------------|-------------------|-----------------|
| Nuclear | 7086            | 7086              | 7086            |
| Coal    | 0               | 1918              | 1918            |
| Gas     | 0               | 0                 | 2165            |
| Oil     | 0               | 0                 | 0               |

**Table 11.4** Profitability of each technology in example 11.3.

|  | Base-load profit $\mu_{i1}$ ($/MWh) | Medium-load profit $\mu_{i2}$ ($/MWh) | Peak load profit $\mu_{i3}$ ($/MWh) | Average $\sum_j \Delta T_j \mu_{ij}$ ($/MWh) | Inv. cost $I_i$ ($/MWh) | Breaks even $\sum_j \Delta T_j \mu_{ij}$ $= I_i$ |
|---|---|---|---|---|---|---|
| Nuclear | 6.06 | 21.02 | 102.7 | 32 | 32 | Yes |
| Coal | 0 | 2.52 | 84.2 | 16 | 16 | Yes |
| Gas | 0 | 0 | 29.2 | 5 | 5 | Yes |
| Oil | 0 | 0 | 0 | 0 | 2 | No |

dispatch of the system. For instance, the price of 12.56 $/MWh during base load periods indeed induces the nuclear technology to produce at full capacity, whereas the remaining technologies have no incentive to produce at all. Table 11.4 presents the profitability of each unit. We note that any units that are actually invested in are exactly recovering their capital cost, whereas any units that are not able to cover their investment cost are not built in the first place.

## 11.3 Market design for generation capacity expansion

Proposition 11.4 demonstrates that the ideal energy-only market can support a long-run expansion plan. This would imply that the deregulation of the electricity sector would take care of inducing investment at the right level and right locations, and the case of regulatory intervention would be closed.

There are a number of real-world complications which invalidate this ideal outcome. These include market power, the absence of reliability markets, risk aversion, and market incompleteness due to bad design choices. We discuss them briefly in turn.

Market power arguments have been used in the literature in order to justify the introduction of price caps in electricity markets. Price caps aim at preventing generators with significant market share from offering their capacity at a price above marginal cost during periods when the system is very stressed,[1] with the goal of profiting from the resulting increase in market clearing prices. Price caps, however, solve one problem while introducing another, since they create **missing money**. Missing money refers to cases where the revenues of generators in the energy market are not sufficient for inducing them to invest. Missing money due to price caps can thus work towards preventing the optimal investment mix from materializing, although the real underlying problem is the absence of perfect competition.

The absence of reliability markets relates to the fact that a significant part of consumers is not responsive to price. Small residential consumers anyway face

[1] This behavior is referred to as **economic withholding**, and should be contrasted to **physical withholding**, whereby generators do not even offer their capacity to the market in the first place.

difficulties in valuing power. Nevertheless, aggregator business models such as those based on priority pricing that are presented in section 10.2 can support the enlisting and active participation of flexible consumers in markets. Although this is possible in theory, it is lagging in practice; therefore the practical reality is that the majority of consumers do not engage in reliability markets. This complicates matters because, when capacity shortages do occur, the curtailment of consumers is random and thus inefficient, but also price discovery is impossible. This issue is discussed further in section 11.3.1. Note that, in the models of the present section, and in contrast to sections 11.1 and 11.2, we assume a fixed demand that is not responsive to price. This reflects the fact that demand does not respond to a wholesale market price, but rather an exogenous retail price that is disconnected from the conditions of the system.

The result of proposition 11.4 does not account for uncertainty. Risk, however, is an important element in electricity markets. The introduction of uncertainty to the capacity expansion model, and its interpretation as a market equilibrium, is straightforward when market agents are risk-neutral. However, the analysis becomes more complex when agents are assumed to be risk averse. Risk is a real issue in electricity markets. Power plants are capital-intensive projects, and a downturn in the economy, an adverse regulatory intervention, or a geopolitical disruption can completely change the landscape of an investment. Fortunately, risk can be mitigated through financial contracts. And there exists theory that generalizes proposition 4.11 to the case of uncertainty, which guarantees that a decentralized system can fully replicate the outcome of an ideal centralized planning process. However, the message of this theory is somewhat pessimistic: in order for a decentralized market system to reproduce the outcome of a centralized planning process it is necessary to establish a complete market for risk, i.e. to define financial contracts that can protect investors from every possible uncertainty. Although this is practically infeasible, it does create a basis for rationalizing regulatory instruments. Concretely, capacity markets can be viewed as an instrument for trading risk, and can thus bring the market closer to the ideal setting of a complete market for risk. This argument has been used by proponents of capacity markets. The intuitive interpretation is simple: capacity markets can help investors hedge their risk, and are thus likely to induce them to invest in generation capacity.

**Complete markets** are markets where all goods are traded. An incomplete market is an institution in which there is a missing market for a good that should be traded. Characteristic examples of incomplete markets in the electricity sector are missing markets for transmission and reserves. Missing markets for transmission are discussed in section 5.3. Zonal pricing creates missing markets because it fails to recognize that electricity produced in different locations is not a homogeneous product, but rather different products. Missing markets for reserve are discussed in section 6.4, where it is pointed out that certain markets have failed to introduce the trading of reserve capacity in real time. The analysis of incomplete markets goes beyond the scope of optimization models that are introduced in this textbook, and the equivalence between centralized optimization and decentralized markets breaks down. This clearly affects investment. A tangible example is the case of Germany in Europe: despite the fact that there is a surplus of renewable supply in Northern Germany, prices are uniform across Germany due to zonal pricing. From the point of view of investors, therefore, the incentive to invest in northern Germany is equally strong as the incentive to invest in southern Germany. Although this is clearly absurd, and prevents the market from guiding towards an optimal location of resources where they are needed most, zonal pricing still prevails in Europe due to political obstacles.

### 11.3.1 VOLL pricing

VOLL pricing refers to the practice of setting prices at an estimate of VOLL in energy markets during periods of capacity shortage. An important challenge with this approach is the fact the VOLL parameter is crucial for investment, but its estimation is notoriously difficult.

The decentralized modeling of VOLL pricing includes the quantity adjustment of generators, the process of setting prices equal to VOLL whenever there is load shedding, and the price adjustment process. We describe these conditions in turn.

Generators adjust investment and production so as to maximize profits in the market. Each generator $i = 1, \ldots, n$ thus solves the following optimization problem:

$$\max_{p,x} \sum_{j=1}^{m} \Delta T_j \cdot (\rho_j - MC_i) \cdot p_{ij} - I_i \cdot x_i$$

$$(\Delta T_j \cdot \mu_{ij}): \quad p_{ij} \leq x_i, j = 1, \ldots, m$$

$$p, x \geq 0.$$

There is no quantity adjustment process for loads, since loads are assumed to be responding to a retail price that is not tracking the price of electricity. Thus, the load can be assumed to be fixed in the market.

The price adjustment process dictates that energy prices adjust at every period $j = 1, \ldots, m$ such that total demand equals total supply:

$$D_j - ls_j - \sum_{i=1}^{n} p_{ij} = 0.$$

Here, $ls_j$ is the amount of load that is shed due to the fact that there is not enough capacity available in the system.

Finally, the following complementarity condition dictates that, for every period $j = 1, \ldots, m$, energy prices are set equal to VOLL whenever there is shortage in the system:

$$0 \leq ls_j \perp VOLL - \rho_j \geq 0.$$

In problem 11.6 we show that this system of market equilibrium conditions is equivalent to the following optimization problem:

$$(VOLLP): \quad \min_{p,x,ls} \sum_{j=1}^{m} \Delta T_j \cdot \left( VOLL \cdot ls_j + \sum_{i=1}^{n} MC_i \cdot p_{ij} \right) + \sum_{i=1}^{n} I_i \cdot x_i$$

$$(\Delta T_j \cdot \rho_j): \quad D_j - l_j - \sum_{i=1}^{n} p_{ij} = 0, j = 1, \ldots, m$$

$$(\Delta T_j \cdot \mu_{ij}): \quad p_{ij} \leq x_i, i = 1, \ldots, n, j = 1, \ldots, m$$

$$p, ls, x \geq 0.$$

This result has two interesting implications. The first implication is that VOLL pricing can replicate the outcome of an ideal centralized planning process, as long as the estimate of VOLL is correct, in the sense that it induces a level of investment that corresponds to the optimal level of load shedding in the system. The second implication is that things can turn out to be very inefficient if the estimate of VOLL is highly inaccurate. This is also a weakness of the mechanism.

---

**Example 11.4** *VOLL pricing.* Consider a market with the load duration curve corresponding to the "original data" label of figure 11.2. The candidate technologies in this system are the ones presented in table 11.1. Suppose that the valuation of loads is equal to 1000 $/MWh. However, let us assume that loads cannot participate in the market, because the necessary demand response technology is not in place. Instead, let us assume that the market clearing price is set equal to VOLL whenever the system runs short of capacity. If the market operator can accurately assess the VOLL, then the mix in the market equilibrium is:

- Oil: 953.5 MW
- Gas: 1417.3 MW
- Coal: 2852.6 MW
- Nuclear: 7353.8 MW

Suppose, instead, that the authorities estimate VOLL to be equal to 3000 $/MWh, rather than the correct value of 1000 $/MWh. Then the equilibrium capacity mix is:

- Oil: 1215.7 MW
- Gas: 1417.3 MW
- Coal: 2852.6 MW
- Nuclear: 7353.8 MW

Note that oil is now over-sized relative to the efficient solution, whereas the mix of all other technologies remains unaffected. This increase in peaking plant capacity is a natural result of the fact that the authorities overestimate the social cost of failing to serve load. The VOLL signal that the market sends is "too strong" and results in building an amount of peaking capacity that is above the optimal level.

---

Despite the potential efficiency of the mechanism if VOLL can be estimated accurately, the estimation of VOLL is difficult and nontransparent. Moreover, VOLL pricing results in highly volatile prices, and the annual duration of load shedding events (which is highly unpredictable) is an important driver of annual producer revenue. Finally, the mechanism can be susceptible to the exercise of market power, since generators with a dominant position can withhold capacity during hours of load shedding, at least in theory.

## 11.3.2 Capacity mechanisms

**Capacity mechanisms** have been proposed as another way for stimulating investment in power generation capacity and mitigating the problem of investment risk.

### General characteristics of capacity mechanisms

The financing of capacity mechanisms can vary between different markets. Inspired by peak load pricing arguments, some have argued that these payments should be collected by the loads that are contributing to the system peak. In other markets, these mechanisms are financed by applying a uniform charge to consumers.

An important challenge of putting in place capacity mechanisms is that of determining the targeted installed capacity. The goal is to target levels of installed capacity at which the average cost of involuntary load shedding approximately equals the investment cost of installing an additional unit of peaking generation capacity. The investment cost of the peaking technology is referred to as the **cost of new entry (CONE)**.

Concretely, let us consider the following subset of KKT conditions of the ($VOLLP$) problem, where $n$ is the peaking technology:

$$0 \leq x_n \perp I_n - \sum_{j=1}^{m} \Delta T_j \cdot \mu_{nj} \geq 0.$$

For those periods during which $ls_j > 0$, we have $0 < p_{nj} = x_n$ (assuming that the peak plant is constructed), and therefore $\rho_j = VOLL$. Thus $\mu_{nj} = VOLL - MC_n$. This means that[2]

$$I_n \simeq \sum_{j=1}^{m} \mathbb{I}[ls_j > 0] \cdot \Delta T_j \cdot \mu_{nj}$$

$$= \sum_{j=1}^{m} \mathbb{I}[ls_j > 0] \cdot \Delta T_j \cdot (VOLL - MC_n)$$

$$= (VOLL - MC_n) \cdot \sum_{j=1}^{m} \mathbb{I}[ls_j > 0] \cdot \Delta T_j$$

$$\simeq VOLL \cdot LOLP.$$

---

[2] In fact, there can be $\mu_{nj} > 0$ even when $ls_j = 0$ (for an hour at which the inelastic demand crosses the supply function at the level of all capacity in the system), thus the precise mathematical statement is that $\sum_{j=1}^{m} \mathbb{I}[ls_j > 0] \cdot \Delta T_j \cdot \mu_{nj} \leq I_n \leq (\sum_{j=1}^{m} \mathbb{I}[ls_j > 0] + 1) \cdot \Delta T_j \cdot \mu_{nj}$.

The function $\mathbb{I}[\cdot]$ in the first line is an indicator function, which evaluates to 1 when its argument is true, and 0 otherwise. The penultimate line is due to the fact that $\Delta T_j$ is unitless in our model (and thus a probability[3]), thus $\sum_{j=1}^{m} \mathbb{I}[ls_j > 0] \cdot \Delta T_j$ indicates the total probability that load is not served, which is by definition $LOLP$. The approximation in the last line is due to the fact that, typically, $VOLL \gg MC_n$.

---

**Example 11.5** We consider the system of example 11.4. The peaking technology $n$ is oil; therefore we have $I_n = 2$ \$/MWh. The marginal cost of the peaking technology is $MC_n = 160$ \$/MWh. We have $VOLL = 1000$ \$/MWh, and at the optimal expansion plan we have that the 20 hours of highest load are not served. Thus $LOLP = \frac{20}{8760} = 0.0023$. We can then readily check that:

$$I_n = 2 \text{ \$/MWh},$$
$$(VOLL - MC_n) \cdot LOLP = (1000 - 160) \cdot 0.0023 = 1.92 \text{ \$/MWh}.$$

Thus, indeed, the two quantities are approximately equal.

---

Determining the level of capacity for which the equality $I_n = VOLL \cdot LOLP$ is achieved requires computing the value of lost load as well as the probability of involuntarily shedding load, or equivalently[4] the **loss of load expectation (LOLE)**. The resource adequacy models that are used for estimating this expectation as a function of installed capacity depend on a host of assumptions, and the agency that performs the resource adequacy computations may have a stake in the process, e.g. an incentive to size the system comfortably. The process of computing targets for capacity mechanisms is sometimes nontransparent and may result in overstated targets. With that being said, it is also interesting to point out that, since the investment cost of peaking capacity is typically lower than that of base-load capacity, oversizing can be more severe in terms of MW than it is in terms of dollars spent.

Another notable challenge of designing capacity mechanisms is that, in their simplest form, they do not differentiate payments by location. Thus, it can be challenging for these mechanisms to steer investment in the right locations. One approach that has been adopted in certain capacity mechanisms, such as the Belgian one, is to introduce a rudimentary network model into the overall process. However, this can raise concerns of discrimination in the design of the mechanism.

Default market designs typically do not foresee capacity mechanisms. Instead, putting in place such mechanisms may require regulatory approval, since they are sometimes viewed as discriminatory out-of-market interventions that involve state intervention. For instance, in European Member States, capacity mechanisms need to be approved by the competition directorate of the European Commission. Member

---

[3]  In the simple model presented here, we assume that different states of the world are represented by different time periods. A generalization of the model presented in this section represents different scenarios and fixes the investment decision over the different scenarios. Loss of load probability and loss of load expectation can then be computed over the time periods of the different scenarios.

[4]  The difference between LOLP and LOLE is that LOLP is a unitless probability, whereas LOLE is expressed in hours: $LOLE = 8760 \cdot LOLP$.

States should thus be in a position to prove that they have quantifiable resource adequacy challenges for upcoming years. Such a resource adequacy computation is undertaken in Europe by the European Network of Transmission System Operators for Electricity (ENTSO-E).

### Types of capacity mechanisms

There are various ways in which capacity mechanisms can be organized. These include capacity auctions, capacity payments, decentralized capacity markets, installed capacity obligations, and strategic reserves.

**Capacity auctions** take place in regular time intervals (e.g. annually) and target a certain delivery period (e.g. future year). Candidate power generation projects (which may - existing plants) submit a price-quantity bid to the auction. The system operator submits a capacity demand curve. The demand curve of the system operator and the supply curve of power generators are typically cleared in a uniform price auction. The power generators are paid the capacity auction price in advance of building their projects, but carry an obligation to actually implement the construction of their project within the foreseen delivery period. Alternatively, capacity payments may be received throughout the year, as an adder to the marginal price of electricity for all units that are cleared in the auction.

A typical shape for a capacity auction demand curve is presented in figure 11.4. The horizontal portion of the demand curve corresponds to a multiple of CONE. The vertical drop occurs at the point of target capacity, and at that point the valuation of the demand curve typically becomes equal to CONE. The demand curve then typically declines linearly up to the point of a multiple of the target capacity, at which point it becomes equal to zero.

The logic of the demand curve is to produce a robust price signal up to the level of target capacity. The linear decline after the level of target capacity is inspired by a desire to mitigate market power in capacity auctions, as well as an intention to avoid binary capacity auction outcomes. Concretely, early designs in the USA with vertical demand curves (i.e. inelastic demands) exhibited significant volatility in the capacity market price, which was deemed undesirable.

---

**Example 11.6** *Capacity auction demand curve.* Consider the system of example 11.4, where we assume that the true VOLL of the system is 1000 $/MWh. The necessary parameters for defining the capacity auction demand curve of this market are the following. (i) The valuation of the horizontal left segment of the demand curve is 1.5 times CONE. (ii) The target capacity is at the level that is determined from the solution of example 11.4. (iii) The capacity demand curve declines linearly and becomes zero at 115% of the target capacity. According to table 11.1, the CONE is the investment cost of the peaking technology, i.e. oil, and is thus equal to 3 $/MWh. The optimal level of capacity in example 11.4 is equal to 12577.1 MW. These values give us the capacity demand curve of figure 11.4.

---

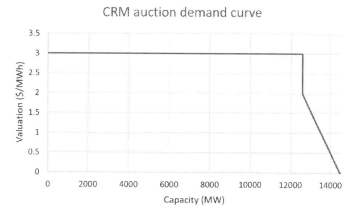

Figure 11.4 The capacity auction demand curve of example 11.6.

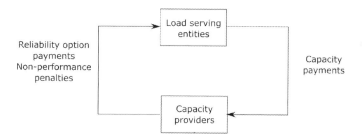

Figure 11.5 Cash flows in capacity auction reliability options.

**Reliability options** have been proposed in a number of markets as a supplement to capacity auctions, with early implementations in the design of the Colombian market. Reliability options are call options that are bundled with each unit that a capacity provider sells to the capacity market. Such reliability options commit the capacity provider to a payment that is equal to the difference between the market price and the strike price of the call option.

Reliability options can be interpreted as risk-sharing arrangements between load-serving entities and capacity providers. Concretely, the reliability options are swapping the risky scarcity rents of capacity providers with the rather predictable capacity payment. Consumers are also protected from price spikes that are traded for the capacity payments. The cash flows of a typical capacity remuneration mechanism are presented in figure 11.5.

Reliability options are essentially call options that are bundled with capacity markets. Call options and their pricing in a risk-neutral market environment are discussed in section 9.3.

Note that the choice of the strike price of the call option is a relevant design parameter in this setup. At one extreme, a strike price of zero would correspond to a mandatory forward contract between load-serving entities and capacity providers, which would arguably interfere excessively with the risk-management practices of stakeholders, who may not be willing to hedge their entire supply/demand in a mandatory forward arrangement. Moreover, the lower the value of the strike price, the larger the value of the reliability option, which creates significant cash flows when the trading of these options becomes mandatory. On the other extreme, an excessively high strike price invalidates the call option, since the option is rarely if ever exercised, and essentially becomes void. The typical practice is to set the strike price at or slightly above the marginal cost of peaking units in the market, so that the cash flow captured by the call option corresponds to the scarcity rents that would be earned by capacity providers when the capacity of the peaking unit is exhausted. It is the intention of the capacity market to hedge these risky cash flows by bundling reliability options with the capacity that is traded in the capacity market.

**Installed capacity obligations**   The idea of **installed capacity (ICAP) obligations** is for a public agency to determine a certain desired level of installed capacity, and to allocate the responsibility of achieving this target to suppliers. Capacity obligations can also be tradable in ICAP mechanisms. In the past, markets that have relied on this approach include PJM, Northwestern Power Pool, and New England.

**Capacity payments** are a price-based alternative to capacity auctions. Rather than targeting a certain quantity through the capacity demand curve, the system operator announces a certain capacity payment, and aims to attract sufficient investment given this payment.

As per the usual price versus quantity dilemma, an important weakness of a price-based mechanism is that it may fail considerably in achieving the desired capacity targets. Even if the planning agency knows the installed capacity target that it wishes to achieve, a miscalculation of the capacity payment that will induce this level of capacity can lead to great errors in achieving the target.

**Decentralized capacity mechanisms**   In **decentralized capacity mechanisms**, suppliers and large consumers are required to furnish capacity guarantees in proportion to their contribution to system peak demand. Capacity guarantees are traded in a decentralized market. Decentralized capacity mechanisms have been implemented in France, the California ISO, and the Southwest Power Pool.

**Strategic reserves** are an indirect form of capacity remuneration that has been employed in the past in certain European countries, including Belgium and certain Nordic countries. The idea is to keep certain mothballed projects available for providing emergency power to the system during periods of peak load (e.g. during the winter, in those systems where heating load dominates). The decision of whether or not to maintain strategic reserve is determined annually.

## Model

A basic capacity market model can be formulated as follows:

$$(CRM): \quad \max_{p,x,xd,ls} \int_{v=0}^{xd} VC(v)dv$$

$$- \sum_{j=1}^{m} \Delta T_j \cdot \left( VOLL \cdot ls_j + \sum_{i=1}^{n} MC_i \cdot p_{ij} \right) - \sum_{i=1}^{n} I_i \cdot x_i$$

$$(\rho C): \quad xd - \sum_{i=1}^{n} x_i = 0$$

$$(\Delta T_j \cdot \rho_j): \quad D_j - ls_j - \sum_{i=1}^{n} p_{ij} = 0, j = 1, \ldots, m$$

$$(\Delta T_j \cdot \mu_{ij}): \quad p_{ij} \le x_i, i = 1, \ldots, n, j = 1, \ldots, m$$

$$x, xd, p, ls \ge 0.$$

The function $VC(\cdot)$ corresponds to the capacity market demand curve discussed previously and presented in figure 11.4. The dual multiplier $\rho C$ corresponds to the price of the capacity market. The equivalence of this model to a market equilibrium is analyzed in problem 11.8.

The model can be generalized straightforwardly in order to accommodate various real-world features. One such feature is **capacity credit**, which is the expected availability of a certain technology during peak periods. For instance, thermal resources may be granted a capacity credit upwards of 90%, whereas renewable resources may earn a credit that is less than 50%. Capacity credit can be introduced by modifying the market clearing condition of the market as follows:

$$xd - \sum_{i=1}^{n} CC_i \cdot x_i = 0,$$

where $CC_i$ is the capacity credit of technology $i$.

Capacity credits can strongly influence the winners and losers of capacity mechanisms, and choosing an appropriate value is far from obvious. A related challenge is the qualification of certain technologies, such as demand response or storage. Even if these technologies are chosen to qualify for a capacity mechanism, choosing an appropriate capacity credit is nontrivial.

The contributions of cross-border resources to capacity mechanisms are also a nontrivial aspect of the design that can have a significant influence on market equilibrium. Concretely, the question is whether resources outside of a system that implements a capacity mechanism should count towards contributing to cover the demand of the system that implements the capacity mechanism. One concern that has been raised in European market design is whether doing so results in Member States that implement capacity mechanisms cross-subsidizing the adequacy of neighboring Member States. Regardless of cross-subsidization effects, there is a nontrivial question of whether the resources of a neighboring system can be counted upon when the system that implements the capacity mechanism is experiencing scarcity, and whether the network can support the delivery of this promised capacity.

Another important determinant of capacity market performance is the shape of the demand curve $VC(\cdot)$. Choosing a curve that is too wide results in excess investment. In problem 11.10, the reader is asked to show that choosing a capacity auction demand curve that evaluates to zero at the level of optimal capacity reproduces the outcome of the $(VOLLP)$ model.

---

**Example 11.7** *A market with a capacity auction.* Consider the system of example 11.4 with a VOLL of 1000 $/MWh, to which we introduce the capacity auction demand curve of example 11.6. The capacities of the model are as follows:

- Oil: 1315.3 MW
- Gas: 1417.3 MW
- Coal: 2852.6 MW
- Nuclear: 7353.8 MW

Comparing to the solution of $(VOLLP)$ in example 11.4, we find that the CRM over-builds peaking oil technology (1315.3 MW instead of 953.5 MW), whereas the capacity of all other technologies remain identical.

---

## 11.3.3 Operating reserve demand curves

Operating reserve demand curves are presented in section 6.4 as a means of introducing price elasticity to the procurement of reserves. ORDCs contribute directly towards covering the reserve requirements of the system, and are thus an instrument for ensuring reliability. Nevertheless, it is important to point out that their original inception is also intended to contribute towards addressing adequacy concerns. In systems with increasing amounts of renewable energy resources, where the low marginal cost of resources such as wind and solar power exerts a downward pressure on energy prices, value shifts from the delivery of energy towards the delivery of reserves, due to the inherent variability, uncertainty, and limited control of these renewable resources. ORDCs are intended to produce robust capacity investment signals, despite this shift to near-zero marginal cost systems.

A basic capacity expansion model with ORDCs can be formulated as follows:

$$(ORDC): \quad \max_{p,r,x,dr,ls} \sum_{j=1}^{m} \Delta T_j \cdot \left( -VOLL \cdot ls_j + \int_{v=0}^{dr_l} VR_j(v)dv - \sum_{i=1}^{n} MC_i \cdot p_{ij} \right)$$

$$- \sum_{i=1}^{n} I_i \cdot x_i$$

$$(\Delta T_j \cdot \rho R_j): \quad dr_j - \sum_{i=1}^{n} r_{ij} = 0, j = 1, \ldots, m$$

$$(\Delta T_j \cdot \rho_j): \quad D_j - ls_j - \sum_{i=1}^{n} p_{ij} = 0, j = 1, \ldots, m$$

$$(\Delta T_j \cdot \mu_{ij}): \quad p_{ij} + r_{ij} \leq x_i, i = 1, \ldots, n, j = 1, \ldots, m$$

$$p, r, x, dr, ls \geq 0.$$

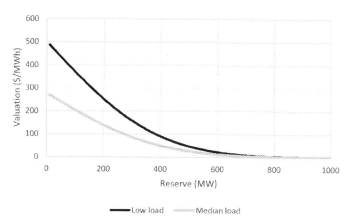

Figure 11.6 ORDCs during periods of median and low load in example 11.8.

The $(ORDC)$ model is equivalent to a market equilibrium (see problem 11.11). Note that the ORDC of period $j$, $VR_j(\cdot)$, can depend on the period in question, meaning that the ORDC can be dynamic and adaptive to system conditions.

---

**Example 11.8** *Adaptive ORDC*. Recall the ORDC formula of equation (6.1):

$$VR_j(x) = \left(VOLL - \widehat{MC_j}\right) \cdot \left(1 - \Phi_{\mu(j),\sigma(j)}(x)\right).$$

Note that the cumulative distribution function used for computing LOLP, $\Phi_{\mu(j),\sigma(j)}$, can depend on the time of year, since the mean and standard deviation of imbalances vary over the year. For instance, the original implementation of scarcity pricing based on ORDC in Texas was based on a distribution function that varied by four-hour blocks of the day and season. Moreover, the approximation of the incremental cost of producing energy, $\widehat{MC_j}$, may also depend on the time period. Let us specifically consider the following formula for $\widehat{MC_j}$:

$$\widehat{MC_j} = \frac{D_j - \min_j D_j}{\max_j D_j - \min_j D_j} VOLL.$$

Let us consider a system with imbalances that have a mean of 0 MW and a standard deviation of 300 MW, as in example 6.13. Let us further assume that the load profile of this system corresponds to that of example 11.4. We present the ORDCs for the periods of median ($j = 4380$, $D_{4380} = 8948.9$ MW, $\widehat{MC}_{4380} = 447.7$ $/MWh) and lowest ($j = 8760$, $D_{8760} = 5600.8$ MW, $\widehat{MC}_{8760} = 0$ $/MWh) load in figure 11.6.

---

The way in which the ORDC mechanism influences adequacy is by virtue of the fact that it can lead to scarcity prices (i.e. prices above the marginal cost of the marginal unit). This happens because the value of reserve is reflected in the reserve demand curve. Since energy and reserve prices are separated only by the marginal cost of the marginal unit (see problem 6.1), an uplift in reserve prices due to ORDCs translates to an uplift of energy prices as well, even if a significant portion of energy supply is covered by near-zero marginal cost renewable resources.

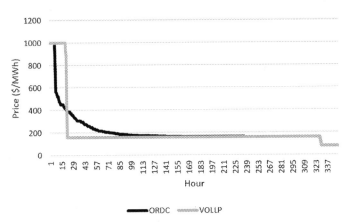

**Figure 11.7** Energy prices in the ORDC design compared to the VOLL pricing design in example 11.9. Zoom in on the 350 periods with highest prices.

---

**Example 11.9** *Capacity investment in a market with an ORDC.* Consider the system of example 11.4 with ORDCs that are given by equation (6.1). The capacity mix resulting from the model is the following:

- Oil: 1150.8 MW
- Gas: 1417.3 MW
- Coal: 2852.6 MW
- Nuclear: 7353.8 MW

As in the case of example 11.7, all technologies other than the peak one have the same capacity as in ($VOLLP$). The peak capacity is greater than the ($VOLLP$) capacity (953.5 MW), but lower than the ($CRM$) capacity (1315.3 MW). The prices resulting from the model are presented in figure 11.7. The figure highlights the periods of peak prices. The effect of the ORDC mechanism in terms of generating scarcity prices that are lower in amplitude but more frequent becomes evident in the figure.

---

## Problems

**11.1** *Ideal energy-only markets lead to optimal long-run capacity expansion.* Show that the energy-only market equilibrium results in an optimal investment and an optimal allocation of power.

**11.2** *Capacity expansion in horizontal slice form.* Formulate the capacity expansion model in horizontal slice form, and confirm that you get the same results as those of the vertical slice form in example 11.2 by comparing your solution to that of problem 1.3. How is the model based on horizontal slices different from the one based on vertical slices?

**11.3** *Adjustment to a natural gas crisis.* Consider the system of example 11.2. You are asked to solve the following using a mathematical programming language.
1. What is the optimal capacity mix, and what are the equilibrium prices?

2. Suppose that the price of natural gas increases to 350 $/MWh. What is the equilibrium price in the short-term market equilibrium (i.e. before capacities are able to adjust)? What are the **windfall profits**[5] of the existing technologies?
3. What is the long-term market equilibrium (i.e. if technologies are adjusted to the new prices)?

**11.4** *The two-market approach.* Suppose that the government decides to split the market of problem 11.3 into two parts, because it aims at containing windfall profits for technologies with low marginal cost. The government specifically decides to assign nuclear units in one market, and remunerates them at the **levelized cost of energy (LCOE)**, which is decided by the government:

$$\lambda_{nuc} = \frac{IC_{nuc} \cdot x_{nuc} + \sum_{j=1}^{m} \Delta T_j \cdot MC_{nuc} \cdot p_{nuc,j}}{\sum_{j=1}^{m} \Delta T_j \cdot p_{nuc,j}}.$$

The government assigns thermal units to the other market. Is there an LCOE for which the long-term equilibrium of the market can support 1000 MW of investment in nuclear units (and thus a totally different capacity mix from the optimal one)? Provide an analytical argument, and support your claim by developing code that detects such an equilibrium solution, in case the answer is yes.

**11.5** *Natural gas subsidies.* Another measure that has been considered in order to cope with the natural gas crisis is to subsidize natural gas units in electricity markets. Consider the following investment model:

$$\min_{p,x} \sum_{g \in G, t \in T} \Delta T_t \cdot MC_g \cdot p_{gt} + 8760 \cdot \sum_{g \in G} IC_g \cdot x_g$$

$$(\mu_{gt}): \quad \Delta T_t \cdot (p_{gt} - x_g) \leq 0, g \in G, t \in T$$

$$(\lambda_t): \quad \Delta T_t \cdot \left( D_t - \sum_{g \in G} p_{gt} \right) = 0, t \in T$$

$$p \geq 0, x \geq 0.$$

The model is populated with the following input data:

- Load duration curve:
  - Base load: $\Delta T_1 = 1260$ hours, $D_1 = 7086$ MW
  - Medium load: $\Delta T_2 = 6000$ hours, $\Delta T_2 = 9008$ MW
  - Peak load: $\Delta T_3 = 1500$ hours, $D_3 = 11169$ MW
- Set of technologies:
  - Nuclear: $MC_{nuc} = 6.5$ $/MWh, $IC_{nuc} = 32$ $/MWh
  - Coal: $MC_{coal} = 25$ $/MWh, $IC_{coal} = 16$ $/MWh
  - Natural gas: $MC_{ng} = 80$ $/MWh, $IC_{ng} = 5$ $/MWh

---

[5] The term windfall profits refers to an increase in profits due to some outside event, as opposed to an increase in profits due to production decisions.

1. Express the KKT conditions of the problem.
2. If you know that the optimal investments and market clearing prices are as follows, find the rest of the problem solution (the primal vector $p$ and the dual vector $\mu$):

   - Prices: $\lambda_1 = 7.62$ \$/MWh, $\lambda_2 = 27.31$ \$/MWh, $\lambda_3 = 109.2$ \$/MWh
   - Investments: $x_{nuc} = 7086$ MW, $x_{coal} = 1922$, $x_{ng} = 2161$ MW

3. Modify the model described in the problem statement in order to express a subsidy on natural gas. Concretely, a subsidy on quantities can be expressed as $S_g$ \$/MWh (where $S_g = 0$ \$/MWh for all units except natural gas), thus it is proportional to the amount of energy production of natural gas units.
4. Suppose that the marginal cost of natural gas changes from 80 \$/MWh (before the natural gas crisis) to 350 \$/MWh (after the natural gas crisis). What should the subsidy be in order for prices to remain identical to those before the crisis?

**11.6** *Relation of VOLL pricing to central planning.* Show that VOLL pricing is equivalent to the optimization problem ($VOLLP$).

**11.7** *VOLL pricing: equilibrium and optimization models.* Implement mathematical programming code that implements the equilibrium conditions of VOLL pricing and thus reproduces the results of example 11.4. Implement mathematical programming code that implements ($VOLLP$). Confirm that the two models give you the same results for $VOLL$ equal to 1000 \$/MWh and 1500 \$/MWh.

**11.8** *Equivalence of ($CRM$) to a market equilibrium.* Prove that the ($CRM$) model is equivalent to a decentralized market equilibrium, where (i) power producers sell power and capacity, (ii) loads are fixed, (iii) the market operator sets the price to VOLL whenever there is involuntary load shedding, and (iv) the system operator buys capacity in the capacity auction.

**11.9** *Mathematical programming model of ($CRM$).* Replicate the results of example 11.7 in a mathematical programming language, where the CRM demand curve consists of 14500 segments with a width of 1 MW each.

**11.10** *Equivalence of ($VOLLP$) and ($CRM$).* Show that ($VOLLP$) and ($CRM$) produce the same investment outcome if $VC(\sum_{i=1}^{n} x_i^\star) = 0$, where $x^\star$ is the vector of optimal capacities of the ($VOLLP$) model.

**11.11** *Equivalence of ($ORDC$) to a market equilibrium.* Prove that ($ORDC$) is equivalent to a decentralized market equilibrium, where (i) power producers sell power and reserve, (ii) loads are fixed, (iii) the market operator sets the price to VOLL whenever there is involuntary load shedding, and (iv) the system operator buys reserve in the reserve auction.

**11.12** *Mathematical programming model of ($ORDC$).* Replicate the results of example 11.9 in a mathematical programming language, where the ORDCs of each period consist of 100 segments with a width of 15 MW each.

**11.13** *Recursive computation of dual multipliers in expansion models.* Use the KKT conditions of the capacity expansion problem in order to derive the dual multiplier values that are reported in table 11.4 for example 11.3.

**11.14** *Reliability options as a hedge for generators.* Consider the oil generator of problem 11.9, which has a marginal cost of 160 \$/MWh and an investment cost of 2 \$/MWh.

1. What is the CRM price of the model?
2. What is the distribution of losses of the oil capacity over a year?
3. If the owner of the oil capacity adheres to a risk criterion of $VaR_{0.01}$, what is the risk-adjusted loss of the plant owner over a year?
4. If a reliability option with a strike price of 300 \$/MWh were traded in the market, what would its price be in a risk-neutral environment (i.e. what would its expected payoff be)?
5. If the owner of the oil plant trades reliability options with a strike price of 300 \$/MWh against their fair market value, what is the distribution of profits of the plant owner over a year?
6. If the owner of the oil plant adheres to a risk criterion of $VaR_{0.01}$, what are the risk-adjusted losses? Is the owner of the oil plant better off with or without the reliability option?

# Bibliography

**Section 11.1**    The material in this section is based on Stoft (2002, §1–3.3) and Birge and Louveaux (2010, §1.3).

**Section 11.2**    The material in this section is based on Stoft (2002, §2–2). The equivalence of centralized planning and energy-only markets in the existence of demand uncertainty and transmission constraints is established by Özdemir (2013).

The two-market approach of problem 11.4 was proposed as a countermeasure to the natural gas crisis of 2022 in Europe. The idea is not new: it has been mobilized in the past in Asian and US markets, and has also resurfaced in recent academic articles Keay and Robinson (2017a,b). The intuition that is advocated in these proposals (Greek delegation, EU Council 2022) is that (i) "nondispatchable" resources (such as wind power), which happen to have a low marginal cost, should not be remunerated at the same prices as thermal resources, but rather at LCOE, because they are not controllable and because they earn unreasonable windfall profits, whereas (ii) thermal "dispatchable" resources (such as natural gas) should compete in a market. Using the model of problem 11.4, we can show that such a design can support the optimal expansion plan. However, we show in problem 11.4 that this design can also support nonoptimal solutions. The intuition of why this is so is that LCOE by construction covers the investment cost of the resources in the "nondispatchable" market segment, but it does not fine-tune the optimal capacity. We can thus find a capacity mix with an arbitrarily low or high level of "nondispatchable" capacity, which defeats the purpose of decentralizing power system operations through markets in order to achieve an optimal long-term capacity mix.

**Section 11.3**  The reader can find related material to this section in paragraph 2 of Stoft (2002). The interplay between market power, price caps, missing money, and capacity markets is analyzed by Fabra (2018). The equivalence between idealized centralized planning and decentralized complete markets for risk is analyzed in a number of publications, including Ralph and Smeers (2015) and Philpott et al. (2016). The connections between risk aversion, capacity markets, and market welfare using the framework of coherent risk measures have been analyzed by Ehrenmann and Smeers (2011b), de Maere d'Aertrycke, Ehrenmann, and Smeers (2017), and Abada et al. (2019). The interplay between incomplete transmission markets and investment is analyzed by Lété et al. (2022).

VOLL pricing is discussed in Stoft (2002). An overview of US capacity mechanisms is provided in Bushnell, Flagg, and Mansur (2017). European capacity mechanisms are reviewed in Papavasiliou (2021b). Strategic reserves in Belgium are discussed in Papavasiliou and Smeers (2017). The shapes of capacity auction demand curves are analyzed in Cramton and Stoft (2005). The bipolar behavior of capacity prices due to vertical demand curves is described in Cramton (2017). The capacity market demand curve described in example 11.6 is inspired by the Irish capacity market design, as documented in European Commission (2017b). Reliability options have been proposed as a supplement to various capacity market designs (Vazquez, Rivier & Perez-Arriaga 2002, Cramton & Stoft 2005, Oren 2005b, Cramton & Stoft 2006, 2008, Cramton, Ockenfels & Stoft 2013). The argument about how ORDCs are intended to generate robust investment signals in near-zero marginal cost systems is developed in Hogan (2005). The use of adaptive cumulative distribution functions in order to compute different ORDCs for different times of the year is documented in ERCOT (2015), and has inspired an analogous proposal in Belgium (Papavasiliou & Smeers 2017) and other European Member States.

The definition of windfall profits in problem 11.3 is due to Varian (2014).

# 12     Beyond electricity

The framework of KKT analysis used throughout the text can be used in broader models of energy economics. In this section we discuss certain classical models from energy sectors beyond electricity, including natural gas, oil, and biofuels. The focus here is on the models, rather than the sectors, and aims at underlining the usefulness of the quantitative framework developed in the text. This is reflected in the organization of the chapter. After a high-level overview of hydrocarbons and biofuels in section 12.1, we proceed to analyze various models that are motivated by these energy resources. In section 12.2 we consider the short- and long-run equilibrium of an industry by considering the world oil market. In section 12.3 we use the example of the Organization of the Petroleum Exporting Countries (OPEC) in order to analyze the monopolistic behavior of a dominant firm, and we discuss cartels on a qualitative basis. Section 12.4 considers the application of tax incidence theory as one of the proposals for coping with the recent natural gas crisis in Europe. The effect of substitutability is analyzed in section 12.5 in the context of the "tortilla crisis," which was triggered by the extensive use of biofuels. Section 12.6 analyzes Hotelling's rule for the evolution of the price of nonrenewable resources.

## 12.1   Hydrocarbons and biofuels

### 12.1.1 Hydrocarbons

Oil and natural gas are formed from organic matter, namely deceased plants and animals. These hydrocarbons take millions of years to form under specific pressure and temperature conditions. For all practical matters, therefore, these resources are nonrenewable.

**Oil**

The oil market is effectively global, since oil is easy to store and transport. It is thus largely impossible to segment the market, which implies that oil prices around the world are approximately equal.

    The precise amount of oil that exists under the surface of the earth, and how much it will be possible to produce in the future, is generally unknown. Reserves are classified as follows: (i) proven reserves correspond to sites where there is a 90–95% probability that commercially recoverable oil exists, (ii) probable reserves correspond to sites with a 50–89% probability, and (iii) possible reserves correspond to sites with a 10–49% probability. The reserves-to-production ratio (R/P ratio) refers to the remaining duration for which the extraction of existing reserves can be sustained, given the current rate at which oil is produced.

**Peak oil** refers to the point in time when half of the global recoverable oil has been extracted. Hubbert's curve, which was presented by Hubbert in 1956, predicted the evolution of US crude oil production. Until 2014, the curve was followed quite closely. However, the US shale revolution overturned this prediction by resulting in a significant uptake of oil after 2014. Given the uncertainty in the amount of oil reserves, it is still unclear whether peak oil has already been reached or not.

Oil production and consumption is typically measured in barrels of oil. Indicatively, the annual global trade of oil corresponded to approximately 35 billion barrels of oil in recent years. The marginal cost of oil suppliers has indicatively ranged from $9 to $45 per barrel in 2016.

### Natural gas

Natural gas is very important for global energy. To the extent that it replaces more polluting fuels, it can improve air quality and limit emissions of carbon dioxide.

In contrast to oil, gas requires a network, import–export facilities, compression, and other processes. There are three major gas markets around the world: (i) the North American market, which covers the USA and Canada, (ii) the East Asian market, where transportation is mainly ensured through LNG ships, and (iii) the European market, which mainly imports from Russia and north Africa, and where transportation also relies largely on pipelines.

Natural gas production and consumption is typically measured in cubic feet, with multiples of thousand cubic feet (mcf) or billion cubic feet (bcf) that are often used in practice. Alternatively, thousand cubic meters and billion cubic meters (bcm) are also used in practice. Since one meter equals approximately 3.281 feet, one cubic meter is equal to approximately 35.315 cubic feet. We also note that 1000 cubic feet of natural gas corresponds to 1 million British thermal units. And since 1 MWh corresponds to 3.4122 MMBtu, this implies that 1 MWh equals 3.411 mcf, or 0.097 thousand cubic meters. This information is summarized in table 3.1.

## 12.1.2 Biofuels

Biofuels are fuels that are produced from organic substances such as corn. Biofuels are dubbed a renewable and sustainable energy source. They are perceived to lead to low greenhouse gas emissions. They are considered somewhat cheap per unit of energy produced. It is argued that there are large amounts of biomass available. Biofuel supporters argue that they enhance energy security, reduce transportation distance, and contribute to local job creation.

There are, however, various points of criticism in what concerns biofuels as well. It has been argued that biofuels are not entirely effective in reducing greenhouse gas emissions when their entire production chain is taken into consideration. Biofuels have also been criticized for leading to high loss of biodiversity, and placing high demands on water. Crucially, they lead to a competition between food and energy. This effect is analyzed subsequently.

## 12.2 Short- and long-term market equilibrium

The world oil market can be considered as a single market, since transportation has a much smaller role in causing price differentiation. Assuming a competitive industry, we can express the world oil market equilibrium using the aggregate marginal cost function $MC_G(p)$ and the aggregate marginal benefit function $MB_L(d)$. Concretely, the equilibrium of the market is simply the intersection of these curves.

Consider a linear aggregate marginal cost function. Suppose, furthermore, that we can estimate the elasticity of supply, which we denote as $\epsilon_S$. Moreover, suppose that we can observe the market clearing price $\lambda_0$ and the market clearing quantity $P_0 = D_0$. Given this information, it is possible to estimate the aggregate marginal cost function. Geometrically, this is intuitive, since we specify the slope of the line through its elasticity, as well as a point through which the line crosses through the historical market equilibrium. Algebraically, we need to solve a system of two equalities in two unknowns, where the unknowns are the parameters of the linear aggregate marginal cost function. An identical argument applies in the case of estimating the aggregate demand function of the industry.

Concretely, let us assume a linear marginal cost function that can be expressed as follows:

$$MC_G(p) = a_S + b_S \cdot p.$$

Here, $a_S$ and $b_S$ are the intercept and the slope of the linear marginal cost function, respectively.

We can then obtain the supply function by solving the above with respect to quantity, as a function of price:

$$P_G(\lambda) = \frac{\lambda - a_S}{b_S}.$$

The elasticity of this supply function is defined as:

$$\epsilon_S = \frac{dP_G(\lambda)/d\lambda}{P_0/\lambda_0} = \frac{\lambda_0}{P_0 \cdot b_S}.$$

This implies that:

$$b_S = \frac{\lambda_0}{P_0 \cdot \epsilon_S}.$$

The historical equilibrium price-quantity pair implies the following equality:

$$\lambda_0 = a_S + b_S \cdot P_0.$$

Substituting $b_S$ into the above equality, we arrive to the following solution for $a_S$:

$$a_S = \lambda_0 - b_S \cdot P_0 = \lambda_0 - \frac{\lambda_0}{\epsilon_S} = \lambda_0 \cdot \frac{\epsilon_S - 1}{\epsilon_S}.$$

The marginal cost function can thus be expressed as follows:

$$MC_G(p) = \lambda_0 \cdot \frac{\epsilon_S - 1}{\epsilon_S} + \frac{\lambda_0}{\epsilon_S} \frac{p}{P_0}.$$

Note that $\epsilon_S > 1$ characterizes an elastic supply side, whereas $0 < \epsilon_S < 1$ characterizes an inelastic supply side. The supply function can be expressed as:

$$P_G(\lambda) = P_0 + \epsilon_S \frac{P_0}{\lambda_0}(\lambda - \lambda_0).$$

Likewise, the aggregate marginal benefit function of the industry can be expressed as:

$$MB_L(d) = \lambda_0 \cdot \frac{\epsilon_D - 1}{\epsilon_D} + \frac{\lambda_0}{\epsilon_D} \frac{d}{D_0}.$$

And the aggregate demand function can be expressed as:

$$D_L(\lambda) = D_0 + \epsilon_D \frac{D_0}{\lambda_0}(\lambda - \lambda_0).$$

Note that $\epsilon_D < -1$ characterizes an elastic demand side, whereas $-1 < \epsilon_D < 0$ characterizes an inelastic demand side.

---

**Example 12.1** *Saudi production cuts.* We revisit the Saudi production cut example of Pindyck (2013). Suppose that the price of oil in the world market is equal to $50 per barrel, and that the world market trades approximately 35 billion barrels per year. We assume that the supply of OPEC corresponds to 12 billion barrels per year, whereas the supply from the remainder of the industry, which equals 23 billion barrels per year, is competitive. Assuming that the production of Saudi Arabia corresponds to 3.6 billion barrels per year, and that Saudi Arabia is a member of OPEC, we would like to compute the short- and long-term equilibrium of the market. The short-run and long-run demand elasticity of the market is assumed to be equal to $-0.05$ and $-0.3$, respectively, whereas the short- and long-run elasticity of the supply side of the market is assumed to be equal to $0.05$ and $0.3$, respectively. Given the available information, the short-run aggregate demand function of the market is expressed as:

$$D_L^{SR}(\lambda) = 36.75 - 0.035 \cdot \lambda.$$

The short-run competitive supply of the market is calibrated from the available data, and is found to be equal to:

$$P_G^{SR, C}(\lambda) = 21.85 + 0.023 \cdot \lambda.$$

Note that, in calibrating this supply function, we have neglected the supply of OPEC from the calculations, meaning that the supply equilibrium quantity against which this supply function is calibrated is equal to 23 billion barrels per year. Assuming that OPEC is price-inelastic and that its quantity is fixed at 12 billion barrels per year, the total supply function of the industry is:

$$P_G^{SR}(\lambda) = 21.85 + 0.023 \cdot \lambda + 12 = 33.85 + 0.023 \cdot \lambda.$$

The same calculations can be used for deriving the long-run supply and demand functions. Concretely, the long-run demand is given as:

$$D_L^{LR}(\lambda) = 45.5 - 0.21 \cdot \lambda.$$

As anticipated by intuition, the long-run demand is more elastic, which suggests that the demand side has alternative ways of substituting oil in the long run if the price of oil increases excessively (e.g. by resorting to electric vehicles). The long-run competitive supply function of the market is calibrated using the same logic as the short-run competitive supply function:

$$P_G^{LR,C}(\lambda) = 16.1 + 0.138 \cdot \lambda.$$

To this we add the inelastic OPEC supply (12 billion barrels), so as to arrive to a total industry supply function:

$$P_G^{LR}(\lambda) = 28.1 + 0.138 \cdot \lambda.$$

As in the case of the demand functions, the long-run supply function has a lower slope than its short-run counterpart. This indicates the ability of suppliers to adapt their production processes to evolving economic conditions over the long term, e.g. by adapting investments in oil production technology, refinery capacity, or other means. Inserting $\lambda_0 = \$50$ per barrel, we confirm that $P_G^{SR}(50) = D_L^{SR}(50) = 35$ billion barrels per year. This means that the short- and long-run equilibrium of the market coincide. The equilibrium is depicted graphically in the left panel of figure 12.1. Let us now assume that Saudi production is fully interrupted, which implies that the total production decreases by 3.6 billion barrels per year. The short- and long-run demand functions of the market remain identical, and the same holds

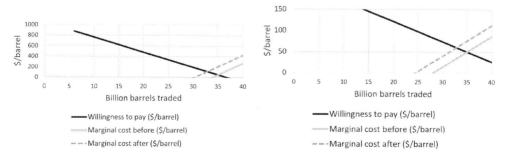

**Figure 12.1** The (Left) short-run and (Right) long-run market equilibrium in example 12.1.

true for the short- and long-run competitive supply of the market. The short-run total supply of the market becomes:

$$P_G^{SR}(\lambda) = 33.85 + 0.023 \cdot \lambda - 3.6 = 30.25 + 0.023 \cdot \lambda.$$

Using a similar reasoning, the long-run total supply of the market becomes:

$$P_G^{LR}(\lambda) = 28.1 + 0.138 \cdot \lambda - 3.6 = 24.5 + 0.138 \cdot \lambda.$$

The intersection of the short-run supply and demand curves yields a short-run equilibrium price of $\lambda = \$112.07$, and the long-run equilibrium price becomes $\lambda = \$60.34$. Notice how the market absorbs the initial price shock in the long run and yields a new price equilibrium that, although higher than the original equilibrium price of \$50, is nevertheless much lower than the short-run equilibrium price. The outcome is depicted in figure 12.1. Note that the figure plots the *inverse* supply function $MC_G(p)$ and *inverse* demand function $MB_L(d)$, whereas we use the supply function and demand function in the actual computations, because this allows us to easily compute the quantity effect of removing Saudi production from the market.

---

The equality of short- and long-run marginal cost at the optimal level of output in example 12.1 is not a coincidence. It is guaranteed by the quantity adjustment process of firms, which aims at minimizing their production cost in the long term, given a profit-maximizing target quantity. The equality of short- and long-run marginal costs for the specific case of investment in power generation capacity is proven in problems 2.6 and 3.9, while the proof of the general result can be found in microeconomics textbooks.

## 12.3   Monopoly, cartel, and the dominant firm

A characteristic of the global oil market is the presence of OPEC. OPEC was established in Iraq in 1960 by five leading producers: Iran, Iraq, Kuwait, Saudi Arabia, and Venezuela. Subsequently, the following countries joined the organization: Qatar in 1961, Indonesia in 1962, Libya also in 1962, the United Arab Emirates in 1967, Algeria in 1969, Nigeria in 1971, Ecuador in 1973, Gabon in 1975, Angola in 2007, Equatorial Guinea in 2017, and Congo in 2018. The organization currently has 13 member countries, since Ecuador withdrew in 2020, Indonesia suspended its membership in 2016, and Qatar terminated its membership in 2019. The stated goal of OPEC is to coordinate petroleum policies of member countries so as to secure a fair and stable return on their investment in the petroleum industry. OPEC accounts for approximately 30–40% of world oil production, 50% of oil trade, and 80% of proven oil reserves, and it is characterized by very low production costs. This implies that it can be in a position to influence the world oil price.

OPEC is an example of a stable cartel. A cartel is a collusion among members that aims at reducing output and increasing prices at a level higher than that of the competitive price. Most of the OPEC members are likely not in a position to influence the market individually (possibly with the exception of Saudi Arabia), nevertheless the collective can influence the market by acting in a coordinated fashion. The members of a cartel have to agree on a market-sharing strategy. In the case of OPEC, the production of each member is a fixed fraction of the total production of OPEC, which is decided during the meetings of the cartel.

The dominant firm model allows us to understand how a monopoly can determine production quantities when it faces a population of **fringe competitors**. Fringe competitors correspond to a population of perfectly competitive firms, which behave as price-takers, and are thus represented by a competitive supply function. When analyzing the strategy of the dominant firm, we effectively consider a monopoly that trades off restricting supply of a product to the market (which results in increased prices) against losing market share. The classical monopoly needs to consider the price elasticity of the market when reaching this decision. In the presence of fringe competitors, these competitors can be interpreted as affecting the price elasticity of the *net demand* that the monopoly faces after some part of the demand has been covered by fringe competitors.

Concretely, we consider a demand function $D(\lambda)$ that represents the price-elastic demand side of the market, as well as a supply function $P_F(\lambda)$ for the fringe competitors in the market. Given a market price $\lambda$, the amount of supply that is made available by fringe competitors amounts to $P_F(\lambda)$, thus $D_N(\lambda) = D(\lambda) - P_F(\lambda)$ corresponds to what is left over for the monopoly to serve. One can then depict the inverse demand function of the net demand as $MB_N(d)$.

The idea of a monopoly manipulating its output decision is that it factors in the effect of its production quantity on the market clearing price, since the price at which the market clears follows the inverse demand function $MB_N(p)$, where $p$ is the amount of output produced by the monopoly, because this net demand must be satisfied by the monopoly. The monopoly thus decides on the amount of output that it wishes to produce by solving the following profit maximization problem:

$$\max_{p \geq 0} MB_N(p) \cdot p - TC(p).$$

The KKT conditions of this problem are expressed as:

$$0 \leq p \perp -MB'_N(p) \cdot p - MB_N(p) + MC(p) \geq 0.$$

For an interior solution ($p > 0$), we thus have:

$$MB'_N(p) \cdot p + MB_N(p) = MC(p).$$

The left-hand side of this equation is the marginal revenue $R'(p)$, where $R(p) = MB_N(p) \cdot p$ for a monopoly supplier. The optimality conditions thus reflect the well-known result of microeconomics whereby firms produce at a level where marginal cost equals marginal revenue, and this applies equally well to competitive firms as well as monopolies.

The two terms of the marginal revenue have a natural interpretation. The first term represents losses in revenue, which result from the fact that additional output from the firm results in a reduction of price, since the market is price-responsive, and this decrease in price impacts the entire quantity sold by the firm. The second term is positive and corresponds to the additional revenue that the firm earns by increasing its output. These two terms are in tension, and in optimizing with respect to quantities the firm seeks the optimal balance.

---

**Example 12.2** *Monopolistic pricing by OPEC.* We revisit example 12.1 in the short-run case, and consider the optimal strategy of the OPEC cartel in this context. The net inverse demand function is the inverse of $D_L^{SR}(\lambda) - P_G^{SR,C}(\lambda) = 36.75 - 0.035 \cdot \lambda - (21.85 + 0.023 \cdot \lambda) = 14.9 - 0.058 \cdot \lambda$, and is thus expressed as:

$$MB_N(d) = 256.897 - 17.241 \cdot d.$$

Note that the slope of the net demand curve decreases relative to the slope of the original demand curve because fringe suppliers effectively act as implicit demand-side elasticity. Let us consider a marginal cost of $10 per barrel.[1] Equating marginal revenue to marginal cost results in the following condition:

$$-17.241 \cdot p + (256.897 - 17.241 \cdot p) = 10.$$

Solving for $p$, we find an optimal output of $p = 7.16$ billion barrels. The total quantity supplied to the market is thus the sum of the 7.16 billion barrels supplied by OPEC plus the supply of fringe suppliers, which means that the following equalities hold in equilibrium:

$$\lambda = MB_L^{SR}(d) = 1050 - 28.571 \cdot d$$
$$d = 7.16 + P_L^{SR,C}(\lambda) = 7.16 + 21.85 + 0.023 \cdot \lambda$$

This results in a global demand of $d = 32.080$ billion barrels and a market price of $\lambda = \$133.46$ per barrel. Given that Saudi Arabia controls 3.6/12 = 30% of the OPEC production, and according to the cartel market sharing rule described above, it is allowed to claim $0.3 \cdot 7.16 = 2.148$ billion barrels, it thus earns a profit of $(133.46 - 10) \cdot 2.148 = \$265.192$ billion per from its participation in the cartel.

---

The dominant firm model can also be interpreted as a **Stackelberg game**. Stackelberg games are leader-follower models, where one player moves first. This player is referred to as the **leader**, and in the case of the dominant firm model it corresponds to the dominant firm. The **follower** then moves by reacting to the decision of the leader. In the context of the dominant firm model, the followers are the fringe producers and the price-responsive demand side of the market.

The Stackelberg game of the dominant firm model can be expressed as a **mathematical program subject to equilibrium constraints (MPEC)**. The idea is to express the profit maximization objective of the dominant firm subject to the KKT conditions that characterize (i) the demand side of the market, which reacts to an exogenous

[1] The marginal cost of $10 per barrel is an approximation of the marginal cost of Iran, Saudi Arabia, and Iraq in 2016, who are among the leading producers in OPEC.

price, (ii) the fringe competitors, which also react to an exogenous price, and (iii) the market-clearing condition, which states that total supply should equal demand. In problem 12.1, the reader is asked to verify the results of example 12.2 by formulating the problem as an MPEC and solving it using a nonlinear programming solver.

It is worth noting that monopoly models with fringe competition have also been used in order to analyze the potential application of market power in electricity markets.

## 12.4  Tax incidence

Models of taxation are interesting from a modeling perspective because they intro-duce a different market clearing price for each side of the market. Concretely, consider the trade of a single good. A tax essentially implies that, if the supply side faces a price of $\lambda_s$, then the demand side faces that same price marked up by a tax: $\lambda_b = \lambda_s + t$. One can then express the resulting market equilibrium problem as follows, where $MC_G$ denotes the aggregate marginal cost and $MB_L$ denotes the aggregate marginal benefit of the market:

- Producers maximize producer surplus:

$$\max_{p \geq 0} \lambda_s \cdot p - \int_{x=0}^{p} MC_G(x) dx$$

- Consumers maximize consumer surplus:

$$\max_{d \geq 0} \int_{x=0}^{d} MB_L(x) dx - \lambda_b \cdot d$$

- The market clears:

$$d - p = 0$$

- Definition of the tax:

$$\lambda_b = \lambda_s + t$$

Collecting the KKT conditions of the producer and consumer surplus maximiza-tion problems, as well as the market clearing condition and the definition of the tax gives us the market equilibrium model as a *complementarity problem*:[2]

$$0 \leq p \perp -\lambda_s + MC_G(p) \geq 0,$$
$$0 \leq d \perp MB_L(d) - \lambda_b \geq 0,$$
$$d - p = 0,$$
$$\lambda_b = \lambda_s + t.$$

---

[2] Complementarity problems are generalizations of optimization problems; see the discussion in section 1.1.2. In certain (well-behaved) cases they are equivalent to optimization problems, although in general this is not the case (and they are therefore harder to solve).

If we substitute out $\lambda_b$ in the above system, we can arrive at the following equivalent set of conditions:

$$0 \leq p \perp -\lambda_s + MC_G(p) \geq 0,$$
$$0 \leq d \perp MB_L(d) - (\lambda_s + t) \geq 0,$$
$$d - p = 0.$$

Note that we can interpret the tax as an equivalent uniform decrease in the marginal benefit of consumers by $t$, which means that the market equilibrium can be expressed equivalently as the following optimization problem:

$$\max_{p \geq 0, d \geq 0} \int_{x=0}^{d} (MB_L(x) - t)dx - \int_{x=0}^{p} MC_G(x)dx$$
$$(\lambda_s): \quad d - p = 0.$$

In this model, the dual variable of the market clearing constraint is the price paid by producers.

An identical argument allows us to express the market equilibrium as an equivalent optimization problem, where the marginal cost of producers is uniformly increased by $t$:

$$\max_{p \geq 0, d \geq 0} \int_{x=0}^{d} MB_L(x)dx - \int_{x=0}^{p} (MC_G(x) + t)dx$$
$$(\lambda_b): \quad d - p = 0.$$

In contrast to the previous model, in this model the dual variable of the market clearing constraint is the price paid by consumers.

Subsidies can be treated identically to taxes, with the difference that buying prices are equal to sell prices *minus* a nonnegative subsidy $s$ (instead of *plus* a nonnegative tax $t$): $\lambda_b = \lambda_s - s$. As in the case of market equilibrium models with taxes, market equilibrium models with subsidies have equivalent optimization models.

Graphically, we can compute this market equilibrium by plotting the aggregate marginal cost and marginal benefit functions of the industry, and scanning the horizontal axis until we reach a point at which the marginal benefit and marginal cost functions are separated by an interval equal to $t$. The intersection of this vertical line with the aggregate marginal benefit function corresponds to the buy price $\lambda_b$, whereas the intersection with the aggregate marginal cost corresponds to the sell price $\lambda_s$. The intersection of this vertical line with the horizontal axis is the market clearing quantity. This is illustrated graphically in figure 12.2.

Tax incidence refers to how the payment of the tax is partitioned between the two sides of the market. This is quantified by comparing how the clearing price after the introduction of the tax ($\lambda_b^{\star}$ on the side of buyers and $\lambda_s^{\star}$ on the side of sellers) compares to the competitive equilibrium price $\lambda^{\star}$ in the absence of a tax. The price increase faced by buyers in equilibrium is $\lambda_b^{\star} - \lambda^{\star}$, whereas the price decrease faced by sellers is $\lambda^{\star} - \lambda_s^{\star}$, and both sum to the tax that needs to be paid, as indicated in the left part of figure 12.2.

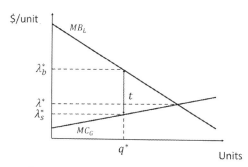
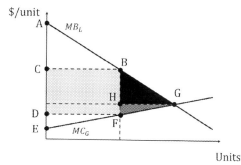

**Figure 12.2** (Left) Graphical solution of the market equilibrium with taxation. In order to find the equilibrium, we seek the level of traded quantity $q^{\star}$ at which the aggregate marginal benefit exceeds the aggregate marginal cost by $t$: $MB_L(q^{\star}) = MC_G(q^{\star}) + t$. Tax incidence refers to how the payment of the tax is split between buyers $(\lambda_b^{\star} - \lambda^{\star})$ and sellers $(\lambda^{\star} - \lambda_s^{\star})$, where $\lambda^{\star}$ is the competitive equilibrium market price without a tax. (Right) Welfare (or deadweight) loss corresponds to the black surface plus the dark gray surface. The light gray surface is the amount of tax collected, and this is *not* lost welfare.

It is interesting to note that the majority of the tax is absorbed by the less elastic side of the market. This is intuitive: the less adaptive a given side of the market is to the introduction of the tax, the less capable it is of avoiding the tax in equilibrium. And agility is measured by the elasticity of demand and supply, respectively, or the slope of the marginal benefit and marginal cost functions, respectively. In the case of figure 12.2, the demand side appears more close to vertical, thus less elastic, and indeed ends up absorbing the majority of the tax. In the extreme of a perfectly inelastic demand side (vertical marginal benefit function), the tax would be absorbed entirely by consumers. Correspondingly, the tax would be entirely absorbed by producers if they were perfectly inelastic (i.e. with a vertical marginal cost curve).

The effect of introducing taxes on welfare can be observed by examining the right panel of figure 12.2. The consumer surplus in the market equilibrium with taxes is measured by the surface of the triangle ABC. And the producer surplus is measured by the surface of the triangle DEF. However, to this surplus we should add the amount of tax collected by the government as a result of introducing the tax, since this surplus returns to the state budget. This corresponds to the surface BCDF. Thus, the total welfare in a market with tax is measured by the surface AEF. This is less than the total welfare generated by the market before the introduction of tax, which corresponds to the surface AEG. The difference of the two, which is referred to as **deadweight loss**, is the sum of the black and dark gray triangles, namely it is equal to the surface of the triangle BFG. It corresponds to loss of economic value as a result of introducing a tax, which precludes potentially beneficial trades between sellers and buyers.

We have pointed out that tax results in deadweight loss, i.e. a loss of economic value. Why, then, would we contemplate introducing a tax? This relates to broader goals of a government. A very interesting case in point is the introduction of tax that has been proposed by Gros (2022) as a means of coping with the energy crisis that occurred in 2022 in Europe as a result of COVID recovery and the invasion of Russia in Ukraine.

---

**Example 12.3** *Taxing Russian natural gas.* The consumption of natural gas in Europe in 2021 was 412 bcm, and the price of natural gas was approximately 78 $/MWh. The equilibrium quantity translates to $D_0 = 14549.78$ bcf, and the equilibrium price translates to $78/3.411 = 22.86$ $/mcf. Assuming a short-term demand elasticity of $\epsilon_D = -0.05$, we can use the results of section 12.2 in order to estimate the following linear demand function:

$$D_T(\lambda) = D_0 + \epsilon_D \frac{D_0}{\lambda_0}(\lambda - \lambda_0) = 15277.3 - 31.82 \cdot \lambda.$$

Let us consider a price-inelastic import of 155 bcm from Russia, which translates to $155 \cdot 35.315 = 5473.825$ bcf; thus the rest ($P_0 = 9075.955$ bcf) is imported from the rest of the world. Assuming a supply elasticity for the rest of the world of $\epsilon_S = 1.1$, and the historical price of $\lambda_0 = 22.86$ $/mcf, we get the following supply function $P_{NR}(\lambda)$ for the rest of the world except Russia:

$$P_{NR}(\lambda) = P_0 + \epsilon_S \frac{P_0}{\lambda_0}(\lambda - \lambda_0) = -907.60 + 436.73 \cdot \lambda.$$

The demand for Russian gas can thus be expressed as:

$$D_R(\lambda) = D_T(\lambda) - P_{NR}(\lambda) = (15277.27 - 31.82 \cdot \lambda) - (-907.60 + 436.73 \cdot \lambda)$$
$$= 16184.87 - 468.55 \cdot \lambda.$$

Let us now consider a scenario whereby the EU imposes a tax on Russian natural gas imports, and let us assume that the supply of Russian natural gas is completely inelastic and equal to the 5473.825 bcf mentioned earlier. It is assumed that, due to pipeline infrastructure, Russia can only sell its output to the European market, namely it is facing the demand function $D_R(\lambda)$, without the option of diverting its sales to Asia or other world markets. The equilibrium outcome is presented in figure 12.3. If the supplier behaves as a perfectly inelastic producer, it ends up

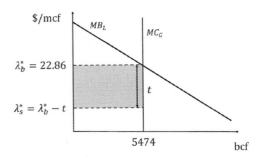

**Figure 12.3** Assuming the Russian supply of gas is inelastic to price in example 12.3, a possible EU tax on Russian natural gas is fully absorbed by Russia.

absorbing the tax entirely, and the equilibrium price remains unaltered at $\lambda_b^\star =$ 22.86 \$/mcf from the point of view of EU consumers. For a tax of 10 \$/mcf, the tax proceeds that are collected from Russia amount to $\$10 \cdot 5473.825 \cdot 10^6$, i.e. approximately \$54 billion per year.

The above example is oversimplified, and ignores the fact that the monopolist would react to the tax that is imposed by further reducing its output. Problem 12.3 recomputes the market equilibrium assuming that the monopolist would adjust its quantity to the imposed tax, and poses the interesting question of whether the resulting loss of consumer surplus for European consumers could be compensated by the tax revenues paid by the monopolist. In Gros (2022), the analysis is further extended to the point of characterizing the optimal level of tax that should be imposed.

## 12.5 One-way substitutability

In this section we analyze the economic effect of one-way substitutability in the context of biofuels. **One-way substitutability** refers to the phenomenon whereby a production factor can be used for covering the needs of two different markets, whereas other production factors cannot.

An example of one-way substitutability in the context of electricity markets is the provision of reserves of different quality: fast generation capacity can be used for covering the demand of system operators for both fast as well as slow reserves, whereas slow generation capacity can only be used for covering needs for slow reserves. The example of one-way substitutability that we focus on in the present section is the use of corn and oil as production factors for both the energy market and the food market. This one-way substitutability is depicted graphically in figure 12.4.

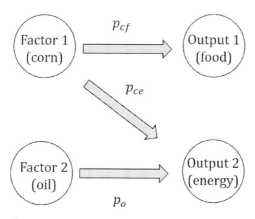

**Figure 12.4** One-way substitutability in the food and energy market.

Substitutability has implications in terms of the market prices of the different goods or services that are being produced. These implications need to be accounted for, both in order to avoid gaming opportunities (recall the design of the California reserve market in example 6.8), as well as in order to anticipate the acceptability of certain energy policies. The latter is profoundly exhibited in the context of the **tortilla crisis**, whereby the one-way substitutability of corn in the food and energy market resulted in a tight coupling of food and energy prices, which led to significant backlash against the use of corn as a means of producing energy.

In order to understand the effect of substitutability, consider the following stylized market model, which aims at capturing the principal effects of the problem:

$$\min_{p \geq 0} C_c \cdot (p_{cf} + p_{ce}) + C_o \cdot p_o$$

$$(\lambda_f): \quad D_f - p_{cf} = 0$$
$$(\lambda_e): \quad D_e - p_{ce} - p_o = 0$$
$$(\mu_c): \quad p_{cf} + p_{ce} \leq P_c^+$$
$$(\mu_o): \quad p_o \leq P_o^+.$$

The amount of corn that is used for producing food and energy is denoted, respectively, by $p_{cf}$ and $p_{ce}$. The objective function aims at minimizing the cost of producing food and energy, where $C_c$ is the marginal cost of corn and $C_o$ is the marginal cost of oil. The first constraint indicates the market clearing condition for the food market, and captures the fact that only corn can be used for covering demand for food, which is assumed inelastic and equal to $D_f$. The market clearing price of the food market is denoted as the dual multiplier $\lambda_f$ of the corresponding constraint. The second constraint corresponds to the market clearing condition of the energy market. It captures the fact that the inelastic demand for energy, which is denoted $D_e$, can be covered by both energy and oil. The total capacity of corn that can be made available to the economy is limited by $P_c^+$, as indicated by the third constraint. The fourth constraint captures the capacity limit $P_o^+$ of oil.

We can express the KKT conditions of the model as follows:

$$D_f - p_{cf} = 0,$$
$$D_e - p_{ce} - p_o = 0,$$
$$0 \leq \mu_c \perp P_c^+ - p_{cf} - p_{ce} \geq 0,$$
$$0 \leq \mu_o \perp P_o^+ - p_o \geq 0,$$
$$0 \leq p_{cf} \perp C_c - \lambda_f + \mu_c \geq 0,$$
$$0 \leq p_{ce} \perp C_c - \lambda_e + \mu_c \geq 0,$$
$$0 \leq p_o \perp C_o - \lambda_e + \mu_o \geq 0.$$

Suppose that we use oil, but we do not run out of it, i.e. $0 < p_o < P_o^+$, while there is not sufficient corn, i.e. $p_{cf} + p_{ce} = P_c^+$. Since there is sufficient oil, we can conclude from the fourth condition that $\mu_o = 0$. And since oil is also used, the last condition implies that $\lambda_e = C_o + \mu_o = C_o$. The dual variable $\mu_o$ can be interpreted as the scarcity rent of oil, and since oil is not scarce, it implies that the profit margin of oil is equal to zero, and thus the price of energy is driven by the marginal cost of oil.

Moreover, since $p_{cf} > 0$, the fifth KKT condition implies that $\mu_c = \lambda_f - C_c$. The dual variable $\mu_c$ can be interpreted as the scarcity rent of corn, and the condition essentially states that the profit margin of corn is equal to the difference between what it can earn in the energy market and its marginal cost. And since $p_{ce} > 0$, the sixth KKT condition implies that $\mu_c = \lambda_e - C_c$, which states that the profit margin of corn is also equal to the difference between what it can earn in the energy market and its marginal cost. Substituting out $\mu_c$ from these two expressions implies that $\lambda_f - C_c = \lambda_e - C_c$. In other words, the profit margin of corn in both the energy and the food market should be equal in equilibrium, so that corn suppliers can be indifferent between supplying their scarce corn in one market or the other. This in turn implies that the price of both markets should be equal: $\lambda_f = \lambda_e$. This becomes the essence of the tortilla crisis: tortillas (made from corn) start tracking the price of energy, due to the one-way substitutability of corn. And if energy prices go up, tortilla prices follow.

---

**Example 12.4** *Tortilla crisis.* Let us consider the above model with the following choice of parameters:

- $C_c = 10$ \$/unit
- $C_o = 20$ \$/unit (note that corn is assumed to be cheaper than oil)
- $D_f = 150$ units
- $D_e = 150$ units
- $P_c^+ = 200$ units (although corn availability can cover demand for food, it cannot also cover the entire demand of the energy market)
- $P_o^+ = 200$ units

In the optimal solution, we have that $p_{cf} = 150$ units are used for covering food demand by corn, and $p_{ce} = 50$ units are used for covering energy demand by corn, which is cheaper than oil. The remainder of the energy demand is covered by oil, thus $p_o = 100$ units. There is no shortage in either of the markets. However, the price of energy is coupled with the price of food due to the scarcity of corn. Concretely, even though the price of corn is only \$10 per unit, the equilibrium price of food becomes \$20, which is also the price of energy, i.e. we have the following relation holding: $\lambda_f = \lambda_e = C_o = \$20$ per unit. This price coupling between the energy and food markets is driven by the one-way substitutability of corn. It has caused a backlash against biofuels, because the energy market starts competing with the food market for the limited factors of production (corn, in this case). And although both needs for energy and food are essential for society, critics of biofuels develop the argument that energy can be produced in a variety of ways, and that corn should not be one of them but rather be left to serve the food market alone.

---

## 12.6 Hotelling's rule

Harold Hotelling was an American mathematical statistician and economic theorist who developed a theory about the evolution of prices for nonrenewable resources over time. Nonrenewable resources, such as oil or natural gas, are particular in that they are finite, and will thus be depleted over a certain time horizon. Assuming that the amount of reserves of nonrenewable resources is known, their market price actually does not follow the marginal cost of their extraction, even in a perfectly competitive economy, but rather obeys **Hotelling's rule** due to the inter-temporal linkage of extraction decisions over time. Hotelling's rule states that the profit that can be generated from the market price of nonrenewable resources should grow at the rate of interest of the economy. It essentially amounts to a no-arbitrage condition.

In order to state the result formally, we introduce the following model:

$$\max_{p\geq0, d\geq0} \sum_{t=1}^{H}(1+r)^{-(t-1)} \cdot \left(\frac{\epsilon}{\epsilon-1}d_t^{\frac{\epsilon-1}{\epsilon}} - C \cdot p_t\right)$$

$$(\lambda_t): \quad (1+r)^{-(t-1)} \cdot (d_t - p_t) = 0, t = 1, \ldots, H$$

$$(\mu): \quad \sum_{t=1}^{H}p_t \leq S.$$

The objective of the economy is to maximize the consumer welfare net of cumulative extraction costs over the horizon of $H$ periods over which the model is considered. Consumer welfare is expressed as an iso-elastic[3] marginal benefit function. The parameter $\epsilon$ expresses the elasticity of demand, while $r$ corresponds to the interest rate of the economy. Decision variables correspond to the amount of the nonrenewable resource that is extracted at every time period $t$, $p_t$, as well as the amount of the nonrenewable resource that is consumed at each time period $t$, $d_t$. The marginal cost of extraction is assumed constant over time and equal to $C$ (although this can be straightforwardly generalized to a time-varying marginal cost of extraction). There are $S$ units of the resource available in total. The first constraint expresses the market clearing condition of each time period, and the dual multipliers $\lambda_t$ correspond to the market price for period $t$. The second constraint expresses the fact that the resource is nonrenewable by stipulating that the total amount of extraction over the full horizon of the model cannot exceed the available reserve of the resource.

Hotelling's result can now be stated formally:

**Proposition 12.1 (Hotelling's rule)**  *For a nonrenewable resource, the marginal profit should grow at the rate of interest, meaning that:*

$$\lambda_{t+1} - C = (1+r) \cdot (\lambda_t - C)$$

*for all $t = 1, \ldots, H-1$.*

---

[3] Concretely, the marginal consumer benefit can be expressed as $MB(d) = d^{-1/\epsilon}$. The elasticity of this inverse demand function is equal to $\epsilon$ for all demand levels $d$.

*Proof*  Unsurprisingly, we use the KKT conditions in order to prove the result. Let us consider an arbitrary time period $t$ and its subsequent period $t + 1$. Complementarity of the dual feasible conditions with respect to $p_t$ implies that:

$$0 \le p_t \perp \frac{1}{(1+r)^{t-1}}(C - \lambda_t + \mu) \ge 0,$$

$$0 \le p_{t+1} \perp \frac{1}{(1+r)^t}(C - \lambda_{t+1}) + \mu \ge 0.$$

Suppose that we are producing throughout the horizon. Then $p_t > 0$ implies that $\mu = \frac{1}{(1+r)^{t-1}}(\lambda_t - C)$. And $p_{t+1} > 0$ implies that $\mu = \frac{1}{(1+r)^t}(\lambda_{t+1} - C)$. Substituting out $\mu$, which can be interpreted as the interest-rate adjusted per-period profit of producers, we arrive at the desired equality:

$$\frac{1}{(1+r)^{t-1}}(\lambda_t - C) = \frac{1}{(1+r)^t}(\lambda_{t+1} - C) \Rightarrow \lambda_{t+1} - C = (1+r) \cdot (\lambda_t - C). \qquad \square$$

---

**Example 12.5** *Hotelling's rule.* Let us consider an instance of the above model with the following choice of parameters: $H = 8$, $S = 1$, $r = 5\%$, $\epsilon = 0.5$, and $C = 2$. The evolution of prices over time is presented in figure 12.5. We indeed observe an increase of prices over time. Note that the actual solution of the model is pinned down by the demand side of the market. Thinking backwards, since prices increase, we can conclude that the amount of the resource extracted at each time step decreases over time, i.e. $d_1 > d_2 > \cdots > d_H$. These quantities cannot evolve arbitrarily, since they must sum to the amount $S$ of the nonrenewable resources, and this is what pins the solution down.

---

It turns out that Hotelling's rule is not entirely confirmed empirically. It would be surprising if it were, since most of the parameters that the model depends on evolve over time, often unpredictably. This is quite clear for the evolution of interest rates $r$, but can also be argued to be the case for the marginal cost of extraction $C$ (e.g. in the case of a technological invention that decreases extraction cost), or even the evolution of the best estimate $S$ for the available reserve of the nonrenewable resource that can be extracted. The following example analyzes the effect of changes in the model parameters on the price trajectory of the nonrenewable resources.

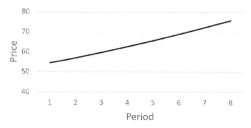

**Figure 12.5**  Evolution of prices over time for the case of example 12.5, where $S = 1$ and $H = 8$.

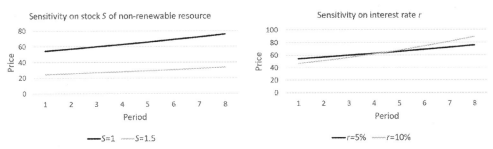

**Figure 12.6** Sensitivity of prices on the amount of reserve $S$ and the interest rate $r$ in the case of example 12.6.

---

**Example 12.6** *Sensitivity of Hotelling's rule to model parameters.* Returning to example 12.5, the left panel of figure 12.6 presents the sensitivity of the price trajectory on the amount of reserve of the resource $S$, in particular when varying the amount of the reserve from $S = 1$ to $S = 1.5$. It is intuitive to observe that an increase in the amount of reserve results in a uniform decrease of the price trajectory. The right panel of figure 12.6 presents the sensitivity of prices to an increase of interest rates from $r = 5\%$ to $r = 10\%$. Although the two price trajectories are not necessarily bounding one another from above or below as in the case of the left panel, one can observe a more accelerated increase of prices in the case of an economy with a higher rate $r$, which is anticipated by Hotelling's rule.

---

The basic model can be generalized in a number of directions. One can consider a backstop technology that can be relied upon after a certain point in time at a given (possibly high) marginal cost. There can be a temporal variation in marginal costs. One can consider capacity constraints. The supply side may correspond to a monopolist, instead of a perfectly competitive collection of suppliers. Each of these elements has interesting effects on the dynamics of price evolution. Nevertheless, the fundamental result of Hotelling's rule is robust to most of these generalizations, and the KKT conditions can be used in order to understand the implication on the temporal evolution of prices.

## Problems

**12.1** *Dominant firm model as an MPEC.* Confirm the results of example 12.2 by formulating it as an MPEC and solving it using a non-convex optimization solver.

**12.2** *Negative intercept of the marginal cost function.* Generalize the analysis of example 12.2 to the case where the marginal cost of the fringe suppliers in the market is no less than a minimum threshold of $12 per barrel by extending the model of problem 12.1.

**12.3** *Russia strikes back.* In example 12.3 we assume that the Russian gas monopoly would not react to the tax imposed by Europe. This is not realistic, and we expand the analysis here.

1.  Assuming that Russia behaves as a monopolistic supplier on the demand for gas $D_R$ derived in example 12.3, and assuming that the monopolist has a constant marginal cost, what is the marginal cost that would lead to a profit maximizing production of 5473.825 bcf, as assumed in the example?

2.  Considering that a tax of 10 \$/mcf, from the point of view of the monopolist, effectively acts as an increase in its marginal cost, how much would the production of the monopolist become after introducing the tax? What would be the resulting equilibrium price of natural gas under this new level of output? What consumer surplus losses would this increase in price imply for European consumers? Is this loss of consumer surplus for EU consumers compensated by the amount of tax collected from the monopolist?

**12.4** *Trading at a single price.* Is the assumption that Russia is limited to selling to the EU market a crucial assumption of the model in example 12.3?

**12.5** *Tax incidence with linear supply and demand functions.* Consider a market with an aggregate supply function $MC_G(p) = p$, and an aggregate demand function $MB_L(d) = 1 - d$. What is the market equilibrium (price and quantity)? What is the market equilibrium if we introduce a tax of $t = 0.5$ (clearing quantity, buy price, sell price)?

**12.6** *Saudi production floods.* Consider a model similar to that of example 12.1, but where Saudi Arabia floods the market instead of withdrawing its production. Concretely, consider the following problem data:

*   Historical price of oil: 60 \$/barrel
*   Global historical consumption/production of oil: 38 billion barrels per year
*   OPEC production: 10 billion barrels per year
*   Competitive (non-OPEC) production: 28 billion barrels per year
*   Saudi production (part of OPEC): 3 billion barrels per year

Suppose that the short-term supply elasticity for the competitive part of the market is 0.04, and that the short-term demand elasticity is −0.04. Suppose that the long-term supply elasticity for the competitive part of the market is 0.2, and that the long-term elasticity is −0.2. Answer the following questions using a mathematical programming language.

1.  Compute the short- and long-term inverse supply and demand function of the market, as well as the short-term and long-term market equilibrium (clearing price and quantity).

2.  What happens to the equilibrium (price and quantity) if Saudi Arabia increases its supply by 2 billion barrels per year?

3.  Use the code that you developed in part 1 in order to confirm the results of example 12.1 before the cuts in Saudi oil production.

4.  Use the code that you developed in part 2 in order to confirm the results of example 12.1 after the cuts in Saudi oil production.

**12.7** *The tortilla crisis as a linear program.* Implement example 12.4 as a linear program in a mathematical programming language.

**12.8** *Hotelling's rule as a linear program.* Implement example 12.5 as a linear program in a mathematical programming language.

**12.9** *Decoupling of prices in the tortilla crisis model.* Consider the tortilla crisis model (example 12.4), where the marginal cost of corn amounts to 30 $/unit. Propose a primal-dual optimal solution for the problem (using the optimization model for finding primal optimal solutions and the KKT conditions for dual optimal solutions). Is there still a coupling between food and energy prices? Why or why not?

**12.10** *Hotelling's rule only holds for scarce resources.* True/false with justification: Hotelling's rule holds, regardless of the leftover amount $S$ of the nonrenewable resource.

# Bibliography

**Section 12.1**  Hubbert's curve first appeared in Hubbert's presentation to the American Petroleum Institute in Hubbert (1956).

**Section 12.2**  Example 12.1 is based on Pindyck (2013). The equality of short- and long-run marginal cost at the level of optimal investment is analyzed in the context of power generation expansion by Boiteux (1960).

**Section 12.3**  The dominant firm model that is presented in this section is developed in Pindyck (2013). Applications of complementarity models in energy markets are covered in Gabriel et al. (2012).

**Section 12.4**  Tax incidence is developed in standard microeconomics textbooks; see for instance Pindyck (2013) and Varian (2014). The proposal of relying on the inelasticity of supply in order to rationalize a tax on Russian gas in order to cope with the 2022 European energy crisis is developed by Gros (2022).

**Section 12.5**  The tortilla crisis is represented in a quantitative model in Hassler and Sinn (2016). The authors use a dynamic general equilibrium model in order to demonstrate how one-way substitutability can lead to substantial food price increases.

**Section 12.6**  Hotelling's rule is discussed in Pindyck (2013). Harold Hotelling argued for regulation in the use of nonrenewable resources in Hotelling (1931). Jevons had already raised concerns about the reliance of England on coal in Jevons (1865). Meadows, Randers, and Behrens III (1972) raised concerns about reliance on oil just before the oil price shocks of the 1970s by pointing out that exhaustible resources are essential inputs to production, and that the long-run prospect of their exhaustion may be worse than stagnation.

A notable application of mathematical programming in energy economics has been the development of **integrated assessment models (IAMs)**. IAMs attempt to quantify the effects of energy policies by representing the entire cycle of energy, the environment, and the economy. Specifically, IAMs capture the loop of interactions whereby fossil fuels generate emissions, with emissions in turn affecting the carbon cycle, which in turn impacts the climate system, which impacts ecosystems and various domains of human activity, which motivates measures for controlling emissions, which feed back into the use of fossil fuels, thereby closing the cycle of IAMs. The Dynamic Integrated Climate Change (DICE) model, and its regional extension, RICE, are two of the most famous examples of IAMs, which have been developed by Nobel laureate William Nordhaus. They rely on a generalization of proposition 4.11 in order to represent the dynamic equilibrium of the global economy as an equivalent nonlinear optimization problem (Nordhaus & Sztorc 2013).

# APPENDIX A

# A brief introduction to linear programming

This chapter provides a brief introduction to the basics of linear programming that are required for advancing in the textbook. Section A.1 introduces mathematical programming models. Linear programming problems, a specific class of mathematical programming problems, are introduced in section A.2. The graphical resolution of linear programs in two dimensions is presented in section A.3, and conveys the central role of extreme points in finding an optimal solution to linear programs. The formulation of linear programs in standard form is presented in section A.4 in order to allow for the definition of bases and to outline the workings of the simplex algorithm at a high level. Linear programming duality, which is used extensively throughout the textbook, is introduced in section A.6. The sensitivity interpretation of dual multipliers is discussed in section A.7. The KKT conditions, which are also used extensively throughout the textbook, are introduced in section 2.3. Section A.9 describes the modeling of piecewise linear functions in linear programs, while section A.10 discusses the dependence of the optimal solution of linear programs to changes in the right-hand side of the model constraints. These two last sections are used in the construction of dynamic programming value functions for hydrothermal planning, as discussed in chapter 8.

## A.1  Mathematical programming models

Mathematical programming models are tools for optimal decision making in complex systems. The prototypical mathematical program amounts to determining a set of *decisions* that optimize a certain *objective* subject to a set of *constraints*.

**Decision variables** are commonly denoted as $x$, and typically lie in a **feasible set** $X$, which is often a subset of $\mathbb{R}^n$. Thus, $x \in X \subseteq \mathbb{R}^n$ denotes the fact that $n$ decisions are being optimized. The objective we aim at optimizing can typically be expressed in functional form, with $f : \mathbb{R}^n \to \mathbb{R}$ an **objective function** that scores the performance of the decision $x$. Mathematical programs either aim at minimizing or maximizing the objective. In the case where the goal is to minimize an objective function, the generic mathematical program can be expressed as:

$$\min_{x} f(x)$$
$$\text{s.t. } x \in X.$$

The subscript under the "min" operator indicates the decision variables of the problem at hand. The "s.t." that precedes the constraints $x \in X$ stands for "subject to," and indicates that the constraints of the problem follow. This is sometimes written as "subject to," or dropped altogether.

Minimization problems can be expressed equivalently as maximization problems, since the minimum of $f(x)$ coincides with the maximum of $-f(x)$. Thus, the above problem can be expressed equivalently as:

$$\max_x g(x)$$
$$\text{s.t. } x \in X,$$

where $g(x) = -f(x)$.

A solution is **feasible** if $x \in X$, meaning that it satisfies all constraints of the problem. A solution $x^\star$ is **optimal** if it is the best among all feasible solutions, meaning that $f(x^\star) \leq f(x)$ for all $x \in X$ in the case of a minimization (or $f(x^\star) \geq f(x)$ for all $x \in X$ in the case of a maximization). A mathematical program is **infeasible** if the feasible set $X$ is empty, $X = \emptyset$. This means that the set of constraints of the problem are not compatible. A mathematical program is **unbounded** if its objective function value can become arbitrarily low in the space of feasible solutions (in the case of a minimization), or arbitrarily large (in the case of a maximization). Infeasible and unbounded mathematical programs are an indication that the underlying problem is ill-defined. We observe that there exists no optimal solution for infeasible and unbounded mathematical programs, i.e. we cannot pinpoint a vector $x^\star$ such that $f(x^\star) \leq f(x)$ for all $x \in X$.

The above format is quite general and can capture special cases where decisions are either discrete or continuous, as we can observe in the following examples.

---

**Example A.1** *Minimizing $f$ is equivalent to maximizing $-f$.* Consider the goal of minimizing $x^2$ over the real line. In this case, the objective function is $f(x) = x^2$, and the constraints of the model are $X = \mathbb{R}$, meaning that $x$ can take on any value in $\mathbb{R}$. The mathematical programming problem can be expressed as $\min_x x^2$, or equivalently as $\max_x -x^2$, as indicated in figure A.1.

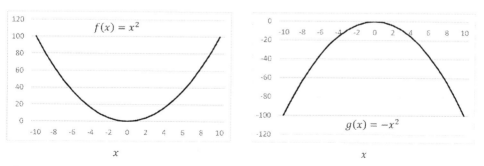

**Figure A.1** Minimizing a function $f(x)$ is equivalent to maximizing $g(x) = -f(x)$. For instance, the minimum of $f(x) = x^2$ in example A.1 is equivalent to the maximum of $g(x) = -f(x)$, and is equal to $x^\star = 0$.

**Example A.2** *Objective and decisions in a discrete problem.* Consider the problem of deciding on buying tickets for an upcoming trip from Athens to Brussels. A round-trip that does not include the weekend costs $600. A round-trip that includes the weekend sells at a discount at $400. One-way tickets cost $350 each way. There are thus three options, which we can enumerate as $X = \{1, 2, 3\}$, with $f(1) = 600, f(2) = 400$, and $f(3) = 700$. If our goal is to minimize the cost of the trip, then we choose to travel with a round-trip that covers the weekend, i.e. $x^\star = 2$ and $f(x^\star) = 400$.

Note that the general mathematical programming problem introduced in this chapter encompasses a very general class of problems. There is no single approach that is best suited for solving general mathematical programs; instead, it is necessary to identify the structure of the specific problem at hand. In the case of example A.1, the optimum can be identified through calculus, namely by identifying the point at which the gradient of the objective function becomes equal to zero. For reasons that are explained in detail in section 2.3, this condition is necessary and sufficient for a point to be optimal in this specific problem. Instead, example A.2 can be solved by enumeration.

Neither of these strategies (enumeration or setting the gradient of the objective function to zero) is fully general or scalable. Scalable strategies exploit the specific structure of mathematical programming problems. One might think that a factor that determines whether a mathematical program is difficult or not is the distinction of whether the objective function $f$ is *linear* or not, and whether the set of constraints $X$ can be expressed as a set of *linear* inequalities or not. It turns out that linearity is not the relevant distinction, but rather *convexity*. Roughly speaking, mathematical programs are computationally much easier to handle if $f$ is a convex function and if $X$ a convex set (convex functions and sets are defined in chapter 2).

## A.2    Linear programming problems

A **linear programming model** is a mathematical programming model where the objective function $f$ can be expressed as a linear function of the decision variables and the constraints $X$ can be expressed as linear equalities of inequalities of the decision variables.

**Example A.3** *Linear programming formulation of economic dispatch.* Consider an economic dispatch problem where two generators are competing to produce electricity in order to cover 100 MWh of demand. Producer 1 is willing to produce up to 60 MWh of energy at 20 $/MWh, whereas producer 2 is willing to produce up to 80 MWh of energy at 50 $/MWh. Our goal is to cover the demand at minimum cost. Decision variables for this problem can be denoted as $p_1$ and $p_2$, and correspond to the amount of energy produced by producer 1 and producer 2, respectively. The economic dispatch problem can then be expressed as follows:

$$\min_{p_1,p_2} 20 \cdot p_1 + 50 \cdot p_2$$

$$p_1 + p_2 \geq 100$$

$$p_1 \leq 60$$

$$p_2 \leq 80$$

$$p_1, p_2 \geq 0.$$

The first constraint expresses the fact that electricity production should at least cover the requested demand. The second and third constraints correspond to the technical limits of the first and second producer, respectively. The last constraint expresses the fact that production cannot be negative. The objective function is $f(p) = 20 \cdot p_1 + 50 \cdot p_2$, which is a linear function of $p$. The set of constraints is expressed as a set of five linear inequalities of the decision variables.

Note that the inequalities of a linear program can be in any of the two senses, i.e. less-than-or-equal ($\leq$) or greater-than-or-equal ($\geq$), and that the two senses can be swapped by multiplying both sides of an inequality by a negative number. Moreover, the decision variables can be constrained to be nonnegative (as in the case of the example above), nonpositive, or free. In the analysis of power systems, examples of nonnegative variables include production and demand, and the level of energy in a hydro reservoir. Examples of free variables include the flow of power over the link of a power network and the amount of power produced by a pumped-hydro unit (which is positive when the unit is in production mode, and negative when the unit is pumping water).

## A.3  Graphical resolution of linear programming problems

The linear programming problem presented in example A.3 can be solved graphically by representing the feasible set and the objective function in the two-dimensional space of the decision variables $p_1$ and $p_2$. This feasible set is indicated as the gray surface in figure A.2.

The gray surface is obtained as the intersection of the inequality constraints of the problem. Concretely, the constraints $p_1 \geq 0$ and $p_2 \geq 0$ imply that the feasible set lies in the first quadrant of the space. The inequality $p_1 + p_2 \geq 100$ corresponds to the half-space in the upper right of the line that crosses the points $(p_1, p_2) = (100, 0)$ and $(p_1, p_2) = (0, 100)$. Similarly, the inequality $p_1 \leq 60$ corresponds to the half-space to the left of the line $p_1 = 60$ and the inequality $p_2 \leq 80$ corresponds to the half-space underneath the line $p_2 = 80$.

Having determined the feasible set of the problem, we now proceed to find the optimal solution. The behavior of the objective function can be understood by drawing different iso-cost curves of the objective function $z = 20 \cdot p_1 + 50 \cdot p_2$. For instance, a cost of \$4400 corresponds to the line $20 \cdot p_1 + 50 \cdot p_2 = \$4400$, which is indicated as the black iso-cost line in figure A.2. There are infinitely many points

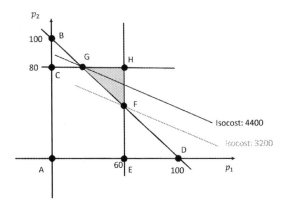

**Figure A.2** The two-dimensional feasible set of example A.3 and different iso-cost lines.

in the feasible set that attain this cost, corresponding to the intersection between the black iso-cost line and the gray surface, including point G. A cost of $3200 is parallel to the iso-cost curve of $4400, and is indicated in gray in figure A.2. It is preferable to the black iso-cost curve because it results in lower cost. An interesting attribute of this iso-cost is that it is as far down and to the left (which is the desirable direction towards which we wish to be moving, given that we aim at minimizing) as possible, while staying in the feasible set. The only point in the gray set that can achieve this objective function value is point F, which turns out to be the optimal solution of the problem. This can be argued geometrically in the figure, since any cost lower than $3200 results in an iso-cost that never intersects the gray feasible set. The coordinates of the point F can then be determined as the intersection of the lines $p_1 = 60$ and $p_1 + p_2 = 100$, namely $(p_1^\star, p_2^\star) = (60, 40)$.

This two-dimensional analysis is quite limited, because we cannot carry over this geometrical idea of pushing for iso-costs at the boundary of the feasible set beyond three dimensions, since we cannot visualize multi-dimensional spaces. But it does convey an important geometric intuition: "corner points" of the feasible set have a special place in linear programs because they qualify as candidates for optimal solutions. We define "corner points" precisely in section A.5, and discuss how this geometric intuition forms the basis of the celebrated simplex algorithm.

## A.4    Linear programming in standard form

A linear programming model in **standard form** with $n$ decision variables and $m$ equality constraints can be described as follows:

$$\min_{x} \sum_{i=1}^{n} c_i x_i$$

$$\text{s.t.} \sum_{i=1}^{n} A_{ij} x_i = b_j, j = 1, \ldots, m$$

$$x_i \geq 0, i = 1, \ldots, n.$$

The parameters $c_i$ can be organized into a vector $c$, the transpose of which, $c^T \in \mathbb{R}^{1 \times n}$, constitutes the objective function coefficients of the linear program. The parameters $A_{ij}$, which are the constraint coefficients of the problem, can be organized into a matrix $A \in \mathbb{R}^{m \times n}$ with $m$ rows (one for each constraint) and $n$ columns (one for each decision variable). The parameters $b_j$ can be organized into a column vector $b \in \mathbb{R}^{m \times 1}$. Using this compact notation, the above problem can be expressed equivalently as:

$$\min_x \; c^T x$$
$$\text{s.t. } Ax = b$$
$$x \geq 0.$$

Any linear program can be converted to standard form through the following operations:

- swapping maximization with minimization
- introducing nonnegative slack variables

Swapping maximization with minimization amounts to exchanging $\max_x c^T x$ with $\min_x -c^T x$.

Introducing nonnegative **slack variables** allows us to convert any linear inequality to an equality by adding or subtracting nonnegative quantities as needed, so as to preserve the equivalence with the original inequality. For instance, the inequality $p_1 \leq 60$ of example A.3 can be expressed equivalently as $p_1 + s = 60$, where $s \geq 0$. This is equivalent to the original condition because if we can find a nonnegative variable $s \geq 0$ such that $p_1 + s = 60$, then it must be the case that $p_1 = 60 - s \leq 60$.

---

**Example A.4** *Economic dispatch in standard form.* Returning to the economic dispatch problem of example A.3, we can express it in standard form as follows:

$$\min_{p_1, p_2, s_1, s_2, s_3} \; 20 \cdot p_1 + 50 \cdot p_2 + 0 \cdot s_1 + 0 \cdot s_2 + 0 \cdot s_3$$
$$p_1 + p_2 - s_1 = 100$$
$$p_1 + s_2 = 60$$
$$p_2 + s_3 = 80$$
$$p_1, p_2, s_1, s_2, s_3 \geq 0.$$

Whereas a nonnegative slack variable is *subtracted* from the left-hand side in the first constraint in order to convert it to an equivalent equality, the second and third inequalities are converted to equalities by *adding* nonnegative slack variables to the left-hand sides of the respective inequalities.

---

The conversion of linear programs into standard form allows us to better understand the relationship between the extreme points of the feasible set and the underlying linear algebra operations that are undertaken by the simplex algorithm for tackling the problem. Although the simplex algorithm is not developed in detail in this textbook since it is not required for understanding the content of the textbook, it is nevertheless briefly introduced in the next section so as to better appreciate the geometry of linear programming problems.

## A.5    Extreme points and the simplex algorithm

Let us consider the linear programming problem in standard form and ignore for a moment the nonnegativity of the decision variables. If there are more constraints ($m$ equalities, as discussed in section A.4) than decision variables ($n$ variables, as discussed in section A.4), then the linear system is infeasible unless there are linearly dependent equality constraints. Therefore, without loss of generality, we focus on the case where there are more variables than constraints, $n \geq m$.

In this case, the set of feasible solutions is of dimension up to $n - m$. A straightforward way to attempt to compute feasible solutions to the original problem is by setting $n - m$ of the variables to zero, and isolating the remaining square $m \times m$ linear system. This is like splitting the original matrix $A$ into two parts, $A = [B, N]$, where $B$ is $m \times m$ and $N$ is $m \times (n - m)$. If the remaining sub-matrix $B$, which is called a **basis**, is invertible, then the linear system is guaranteed to have a unique solution. This is referred to as a **basic solution**. The $n - m$ variables that are set to zero at the outset of this process are referred to as **nonbasic variables**. The remaining $m$ variables that participate in the linear system are called **basic variables**. If the unique solution of the $m \times m$ system is nonnegative, then we have a so-called **basic feasible solution** to the original problem. It is interesting to note that such a basic feasible solution corresponds to an **extreme point** of the original feasible set, i.e. a point that cannot be expressed as the convex combination of any two other distinct points in the feasible set. Geometrically, extreme points correspond to "corners" of the feasible set.

---

**Example A.5** *Computing a basic solution for the economic dispatch problem.* Let us consider the economic dispatch problem of example A.4. The constraint matrix of the problem can be expressed as:

$$A = \begin{pmatrix} 1 & 1 & -1 & 0 & 0 \\ 1 & 0 & 0 & 1 & 0 \\ 0 & 1 & 0 & 0 & 1 \end{pmatrix},$$

where the columns of the matrix correspond to the variables $(p_1, p_2, s_1, s_2, s_3)$ respectively. The matrix has rank 3, which means that all constraints are linearly independent of each other. Choosing $(s_1, s_3)$ as the nonbasic variables that are set to zero, the remaining linear system, which consists of only basic variables $(p_1, p_2, s_2)$, can be expressed as follows:

$$\begin{pmatrix} 1 & 1 & 0 \\ 1 & 0 & 1 \\ 0 & 1 & 0 \end{pmatrix} \begin{pmatrix} p_1 \\ p_2 \\ s_2 \end{pmatrix} = \begin{pmatrix} 100 \\ 60 \\ 80 \end{pmatrix}.$$

This essentially amounts to isolating columns 1, 2, and 4 of the original matrix $A$. The resulting $3 \times 3$ matrix is invertible because it too has rank 3. Inverting the matrix allows us to obtain the basic solution corresponding to basic variables $(p_1, p_2, s_2)$. Algebraically, this corresponds to solving the linear system of three equations over three unknowns, thus arriving to the following basic solution:

$$
\begin{pmatrix} p_1 \\ p_2 \\ s_2 \end{pmatrix} = \begin{pmatrix} 1 & 1 & 0 \\ 1 & 0 & 1 \\ 0 & 1 & 0 \end{pmatrix}^{-1} \begin{pmatrix} 100 \\ 60 \\ 80 \end{pmatrix} = \begin{pmatrix} 20 \\ 80 \\ 40 \end{pmatrix}.
$$

Note that the basic variables are nonnegative, so this linear subsystem happens to produce a basic solution that is also feasible. It corresponds to the extreme point G in figure A.2. The slack variable $s_3$ being equal to zero should be interpreted as indicating that the third constraint of the original problem holds as an exact equality, which means that point G lies exactly on the line $p_2 = 80$, which is indeed confirmed visually in the figure.

One of the key ideas behind the simplex algorithm is that, if the original problem has an optimal solution (i.e. in the case that it is neither infeasible nor unbounded[1]), then one of them is to be found in the extreme points of the feasible set. To put it differently, if we could enumerate all extreme points of the feasible set, then picking the one with the lowest objective function value would give us an optimal solution to the original problem. This is not to say that there cannot exist optimal solutions that are not extreme points, but at least one optimal solution, if it exists, is an extreme point. Now, of course, enumerating all extreme points is computationally prohibitive $\left(\text{there are up to } \binom{n}{n-m} \text{ of them}\right)$, but at least it gives us a way to navigate a finite set of points instead of infinitely many.

**Example A.6** *Basic solutions of the economic dispatch problem.* We return to example A.4 and describe the set of basic solutions in table A.1. We do this by considering two nonbasic variables at a time, so that we are left with a system of three variables in three constraints. There are $\binom{5}{2} = 10$ ways in which we can choose nonbasic variables. There are eight choices that result in invertible sub-matrices, i.e. there exist eight basic solutions to the problem. The choices that result in non-invertible $3 \times 3$ sub-matrices are indicated in gray font in the corresponding row of the matrix, and no basic solution can be produced by this choice of nonbasic variables. Geometrically, these correspond to parallel lines that never intersect and thus never produce a basic solution. The table presents the resulting basic solutions, whether they are feasible or not, and their corresponding objective function values. Note that the basic feasible solutions exactly correspond to the corners of the feasible set in figure A.2, and that the optimal solution of the problem corresponds to one of these extreme points. In particular, the optimal solution corresponds to point F, which attains a cost of $3200. This confirms the result of the graphical solution in section A.3.

---

[1] There exist mathematical programs that have no optimal solutions even if they are neither infeasible nor unbounded. For instance, this can occur when the feasible set is an open set. The feasible sets of linear programs are closed sets.

**Table A.1** Basic solutions of example A.6. The gray font in the corresponding row indicates choices of nonbasic variables that fail to produce a basic solution, since the resulting $3 \times 3$ system is not invertible.

| Non-basic variables | Basic variables | Basic solution | Extreme point | Feasible? $(\geq 0)$ | Objective function |
|---|---|---|---|---|---|
| $(p_1, p_2)$ | $(s_1, s_2, s_3)$ | $(-100, 60, 80)$ | A | No | - |
| $(p_1, s_1)$ | $(p_2, s_2, s_3)$ | $(100, 60, -20)$ | B | No | - |
| $(p_1, s_2)$ | $(p_2, s_1, s_3)$ | - | - | - | - |
| $(p_1, s_3)$ | $(p_2, s_1, s_2)$ | $(80, -20, 60)$ | C | No | - |
| $(p_2, s_1)$ | $(p_1, s_2, s_3)$ | $(100, -40, 80)$ | D | No | - |
| $(p_2, s_2)$ | $(p_1, s_1, s_3)$ | $(60, -40, 80)$ | E | No | - |
| $(p_2, s_3)$ | $(p_1, s_1, s_2)$ | - | - | - | - |
| $(s_1, s_2)$ | $(p_1, p_2, s_3)$ | $(60, 40, 40)$ | F | Yes | 3200 |
| $(s_1, s_3)$ | $(p_1, p_2, s_2)$ | $(20, 80, 40)$ | G | Yes | 4400 |
| $(s_2, s_3)$ | $(p_1, p_2, s_1)$ | $(60, 80, 40)$ | H | Yes | 5200 |

Instead of enumerating all extreme points, the simplex algorithm navigates by performing one **pivot** at a time. Algebraically, this corresponds to swapping one nonbasic variable for one basic variable in the $m \times m$ linear system. Geometrically, it corresponds to traversing to a neighboring extreme point. At any given iteration (i.e. at every basic solution computed along the way), there is well-established theory for performing numerical checks that inform us whether the current solution is optimal (see also section A.10). If the current basic solution is not optimal, there are well-defined procedures for choosing which nonbasic variable should exit, such that we are guaranteed that the next iteration will produce a solution that is at least as good as the present solution. This property also explains the computational savings relative to enumeration, as well as the finite convergence of the algorithm: by guaranteeing that the new solution is no worse than the previous one, and as long as we have a procedure for avoiding cycling between solutions that attain the same objective function value, we are guaranteed to terminate the procedure in a finite number of steps, and can avoid exploring a (potentially massive) set of points that are not promising compared to the current iterate.

## A.6  Linear programming duality

Duality theory is a theoretical area of mathematical programming that concerns all classes of mathematical programs. It has both important implications in terms of algorithms that can be used for solving large-scale problems as well as the economic interpretation of mathematical programming models. In this section we focus on the application of this theory to linear programming. General results are developed in detail in section 2.2. The focus in the present section is on the application of these results to linear programs, whereas section 2.2 explains why these results hold in the first place.

**Table A.2 Linear programming duality mnemonic table.**

| Primal | Minimize | Maximize | Dual |
|---|---|---|---|
| Constraints | $\geq b_i$ | $\geq 0$ | Variables |
| | $\leq b_i$ | $\leq 0$ | |
| | $= b_i$ | Free | |
| Variables | $\geq 0$ | $\leq c_j$ | Constraints |
| | $\leq 0$ | $\geq c_j$ | |
| | Free | $= c_j$ | |

Every linear programming problem has a **dual problem**. If the original problem, which is referred to as the **primal problem**, is a minimization problem then the dual problem is a maximization, whereas if the primal problem is a maximization then the dual problem is a minimization. The decision variables of the dual problem can be derived by assigning a **dual variable** to every constraint of the primal problem. Thus, for every constraint of the primal problem there exists a dual variable. Moreover, for every variable of the primal problem there exists a dual constraint. The rule by which we can derive a dual problem from a primal problem is summarized in table A.2, and is also repeated in table 2.1, where its validity is proven mathematically.

The way in which table A.2 can be read is as follows:

- Suppose that we are dealing with a primal problem that is a minimization. Then we notice from row one of the table that we are in the left side of the table because a minimization corresponds to the second column of the table, not the third.
- For every primal constraint of the type ($\geq$) (row two, column two of the table), there corresponds a nonnegative dual variable (row two, column three of the table). Similarly, from row two we read that for every primal constraint of the type ($\leq$) there corresponds a nonpositive dual variable, and from row three we read that for every equality constraint of the primal problem there corresponds a free dual variable.
- For every nonnegative primal variable (row four, column two) there corresponds a dual constraint of the type ($\leq$) (row four, column three). Similarly, row five indicates that for every nonpositive primal variable there corresponds a dual constraint of the type ($\geq$), and row six indicates that for every free primal variable there corresponds a free dual variable.
- If we are dealing with a primal maximization problem we follow the mapping of the table from right to left, instead of from left to right.

Now that we have determined the sign of the dual variables and the sense of the dual constraints, it remains to populate the dual problem with coefficients. Every dual variable in the dual problem objective function is multiplied by the right hand-side parameter $b_j$ of the constraint that it corresponds to. Analogously, the right-hand side parameter of a dual constraint is equal to the objective function coefficient $c_i$ of the corresponding primal variable. Finally, the left-hand side of the dual constraints is determined by scanning the primal constraints, multiplying the coefficient of the primal variable that the dual constraint corresponds to by its corresponding dual variable, and summing up the result. We illustrate this procedure in a simple example, and then proceed to demonstrate it on the running economic dispatch example of this section.

---

**Example A.7** *Dual of a linear program.* Consider the following linear program:

$$\min_x 3x_1 + 2x_2 + x_3$$

$$(\pi_1): \quad x_1 + x_2 + x_3 \leq 3$$
$$(\pi_2): \quad 2x_1 + 2x_2 + 2x_3 \geq 0$$
$$(\pi_3): \quad x_3 = 5$$
$$x_1 \geq 0, x_2 \leq 0.$$

The dual variables corresponding to each primal constraint are indicated in the left of the constraints, within parentheses, followed by a colon. The dual problem can be expressed as follows:

$$\max_\pi 3\pi_1 + 0\pi_2 + 5\pi_3$$

$$(x_1): \quad \pi_1 + 2\pi_2 \leq 3$$
$$(x_2): \quad \pi_1 + 2\pi_2 \geq 2$$
$$(x_3): \quad \pi_1 + 2\pi_2 = 1$$
$$\pi_1 \leq 0, \pi_2 \geq 0.$$

**Example A.8** *Dual of the economic dispatch problem.* We return to the economic dispatch problem of example A.3 and rewrite it with corresponding dual variables indicated in the left of the constraints and with a slight rearrangement of the objective function and the first inequality constraint:

$$\max_{p_1, p_2} -20 \cdot p_1 - 50 \cdot p_2$$

$$(\lambda): \quad -p_1 - p_2 \leq -100$$
$$(\mu_1): \quad p_1 \leq 60$$
$$(\mu_2): \quad p_2 \leq 80$$
$$p_1, p_2 \geq 0.$$

Since the optimal objective function value of the original problem of example A.3 is 3200, the optimal objective function value of the maximization version of the problem is $-3200$. The dual problem can be written as follows:

$$\max_{\lambda, \mu} -100\lambda + 60\mu_1 + 80\mu_2$$

$$(p_1): \quad -\lambda + \mu_1 \geq -20$$
$$(p_2): \quad -\lambda + \mu_2 \geq -50$$
$$\lambda \geq 0, \mu_1 \geq 0, \mu_2 \geq 0.$$

The optimal solution of the dual problem is $\lambda^\star = 50$, $\mu_1^\star = 30$, $\mu_2^\star = 0$. The optimal objective function value of the dual problem is thus $-3200$. Note that the objective function value of the dual problem is exactly equal to that of the primal problem. This is not coincidental, and is referred to as strong duality. We return to strong duality shortly.

We note that the procedure described above verbally can be summarized succinctly using the mathematical notation introduced earlier. For instance, the dual of a primal problem in canonical form,

$$\min_x c^T x$$
$$(\pi): \quad Ax = b$$
$$x \geq 0,$$

can be expressed as follows:

$$\max b^T \pi$$
$$\pi^T A \leq c^T.$$

A **relaxation** of a mathematical program is a version of the original mathematical program in which the feasible set is enlarged, for instance some constraint is ignored. Thus, a relaxation of a minimization problem gives a solution that results in an objective function value that is less than or equal to the objective function value of the original minimization problem. Likewise, a relaxation of a maximization problem gives a solution that results in an objective function value that is greater than or equal to the objective function value of the original maximization problem.

A very general result in mathematical programming theory states that dual problems bound their corresponding primal problems because they originate from *relaxations* of the original primal problems. This is referred to as **weak duality**. The result is essentially proposition 2.1.

A stronger result in mathematical programming states that, for certain classes of mathematical programs, the objective function value of the dual problem is *equal* to the objective function value of the primal problem. This is referred to as **strong duality**. Strong duality holds in the case of linear programs. It is commonly the case that mathematical programs for which strong duality holds are computationally tractable.

Strong duality in linear programming can be stated in a more nuanced way so as to account for cases of infeasible and unbounded primal linear programs. Concretely:

- If a primal problem has an optimal solution, then its dual has an optimal solution with an objective function value that is equal to that of the primal problem.
- If a primal problem is unbounded, then its dual is infeasible.
- If a primal problem is infeasible, then its dual may be infeasible or unbounded.

Duality is often exploited in optimization algorithms of large-scale problems. The general principle is to tackle the dual problem, which is often easier to handle computationally than the primal problem. This leads to a bound of the original problem in the case of weak duality (from which feasible solutions of high quality can be recovered), or an optimal solution of the original problem in the case of strong duality.

Finally, it is interesting to note that the dual problem of a dual problem corresponds to the original primal problem.

## A.7  Sensitivity

Dual variables carry important information about the behavior of the objective function value in the neighborhood of the optimal solution. Concretely, a dual variable quantifies the sensitivity of the objective function value of the original problem with respect to a change in the right-hand side of its corresponding constraint. This result is proven formally in proposition 2.8. We illustrate this by revisiting the familiar running example of economic dispatch.

---

**Example A.9** *Sensitivity in the economic dispatch model.* Consider the reformulation of the economic dispatch problem in example A.8, and suppose that we increase the right-hand side of the first constraint $(-p_1 - p_2 \leq -100)$ by one unit, thus making the right-hand side equal to $-99$ instead of $-100$:

$$\max_{p_1, p_2} -20 \cdot p_1 - 50 \cdot p_2$$

$$-p_1 - p_2 \leq -99$$

$$p_1 \leq 60$$

$$p_2 \leq 80$$

$$p_1, p_2 \geq 0.$$

We can re-solve the problem using the graphical approach. The solution is indicated in figure A.3, where the gray dot $F$ indicates the original optimal solution and the gray

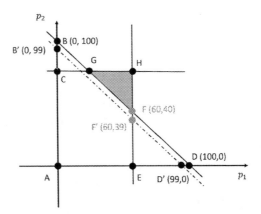

**Figure A.3** Sensitivity of the optimal solution of the economic dispatch problem in example A.9.

dot $F'$ indicates the new optimal solution. Note that the feasible set has increased, since the constraint $p_1 + p_2 \geq 100$ has been moved down and to the left, by being reexpressed as $p_1 + p_2 \geq 99$. This constraint can easily be identified by the line crossing from point B' at coordinates (0, 99) and point D' at coordinates (99,0). The expansion of the feasible set allows us to identify a new and better iso-cost that passes through the point F' at coordinates (60, 39). The resulting cost of this solution is 3150, thus the optimal objective function value of the problem expressed in maximization form is $-3150$. The change in the optimal objective function value is thus equal to $-3150-(-3200) = 50$. Notice that this is exactly the optimal value of $\lambda^\star$, which is the dual variable of the constraint that we have relaxed. Thus, we observe that the dual optimal variable already carries this information of the sensitivity of the objective function value to a change in the right-hand side of the corresponding constraint.

## A.8 KKT conditions for linear programs

The **Karush–Kuhn–Tucker (KKT) conditions** are a set of mathematical conditions that characterize the optimal solution of a primal problem and its dual for certain classes of mathematical programs, including linear programs. The KKT conditions are a collection of equalities, inequalities, and *complementarity conditions* that implicate both primal and dual variables, and that serve as certificates of the optimality of a candidate primal and dual solution. They are both necessary and sufficient for the optimality of certain classes of problems (including linear programs), which means that (i) if a primal-dual vector $(x, \pi)$ is claimed to be optimal then this can be checked by verifying that it satisfies the KKT conditions, and (ii) any optimal solution $x$ of the primal problem and $\pi$ of the dual problem must satisfy these conditions.

The complementarity operator is denoted as $\perp$ and it is a binary operator. The expression $a \perp b$ implies that $a \cdot b = 0$. We also use the condensed notation $0 \leq a \perp b \geq 0$ to indicate that the following three conditions must hold simultaneously: $a \geq 0, b \geq 0$, and $a \cdot b = 0$. Thus, $0 \leq a \perp b \geq 0$ means that either $a = 0, b = 0$, or both, but both cannot be positive at the same time, thus if one is positive the other must be zero.

In the case of linear programming, the KKT conditions are necessary and sufficient for a solution to be optimal. Concretely, the following result holds, which is a restatement of proposition 2.5 for the case of a linear objective function:

**Proposition A.1**  *Consider the following linear program:*

$$\max c_x^T x + c_y^T y$$
$$(\lambda): \quad Ax + By \leq b$$
$$(\mu): \quad Cx + Dy = d$$
$$(\lambda_2): \quad x \geq 0.$$

*Then the KKT conditions have the following form:*

$$Cx + Dy - d = 0,$$
$$0 \le \lambda \perp Ax + By - b \le 0,$$
$$0 \le x \perp \lambda^T A + \mu^T C - c_x^T \ge 0,$$
$$\lambda^T B + \mu^T D - c_y^T = 0,$$

*and are necessary and sufficient for optimality.*

These KKT conditions can be summarized in the following mnemonic:

- The first set of KKT conditions are the original equality constraints of the primal problem.
- The second set of KKT conditions are the inequality constraints of the primal problem, which are complementary to nonnegative dual variables.
- The third set of KKT conditions are nonnegative primal decision variables that are complementary to dual inequalities of the $\le$ type. The dual inequalities are obtained as the following expression: minus[2] the objective function coefficient of the corresponding primal variable $\left(-c_x^T\right)$ plus the product of the coefficient of the corresponding primal variable wherever it appears in a primal constraint with the corresponding dual multiplier $\left(\lambda^T A + \mu^T C\right)$.
- The fourth set of KKT conditions are dual equalities corresponding to every free primal variable. The dual equalities are obtained as the following expression: minus the objective function coefficient of the corresponding primal variable $\left(-c_y^T\right)$ plus the product of the coefficient of the corresponding primal variable wherever it appears in a primal constraint with the corresponding dual multiplier $\left(\lambda^T B + \mu^T D\right)$.

We proceed to apply this procedure to the running economic dispatch example.

---

**Example A.10** *KKT conditions of the economic dispatch problem.* Consider the reformulation of the economic dispatch problem in maximization form in example A.8:

$$\max_{p_1,p_2} -20 \cdot p_1 - 50 \cdot p_2$$
$$(\lambda): \quad -p_1 - p_2 \le -100$$
$$(\mu_1): \quad p_1 \le 60$$
$$(\mu_2): \quad p_2 \le 80$$
$$p_1, p_2 \ge 0.$$

Note that the primal problem is directly cast in the format of proposition A.1. The KKT conditions of the problem can be summarized as follows:

---

[2] Note that, for minimization problems, the dual inequality and equality constraints are obtained by using the objective function coefficient of the corresponding primal variable $\left(c_x^T\right)$ instead of minus the coefficient, which is consistent with the fact that the minimization of a function $f$ is equivalent to the maximization of $-f$.

$$0 \leq \lambda \perp p_1 + p_2 - 100 \geq 0,$$
$$0 \leq \mu_1 \perp 60 - p_1 \geq 0,$$
$$0 \leq \mu_2 \perp 80 - p_2 \geq 0,$$
$$0 \leq p_1 \perp 20 - \lambda + \mu_1 \geq 0,$$
$$0 \leq p_2 \perp 50 - \lambda + \mu_2 \geq 0.$$

Note that there are no free primal variables, thus no dual equality constraints. We claimed earlier that $\lambda^\star = 50, \mu_1^\star = 30, \mu_2^\star = 0$ is an optimal solution to the dual problem without formally proving it. This can be verified by checking that this claimed dual optimal solution also satisfies the KKT conditions of the economic dispatch problem when combined with the primal solution $p_1^\star = 60, p_2^\star = 40$, which has indeed graphically been verified to be the optimal solution of the primal problem.

---

The KKT conditions are recurrent throughout this textbook because they allow us to draw formal conclusions regarding the properties of perfect competition economic models. Economic models of perfect competition are a starting basis for pricing in electricity market models. The KKT conditions thus characterize how prices should behave in a context of perfect competition and illuminate the behavior in a range of models, including the pricing of energy in economic dispatch models, the pricing of transmission in optimal power flow models, the pricing of reserves in energy and reserves co-optimization models, the pricing of scarce capacity in long-term investment models, the pricing of energy in ramp-constrained models, the effect of storage on prices, the effect of substitutability on energy prices, and a number of other interesting insights. The KKT conditions are thus indispensable to this textbook and revisited throughout the text.

---

**Example A.11** *The economic dispatch problem embeds the profit maximization objectives of generators.* In order to appreciate why dual multipliers can be interpreted as economic indicators, we return to the economic dispatch example and analyze its KKT conditions more closely. Suppose that, instead of being instructed about how to produce its power, generator 1 were responding to a market price, which we denote by $\lambda$. As the reader will soon discover, the choice of notation is not coincidental. Given an exogenous price $\lambda$, the selfish profit maximization problem of the producer can be expressed as follows:

$$\max_{p_1 \geq 0} \lambda \cdot p_1 - 20 \cdot p_1$$
$$(\mu_1): \quad p_1 \leq 60.$$

Note that the generator produces by maximizing its profit, but while respecting its private operating constraints: $0 \leq p_1 \leq 60$. The dual variable corresponding to the constraint $p_1 \leq 60$ is denoted as $\mu_1$. Here, too, the choice of notation is not coincidental. This constraint is analogous to the second constraint of the economic dispatch problem of example A.10, but applies to a different optimization problem: that of the private profit maximization of generator 1, not the overall

**Table A.3 The generator data of example A.12.**

| Unit | Marg. cost ($/MWh) | Ramp limit (MW) | Capacity (MW) |
|------|-------------------|-----------------|---------------|
| 1 | 20 | 60 | $+\infty$ |
| 2 | 50 | $+\infty$ | $+\infty$ |

system economic dispatch problem. The KKT conditions of this profit maximization problem can be expressed as follows:

$$0 \leq p_1 \perp 20 - \lambda + \mu_1 \geq 0,$$
$$0 \leq \mu_1 \perp 60 - p_1 \geq 0.$$

Note that these conditions are *identical* to the second and fourth expression of the KKT conditions of the original economic dispatch problem of example A.10 (and this explains why we choose this specific notation for the dual variables of the profit maximization problem). This is an important observation: it means that the optimal solution of the economic dispatch problem already *embeds* within it the profit maximization goal of generator 1. In other words, it means that the primal-dual vector $(\lambda^\star, \mu_1^\star, p_1^\star)$ generated by the economic dispatch model also solves the profit maximization problem of generator 1, as long as the dual variable $\lambda$ assumes the role of a *market price* for trading energy in the context of the decentralized profit maximization setting. The reader can readily verify, through an identical analysis, that the profit maximization problem of generator 2 is embedded in the second and fifth condition of the KKT conditions of example A.10. This leaves us with the interpretation of the first KKT condition of the economic dispatch problem: $0 \leq \lambda \perp p_1 + p_2 - 100 \geq 0$. We revisit this constraint in detail in chapter 4, but can already state that it can be interpreted as a *market clearing condition*: given a nonzero market price ($\lambda > 0$), the supply of generators should match the demand in the market ($p_1 + p_2 = 100$) unless there is excess production in the market ($p_1 + p_2 > 100$), in which case the market price of energy is zero ($\lambda = 0$) because energy is abundant. Thus, the KKT conditions of the centralized economic dispatch problem can be re-expressed as embedding the following information:

- The primal-dual solution should be such that the profit of generator 1 is maximized.
- The primal-dual solution should be such that the profit of generator 2 is maximized.
- The market should clear.

This analysis is formalized in chapter 4. Nevertheless, the reader can already appreciate the economic interpretation of the dual variable $\lambda$ as a market price, i.e. a price signal that can coordinate decentralized surplus-maximizing agents to react in a way that maximizes the overall performance of the system (i.e. that minimizes the cost of production, in the context of this example).

**Example A.12** *Pricing and mis-pricing in multi-period economic dispatch with ramps.* Consider a two-period economic dispatch in a system that consists of two units. The technical characteristics of the units are summarized in table A.3. The system demand over the two periods of the optimization horizon is:

- Demand in period 1: 100 MWh
- Demand in period 2: 200 MWh

The linear program that describes the optimal dispatch problem can be expressed as follows:

$$\min_{p \geq 0} 20 \cdot (p_{11} + p_{12}) + 50 \cdot (p_{21} + p_{22})$$

$$(\lambda_1): \quad 100 - p_{11} - p_{21} = 0$$
$$(\lambda_2): \quad 200 - p_{12} - p_{22} = 0$$
$$(\delta^+): \quad p_{12} - p_{11} \leq 60$$
$$(\delta^-): \quad p_{11} - p_{12} \leq 60.$$

Here, $p_{it}$ indicates the production of unit $i$ in period $t$, thus the first index indicates the unit whereas the second index indicates the time period. The optimal dispatch solution is $p_{11}^\star = 100$ MW, $p_{21}^\star = 160$ MW, $p_{22}^\star = 40$ MW, i.e. the system sources 100 MWh from generator 1 and ramps this unit up to 160 MWh in period 2, making up the difference to cover the remaining demand with unit 2. The KKT conditions of the problem are described as follows:

$$100 - p_{11} - p_{21} = 0,$$
$$200 - p_{12} - p_{22} = 0,$$
$$0 \leq 60 - p_{12} + p_{11} \perp \delta^+ \geq 0,$$
$$0 \leq 60 - p_{11} + p_{12} \perp \delta^- \geq 0,$$
$$0 \leq p_{11} \perp 20 - \lambda_1 - \delta^+ + \delta^- \geq 0,$$
$$0 \leq p_{12} \perp 20 - \lambda_2 + \delta^+ - \delta^- \geq 0,$$
$$0 \leq p_{21} \perp 50 - \lambda_1 \geq 0,$$
$$0 \leq p_{22} \perp 50 - \lambda_2 \geq 0.$$

By analogy to example A.11, the dual variables $\lambda_1$ and $\lambda_2$ can be interpreted as market prices in a decentralized setting. Concretely, the third, fourth, fifth, and sixth KKT condition are equivalent to the following profit maximization problem for generator 1:

$$\max_{p \geq 0} (\lambda_1 - 20) \cdot p_{11} + (\lambda_2 - 20) \cdot p_{12}$$

$$(\delta^+): \quad p_{12} - p_{11} \leq 60$$
$$(\delta^-): \quad p_{11} - p_{12} \leq 60.$$

Similarly, the seventh and eighth KKT condition are equivalent to the following profit maximization problem for generator 2:

$$\max_{p \geq 0} (\lambda_1 - 50) \cdot p_{21} + (\lambda_2 - 50) \cdot p_{22}.$$

As in the case of example A.11, the KKT conditions of the centralized economic dispatch problem embed the profit maximization conditions of individual agents. This implies that the pair of prices and dispatch instructions produced by the centralized problem is consistent with the profit maximization objectives of agents, and thus the prices are a suitable decentralized control signal. This observation can be used in order to understand the prices that are produced by the model. In particular, since unit 2 is asked to produce a positive quantity in period 2, the only price that can be used to induce this unit to produce a nonzero but non-infinite quantity in period 2 is its marginal cost, since a price that is greater than its marginal cost would induce the unit to produce its maximum (i.e. infinity) whereas a price signal that is lower than its marginal cost would induce the unit to produce zero. In contrast, we cannot conclude what the price is in period 1 by observing generator 2: since the generator is not producing, the price $\lambda_1$ has to be less than or equal to its marginal cost, but determining its specific value requires an examination of the profit maximization problem of generator 1. The profit maximization problem of generator 1 has to be considered over both periods because the ramp constraints couple the two periods. If the average price over both periods exceeds 20 \$/MWh, then the unit has an interest to produce an arbitrarily large quantity in period 1, and that same quantity plus 60 MWh (its ramp constraint) in period 2. Instead, if the average price over both periods is below 20 \$/MWh, then the unit has an interest to produce 0 MWh in both periods. Since neither of these two extremes is occurring in the optimal solution, the average price over both periods must be 20 \$/MWh, so that the unit can be induced to produce 100 MWh in period 1 and 160 MWh in period 2, which implies that the price of period 1, $\lambda_1$, must be $-10$ \$/MWh. A negative price may seem exotic, but it does occur in electricity markets. It implies that consumers are *paid*, instead of paying, for consuming energy. The value of $\lambda_1$ can also be derived from the KKT conditions of the problem. Concretely, since $p_{11} > 0$, the fifth KKT condition of the centralized problem implies that:

$$\lambda_1 = 20 - \delta^+ + \delta^-.$$

And since $p_{12} > 0$, the sixth KKT condition of the centralized problem implies that:

$$\lambda_2 = 20 + \delta^+ - \delta^-.$$

We already know that $\lambda_2 = 50$. This implies that $\delta^+ - \delta^- = 30$, which in turn implies that $\lambda_1 = -10$. It is worth contrasting this price solution to an ad hoc heuristic that has been used by certain transmission system operators in practice: setting the price equal to the marginal cost of the cheapest unit that is producing a nonzero quantity that is lower than the nominal capacity of a unit. Applying this heuristic ad hoc pricing method to the optimal dispatch solution derived earlier would imply a market price of 20 \$/MWh in both period 1 and period 2 (since unit 1 is producing a nonzero quantity in both periods, and is the cheapest unit in the market). But this ad hoc pricing method is *not* profit-maximizing for the generators, and thus not consistent with their private incentives.

**Example A.13** *Pricing and mis-pricing in multi-period economic dispatch with hydro.* Hydro resources allow the system operator to move energy consumption from periods when the system is more loaded to periods when the system is more comfortable. The general effect of this shifting, which is economically efficient, is to produce prices that are more level across time periods. In the limit case where hydro has no operational limits (pump capacity, production capacity, energy storage capacity), energy prices are guaranteed to be equal throughout the market clearing horizon. We demonstrate this phenomenon in a two-period context using a simple linear programming formulation. Consider the following two-period model:

$$\min_{p \geq 0, p^H} \sum_{t=1}^{2} \sum_{g \in G} MC_g \cdot p_{gt}$$

$$(\lambda_t): \quad D_t - \sum_{g \in G} p_{gt} - p_t^H = 0, t \in T$$

$$(\mu_{gt}): \quad p_{gt} \leq P_g, g \in G, t \in T$$

$$(\delta): \quad \sum_{t=1}^{2} p_t^H = 0.$$

The set of thermal generators is indicated as $G$. The objective function of the model states that the goal is to minimize cost, which is the total cost over the optimization horizon of $T = \{1, 2\}$ periods. The parameter $MC_g$ corresponds to the marginal cost of each thermal generator $g$, while $p_{gt}$ is the production of thermal unit $g$ during period $t$. The first constraint requires that the supply of power equals the demand of power, where $p_t^H$ is the amount of hydro power that is produced at each period $t$ and $D_t$ is the load of each period. Note that $p_t^H$ is unrestricted in sign, since a hydro unit can both pump power in order to store water (negative production) or release water to produce electricity (positive production). Production limits of thermal units are imposed in the second constraint, with $P_g$ corresponding to the nominal capacity of thermal unit $g$. The last constraint requires that the hydro resource produce as much power as it consumes throughout the optimization horizon. The KKT conditions of the model can be expressed as follows:

$$D_t - \sum_{g \in G} p_{gt} - p_t^H = 0, t \in T$$

$$\sum_{t=1}^{2} p_t^H = 0$$

$$0 \leq p_{gt} \perp MC_g - \lambda_t + \mu_{gt} \geq 0, g \in G, t \in T$$

$$0 \leq \mu_{gt} \perp P_g - p_{gt} \geq 0, g \in G, t \in T$$

$$\left(p_t^H\right): \quad \delta - \lambda_t = 0, t \in T.$$

The dual multiplier $\lambda_t$ corresponds to the price of electricity in period $t$. The last KKT condition implies that prices are equal across all time periods. Let us concretely consider an example of a system with two thermal units, and one hydro unit without operational limits. The two units correspond to a cheap generator, denoted G1, which can produce up to 60 MW at a marginal cost of 10 \$/MWh, and an expensive generator, denoted G2, which can produce up to 100 MW at a marginal cost of 50 \$/MWh. We further assume that the load of the system is equal to 50 MWh in hour 1 and 100 MWh in hour 2. The optimal dispatch in this example is as follows:

- In period 1, we have $p_{G_1,1} = 60$ MWh, $p_{G_2,1} = 30$ MWh, and $p_1^H = -40$ MWh.
- In period 2, we have $p_{G_1,2} = 60$ MWh, $p_{G_2,2} = 0$ MWh, and $p_2^H = 40$ MWh.

Applying the heuristic pricing method of example A.12, whereby the cheapest dispatched unit producing below its technical limit sets the price, the price in period 1 according to this method would equal 50 \$/MWh and the price in period 2 would equal 10 \$/MWh. However, this is not an equilibrium price. Instead, the equilibrium prices of this dispatch model correspond to 50 \$/MWh during both periods. The price equality result can be understood intuitively as follows: if prices are different between time periods, the hydro resource has an incentive to pump an arbitrary amount of energy during the low-price period and produce an arbitrary amount of energy during the high-price period. The only way in which the resource can be indifferent about producing and consuming a finite amount of energy is if the price in both periods is equal. There is an interesting analogy between this result and the single-period optimal dispatch, where the optimal solution strives to equalize marginal costs across generators. The intuition is the same over multiple periods: since hydro effectively acts as a means of shifting demand, the equalization of marginal costs now occurs not only across generators but also between time periods.

**Example A.14** *Cap-and-trade.* Consider a collection of firms $G$ that face a marginal cost $MC_g(x_g), g \in G$, for reducing CO2 emissions, where $x_g$ corresponds to the amount of CO2 emissions abatement of firm $g$. The **cap and trade** system is a market-based mechanism for limiting CO2 emissions that has been implemented in the European Union as well as other worldwide markets. It works as follows:

- The government issues permits for annual CO2 emissions. A firm can only emit CO2 as long as it holds permits. The permits can be **grandfathered** (i.e. handed out for free to firms based on past emissions), or auctioned off on a regular basis (e.g. annually). This is the *cap* portion of the cap-and-trade mechanism.
- Firms are allowed to trade permits depending on the extent to which they intend to decrease CO2 emissions. This corresponds to the *trade* portion of the cap-and-trade mechanism.

Suppose that the government introduces a cap $E$ on the amount of CO2 emissions that the economy can produce on an annual basis. This implies that the total demand for abatement can be expressed as $D = E_0 - E$, where $E_0$ is the business-as-usual level of CO2 emissions in the economy. Consider an example with one hundred firms,

where firm $g \in \{1, \ldots, 100\}$ has a marginal cost of abatement of $MC_g = g$ \$/ton
CO2 and can abate up to $X_g = 25$ million tons of CO2. Suppose that $E_0 = 2.73$
billion tons (which corresponds to the CO2 emissions level of Europe in 2021), and
suppose that the goal for the following year is to emit 2.2 billion tons. The problem
of reducing emissions at minimum cost can be expressed as:

$$\min_x \sum_{g \in G} MC_g \cdot x_g$$

$$(\lambda): \quad D - \sum_{g \in G} x_g = 0$$

$$(\mu_g): \quad x_g \leq X_g, g \in G$$

$$x \geq 0.$$

The objective function expresses the social goal of minimizing the cost of reducing
emissions. The first constraint requires that the total demand for abatement should
be supplied from firms that are reducing emissions in their individual operations. The
second constraint expresses the upper limit on the amount of abatement that each
firm can achieve. The third constraint requires that no firm increases its emissions.
The optimal solution to this problem can be determined by noting that we abate from
the most expensive firms first, until we achieve the target abatement. Concretely, since
we need to abate by 2730–2200 = 530 million tons, and since each firm can abate by
25 million tons, we conclude that 21 firms need to abate fully, and the 22nd firm needs
to provide 5 million tons of abatement. The KKT conditions of the problem can be
expressed as follows:

$$D - \sum_{g \in G} x_g = 0,$$

$$0 \leq \mu_g \perp X_g - x_g \geq 0, g \in G,$$

$$0 \leq x_g \perp MC_g - \lambda + \mu_g, g \in G.$$

Consider the KKT conditions of the 22nd firm. Since $0 < x_{22} < X_{22}$, we can
conclude from the KKT conditions that $\mu_{22} = 0$ and $\lambda = MC_{22} + \mu_{22} = 22$ \$/ton.
In problem A.3, the reader is asked to verify that this is the equilibrium price of cap-
and-trade permits. In other words, if emissions permits are traded at this price $\lambda$,
then firms for which abatement is too expensive choose to procure emissions permits
instead of reducing emissions. Note how the cap-and-trade system thus allows firms
to self-organize such that the total emissions goal is achieved at minimum cost. This
is a general property of market-based systems, and the formal result is stated and
proven in chapter 4 and invoked repeatedly throughout the textbook.

## A.9    Representing piecewise linear functions

A useful modeling tool that is often used in linear programming is the representation
of piecewise linear functions using linear constraints. Although piecewise linear

functions are not linear, if their curvature is in the right "direction" then they can be represented through linear inequalities.

Specifically, piecewise linear convex functions can be represented with linear constraints in minimization problems, and piecewise linear concave functions can be represented with linear constraints in maximization problems. The crucial observation is that a piecewise linear convex function is the upper envelope of the "lines"[3] that define it, and a piecewise linear concave function is the lower envelope of the "lines" (i.e. hyperplanes) that define it.

Concretely, let us define $f(x) = \max_{i=1}^{k} \left( a_i^T x + b_i \right)$ as a piecewise linear convex function. Then we can express the minimization $\min_x \left( c^T x + f(x) \right)$ equivalently as follows:

$$\min_{x,\theta} c^T x + \theta$$
$$\theta \geq a_i^T x + b_i, i = 1, \ldots, n.$$

Similarly, let us define $g(x) = \min_{i=1}^{k} \left( a_i^T x + b_i \right)$ as a piecewise linear concave function. Then we can express the maximization $\max_x \left( c^T x + g(x) \right)$ equivalently as follows:

$$\max_{x,\theta} c^T x + \theta$$
$$\theta \leq a_i^T x + b_i, i = 1, \ldots, n.$$

---

**Example A.15** *Minimization with piecewise linear convex functions.* Consider the following nonlinear optimization problem:

$$\min_x 2 \cdot x + |x| + 2.5 \cdot |x - 2|.$$

The optimal solution of this model is $x = 2$. Intuitively, the slope of the objective function is $-1.5$ for $x < 0$, it becomes $-0.5$ for $0 \leq x < 2$, and finally becomes $5.5$ for $x > 2$. If our goal is to minimize, we should move along $x$ while the slope is negative, but no further than that, i.e. we should stop at $x = 2$. Observe that $|x|$ and $|x - 2|$ are both piecewise linear. Concretely, $|x| = \max(x, -x)$ and $|x - 2| = \max(x - 2, 2 - x)$. The model can be expressed equivalently as a linear program:

$$\min_{x,\theta_1,\theta_2} 2 \cdot x + \theta_1 + 2.5 \cdot \theta_2$$
$$\theta_1 \geq x$$
$$\theta_1 \geq -x$$
$$\theta_2 \geq x - 2$$
$$\theta_2 \geq 2 - x.$$

---

[3] Lines in higher dimensions are generalized as **hyperplanes**. A hyperplane in $\mathbb{R}^n$ is a set of points that can be described as $\left\{ x \in \mathbb{R}^n | a^T x = b \right\}$, where $a \in \mathbb{R}^n$ and $b \in \mathbb{R}$.

Piecewise linear functions emerge in the context of hydrothermal planning. Specifically, the dynamic programming **value functions** that map the level of water in hydro reservoirs to their expected future benefit are piecewise linear. The above modeling methods are used in decomposition algorithms that are employed for tackling multistage stochastic linear programs. This topic is discussed further in section 8.2.2.

## A.10 Sensitivity analysis

Recall that a **basis** is a choice of $m$ linearly independent columns of $A$, with $A = [B, N]$. As discussed previously, each basis has an associated **basic solution** $\begin{bmatrix} x_B \\ x_N \end{bmatrix}$ with $x_B = B^{-1}b$ and $x_N = 0$. The corresponding objective function value is computed as $z = c_B^T B^{-1} b$. Basic feasible solutions correspond to extreme points of the feasible region $\{x | Ax = b, x \geq 0\}$. A basis is **feasible** if $B^{-1}b \geq 0$ and it is **optimal** if, in addition, $c_N^T - c_B^T B^{-1} N \geq 0$, as shown in the following proposition.

**Proposition A.2** *A basic solution is optimal if $B^{-1}b \geq 0$ and $c_N^T - c_B^T B^{-1} N \geq 0$.*

*Proof* The feasibility of the basic solution follows immediately from the fact that $x_B = B^{-1}b \geq 0$ and $x_N = 0$. Consider how the objective function value changes in the neighborhood of the current solution. For this purpose, we can substitute basic variables for nonbasic variables in the objective function $c^T x$. Note that

$$[B \quad N] \begin{bmatrix} x_B \\ x_N \end{bmatrix} = b$$
$$\Leftrightarrow Bx_B + Nx_N = b$$
$$\Leftrightarrow x_B = B^{-1}(b - Nx_N). \tag{A.1}$$

Substituting equation (A.1) into the objective function,

$$c^T x = c_B^T x_B + c_N^T x_N$$
$$= c_B^T B^{-1} b + (c_N^T - c_B^T B^{-1} N)x_N. \tag{A.2}$$

Note that the first term is a constant. Since nonbasic variables can only increase when moving away from the current solution while remaining feasible, the second term of equation (A.2) cannot decrease the objective function when moving to *any* feasible solution. It therefore follows that the solution $(x_B^T, x_N^T)$ is optimal. $\square$

The vector $c_N^T - c_B^T B^{-1} N \in \mathbb{R}^{n-m}$ corresponds to a specific basis $B$ and is called the **reduced cost** of the basis. The result of the above proposition therefore provides sufficient conditions for the optimality of a candidate solution: if the reduced cost of the basis is nonnegative, then the basis is optimal.

The nonnegativity of reduced cost is also necessary for guaranteeing the optimality of a solution, provided the basic solution is nondegenerate, meaning that none of the basic variables are equal to zero (see theorem 3.1 in Bertsimas and Tsitsiklis [1997]).

For some of the material that is developed in the textbook (e.g. the discussion on value functions for hydrothermal problems), it is useful to understand how the optimal value of a linear program depends on the right-hand side parameters $b$. Insights can be gained by examining both the primal linear program, as well as its dual. First, consider the following example.

---

**Example A.16** *Feasible and optimal bases for the diet problem.* Consider the following instance of the diet problem. A least-cost diet is sought that consists of three dishes, $x_1$, $x_2$, and $x_3$, and includes two types of nutrients in quantities $b_1$ and $b_2$. The first, second, and third dish cost \$1, \$2, and \$1, respectively. The amount of each nutrient contained in each dish is presented in table A.4. The problem can be expressed as follows, parametrically in $b$:

$$\min_x x_1 + 2 \cdot x_2 + x_3$$
$$\text{s.t.} \quad 0.5 \cdot x_1 + 4 \cdot x_2 + x_3 = b_1$$
$$2 \cdot x_1 + x_2 + 2 \cdot x_3 = b_2$$
$$x_1, x_2, x_3 \geq 0.$$

Note that the problem is already expressed in standard form. Since the columns of the constraint matrix $A$ are linearly independent, three possible bases can be constructed. They are as follows:

$$B_1 = \begin{bmatrix} 0.5 & 4 \\ 2 & 1 \end{bmatrix}, B_2 = \begin{bmatrix} 0.5 & 1 \\ 2 & 2 \end{bmatrix}, B_3 = \begin{bmatrix} 4 & 1 \\ 1 & 2 \end{bmatrix}.$$

Note that the reduced cost corresponding to bases $B_1$, $B_2$, and $B_3$ is $-0.2$, $1.5$, and $0.2143$, respectively. Therefore, unless any basic solutions are degenerate, only the second and third bases stand a chance at being optimal. The bases lead to the following basic solutions, parametrized with respect to $(b_1, b_2)$:

$$x_{B_1} = \begin{bmatrix} -0.1333 \cdot b_1 + 0.5333 \cdot b_2 \\ 0.2667 \cdot b_1 - 0.0667 \cdot b_2 \end{bmatrix},$$

$$x_{B_2} = \begin{bmatrix} -2 \cdot b_1 + b_2 \\ 2 \cdot b_1 - 0.5 \cdot b_2 \end{bmatrix},$$

$$x_{B_3} = \begin{bmatrix} 0.2857 \cdot b_1 - 0.1429 \cdot b_2 \\ -0.1429 \cdot b_1 + 0.5714 \cdot b_2 \end{bmatrix}.$$

**Table A.4  The amount of nutrients in each dish for the diet problem of example A.16.**

|  | Dish 1 | Dish 2 | Dish 3 |
|---|---|---|---|
| Nutrient 1 | 0.5 | 4 | 1 |
| Nutrient 2 | 2 | 1 | 2 |

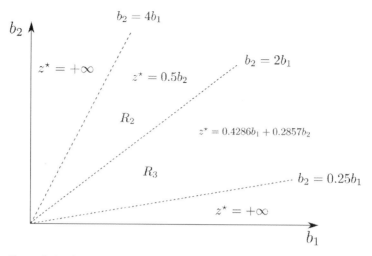

**Figure A.4** The space of right-hand side parameters $b$ of the diet problem of example A.4 can be split into subsets where different bases are optimal. The optimal value $z^\star$ of the diet problem is piecewise linear with respect to $b$.

In order for a basic solution to be feasible, it is necessary that $b$ be such that $x_B \geq 0$. Denoting $R_i = \{(b_1, b_2) : x_{B_i} \geq 0\}$, the following polyhedra are obtained:

$$R_1 = \{0.25 \cdot b_1 \leq b_2 \leq 4 \cdot b_1\},$$
$$R_2 = \{2 \cdot b_1 \leq b_2 \leq 4 \cdot b_1\},$$
$$R_3 = \{0.25 \cdot b_1 \leq b_2 \leq 2 \cdot b_1\}.$$

The cost of each basic solution, parametrized with respect to $(b_1, b_2)$, can be expressed as follows:

$$c_{B_1}^T x_{B_1} = 0.4 \cdot b_1 + 0.4 \cdot b_2,$$
$$c_{B_2}^T x_{B_2} = 0.5 \cdot b_2,$$
$$c_{B_3}^T x_{B_3} = 0.4286 \cdot b_1 + 0.2857 \cdot b_2.$$

It is shown in proposition A.2 that if a basis is feasible and results in nonnegative reduced cost, then the corresponding basic solution is optimal. From this it can be concluded that $x_{B_2}$ is optimal in region $R_2$, and $x_{B_3}$ is optimal in region $R_3$, as shown in figure A.4. Note that the optimal value of the problem is a piecewise linear function of the right-hand side vector $b$. Region $R_2$ corresponds to a mixture of dishes 1 and 3. Region $R_3$ corresponds to a mixture of dishes 2 and 3.

---

In example A.16 we show that the optimal objective function value of the diet problem is a piecewise linear function of the right-hand side parameters $b_1, b_2$. We now proceed to show the same result using duality.

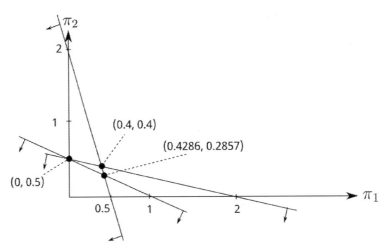

**Figure A.5** The dual feasible region of example A.17. Each black dot is a basic solution of the dual feasible region and corresponds to a basis of the primal problem in standard form.

As mentioned previously, the dual of the primal problem in standard form is the following linear program:

$$(D): \quad \max \pi^T b$$
$$\text{s.t. } \pi^T A \leq c^T.$$

---

**Example A.17** *Feasible set of the dual of the diet problem.* Consider again the diet problem of example A.16. Using the mnemonic of table 2.1, the dual problem is obtained as follows:

$$\max_\pi b_1 \cdot \pi_1 + b_2 \cdot \pi_2$$
$$\text{s.t. } \quad 0.5 \cdot \pi_1 + 2 \cdot \pi_2 \leq 1$$
$$4 \cdot \pi_1 + \pi_2 \leq 2$$
$$\pi_1 + 2 \cdot \pi_2 \leq 1.$$

The feasible region is shown in figure A.5.

---

A **basic solution** of a polyhedron $P \subset \mathbb{R}^n$ (which is not necessarily expressed in standard form) that is defined by linear equalities and inequalities is defined as a vector $x$ such that (i) all equality constraints are active and (ii) out of the constraints that are active at $x$, $n$ are linearly independent.

For linear programs in standard form, each basis $B$ of the primal coefficient matrix $A$ corresponds to a basic solution of the dual feasible set, according to the relationship $\pi^T = c_B^T B^{-1}$. Conversely, for every basic solution of the dual problem there exists a basis of the primal matrix such that the previous relationship holds.[4] If a

---

[4] These are the $m$ columns of $A$ that correspond to the $m$ active constraints of the dual problem.

dual optimal solution exists, then one dual basic solution must be optimal. Therefore, the dual problem can be written equivalently as

$$\max_{i=1,\dots,r} \pi_i^T b,$$

where $r$ indicates the number of basic feasible solutions of $(D)$ (there are finitely many of them). It can therefore be concluded that if the optimal value of the linear program, $z(b)$, is parametrized with respect to $b$, then it is a piecewise linear function of $b$. One can argue similarly for primal linear programs in nonstandard form.

---

**Example A.18** *Basic solutions of the dual of the diet problem.* Recall from example A.16 that three bases correspond to the primal problem. From these bases, the following basic solutions can be computed for the dual feasible region, according to the relation $\pi^T = c_B^T B^{-1}$:

$$(\pi_1, \pi_2) = (0.4, 0.4), (\pi_1, \pi_2) = (0, 0.5), (\pi_1, \pi_2) = (0.4286, 0.2857).$$

The dual basic solutions are indicated in figure A.5.

---

## Problems

**A.1** *Hydrothermal optimization as a linear program.* Implement example A.13 as a linear program and reproduce the results of the example (thermal production, hydro dispatch, prices).

**A.2** *Hydro with efficiency.* Extend the analysis of example A.13 to the case where the hydro unit has an efficiency factor $\eta$, meaning that 1 MWh of electric energy produced by the hydro unit requires $1/\eta$ MWh of hydro energy. Concretely:

1. Formulate the model as a linear program.
2. Implement the model as linear programming code on the instance of example A.13 with $\eta = 0.8$, and report the resulting dispatch solution and market clearing price.
3. Provide an interpretation about the market clearing prices as equilibrium prices for the hydro plant.
4. Compare the equilibrium price to that of the heuristic (mis-)pricing method described in the main text whereby the cheapest unit producing below its nominal capacity is setting the price.

**A.3** *Decentralized model of cap-and-trade.* Consider the profit maximization problem in example A.14 for a firm $g$ with a marginal abatement cost $MC_g$, a maximum abatement possibility of $X_g = 25$ million tons, and an initial *net* allocation of $E_g$ permits (net refers to the baseline emissions of the firm minus the amount of permits that are grandfathered). For permit price $\lambda$, the profit maximization problem of firm $g$ can be expressed as:

$$\max_x \lambda \cdot (E_g + x_g) - MC_g \cdot x_g$$
$$x_g \leq X_g$$
$$x_g \geq 0.$$

Express the meaning of the decision variable, objective function, and each constraint. Argue that the problem can be expressed equivalently as

$$\max_x \lambda \cdot x_g - MC_g \cdot x_g$$
$$x_g \leq X_g$$
$$x_g \geq 0.$$

For the numerical values used in example A.14, what is the optimal decision of firms 1 through 21, firm 22, and firms 23 to 100? How do the optimal selfish decisions compare the socially optimal allocation of abatement in example A.14? What is the profit of each firm if we assume that firms are emitting equal quantities in the reference year, if all firms receive an equal share of permits, and if the cost of not abating is 0?

**A.4** *KKT conditions.* Consider the following optimization problem:

$$\min_{x_1,x_2,z_1,z_2} 5x_1 + x_2 - z_1 + 0.5z_2$$
$$(\lambda_1): \quad -x_1 + x_2 \leq -3$$
$$(\lambda_2): \quad z_1 \leq 0$$
$$(\mu): \quad -z_1 + z_2 = 3$$
$$x_2 \geq 0.$$

1. State the KKT conditions of the problem.
2. Propose a primal-dual optimal pair of solutions, and prove that it is optimal by verifying that it satisfies the KKT conditions.

**A.5** *Welfare maximization is equivalent to cost minimization when demand is inelastic.* Consider the following economic dispatch problem:

$$\max_{p,d} V \cdot d - \sum_{g \in G} MC_g \cdot p_g$$

$$\sum_{g \in G} p_g = d$$
$$d \leq D$$
$$p_g \leq P_g, g \in G$$
$$p, d \geq 0,$$

where $G$ is the set of generators, $p$ and $d$ are the production of generators and consumption, respectively, $V$ and $D$ are the valuation and load of consumers, respectively, and $MC_g$ and $P_g$ are the marginal cost and production capacity of generator $g$, respectively. Prove that, if $V$ is arbitrarily high, then the above problem is equivalent to:

$$\min_p \sum_{g \in G} MC_g \cdot p_g$$

$$\sum_{g \in G} p_g = D$$

$$p_g \le P_g, g \in G$$

$$p \ge 0.$$

**A.6** *Strong duality*. True or false: Consider the following linear program:

$$\min_{x_1, x_2} 2x_1 + 5x_2$$

$$x_1 - 3x_2 = 5$$

$$x_1 \ge 0, x_2 \ge 0.$$

The optimal value of the objective function of the dual linear program is less than 5.

## Bibliography

The organization of the exposition in sections A.2–A.8 is inspired by Taha (2017).

**Section A.6**  The mnemonic table A.2 is sourced from Bertsimas and Tsitsiklis (1997).

**Section A.8**  The cap-and-trade example (example A.14 and problem A.3) is inspired by the cap-and-trade discussion in Varian (2014).

**Section A.10**  The material is inspired by Birge and Louveaux (2010) and Bertsimas and Tsitsiklis (1997).

# APPENDIX B

# The direct current power flow

This chapter provides an overview of the physical laws that govern the flow of power over an electric power network. The ultimate goal of the chapter is to arrive to a linear approximation of the power flow equations used in chapter 5. To get there, we introduce graph Laplacians in section B.1, which allow us to understand why injections of power at nodes imply flows of power over lines in a unique way, in the special case of electric power networks (which is typically not true in other networks). Section B.2 introduces basic notions in circuit theory. Section B.3 derives the power flow equations. A linear mapping of real power injections in buses to bus voltage angles is derived in section B.4. A linear mapping from voltage angles on buses to power flows on lines is presented in section B.5. The composition of these two mappings (i.e. a mapping from real power injections to line flows) leads to the definition of power transfer distribution factors, as indicated in figure B.1. The linear approximation of losses is described in section B.6.

## B.1    Graph Laplacian

Given a graph $G = (N, E)$, where $N$ is the set of nodes in the graph and $E$ is the set of edges, the **incidence matrix** of the graph is defined as $A = (A_{ij}), i, j \in N$, where

$$A_{ij} = \begin{cases} 1, & \text{if } (i,j) \in E \text{ or } (j,i) \in E, \\ 0, & \text{otherwise.} \end{cases}$$

The **degree matrix** is defined as $D = (D_{ij}), i, j \in N$, where

$$D_{ij} = \begin{cases} d_i, & \text{if } (i = j), \\ 0, & \text{otherwise.} \end{cases}$$

Here, $d_i$ is the degree of node $i$, which is the number of edges that are incident to the node.[1] The **Laplacian** of the graph is then defined as

$$L = D - A.$$

---

**Example B.1** *Laplacian of a three-node network.* Consider the 3-node graph of figure B.2. The incidence matrix is

---

[1] For weighted graphs the degree generalizes to the sum of the weights of the edges that are incident to the node.

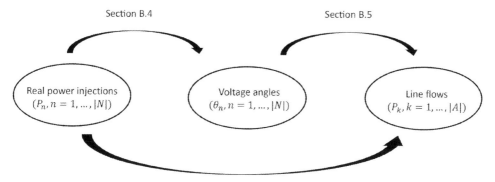

Power transfer distribution factors (PTDFs)

**Figure B.1** Power transfer distribution factors constitute a mapping from real power injections in buses to real power flows in lines in a power network. They are the composition of a linear mapping from real power injections to node bus angles (described in section B.4) and bus angles to line flows (described in section B.5).

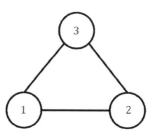

**Figure B.2** The 3-node graph of example B.1.

$$A = \begin{bmatrix} 0 & 1 & 1 \\ 1 & 0 & 1 \\ 1 & 1 & 0 \end{bmatrix},$$

and the degree matrix is

$$D = \begin{bmatrix} 2 & 0 & 0 \\ 0 & 2 & 0 \\ 0 & 0 & 2 \end{bmatrix},$$

from which the Laplacian is derived as

$$L = \begin{bmatrix} 2 & -1 & -1 \\ -1 & 2 & -1 \\ -1 & -1 & 2 \end{bmatrix}.$$

A matrix $A \in \mathbb{R}^{n \times n}$ is **positive semidefinite** if, for any nonzero vector $f \in \mathbb{R}^n$, $f^T A f \geq 0$. The Laplacian of a graph is positive semidefinite. To see this, pick any vector $f \in \mathbb{R}^n$, where $n = |N|$. The $i$th component of $Lf$ is $(Lf)_i = \sum_{j \in N | (i,j) \in E} (f_i - f_j)$. Then,

$$f^T L f = \sum_{i \in N} f_i \sum_{j \in N | (i,j) \in E} (f_i - f_j)$$

$$= \sum_{(i,j) \in E} f_i(f_i - f_j) + \sum_{(i,j) \in E} f_j(f_j - f_i)$$

$$= \sum_{(i,j) \in E} (f_i(f_i - f_j) - f_j(f_i - f_j))$$

$$= \sum_{(i,j) \in E} (f_i - f_j)^2,$$

which is nonnegative. An intuitive interpretation is that, if $f$ is a vector that assigns a value $f_i$ to each node, the quadratic form $f^T L f$ measures the variance of $f$ across the edges of the graph.

---

**Example B.2** *Positive definiteness of the Laplacian of a three-node graph.* Consider the graph of example B.1 and any nonzero vector $f = (f_1, f_2, f_3)$. Then,

$$L f = \begin{bmatrix} 2f_1 - f_2 - f_3 \\ 2f_2 - f_1 - f_3 \\ 2f_3 - f_1 - f_2 \end{bmatrix} = \begin{bmatrix} (f_1 - f_2) + (f_1 - f_3) \\ (f_2 - f_1) + (f_2 - f_3) \\ (f_3 - f_1) + (f_3 - f_2) \end{bmatrix}.$$

Left-multiplying by $f$,

$$f^T L f = f_1(f_1 - f_2) + f_1(f_1 - f_3)$$
$$+ f_2(f_2 - f_1) + f_2(f_2 - f_3)$$
$$+ f_3(f_3 - f_1) + f_3(f_3 - f_2)$$
$$= (f_1 - f_2)^2 + (f_2 - f_3)^2 + (f_1 - f_3)^2.$$

---

**Proposition B.1** *The multiplicity of the eigenvalue $\lambda = 0$ in the Laplacian of a graph is equal to the number of connected components of the graph.*

*Proof* The vector $e = (1, \ldots, 1)$ is an eigenvector of $L$ with eigenvalue 0, since $(Lf)_i = \sum_{j \in N | (i,j) \in E}(1 - 1) = 0$. By the same reasoning, if $G_1, \ldots, G_r$ are the connected components of $G$ with nodes $N_j$ being the set of nodes for $G_j$, then the vector

$$e^j = \begin{cases} 1, & \text{if } i \in N_j, \\ 0, & \text{otherwise,} \end{cases} \quad j = 1, \ldots, r,$$

is an eigenvector with eigenvalue of zero for $L$, and $e^j$ are orthogonal to each other since they have mutually exclusive supports. Therefore, the multiplicity of the zero eigenvalue is at least $r$.

To show that the multiplicity of $\lambda = 0$ is no greater than $r$, it needs to be shown that it is impossible to identify an eigenvector that is linearly independent of $e^1, \ldots, e^r$. Therefore, it needs to be shown that any eigenvector $v$ for 0 lies in the span of $e^1, \ldots, e^r$. Note that

$$v^T L v = \sum_{(i,j)\in E} (v_i - v_j)^2.$$

Since $v$ is the eigenvector corresponding to an eigenvalue of zero, $Lv = 0$. Therefore the quadratic form above has to be zero. This is only possible if $v_i = v_j$ for all $(i,j) \in E$, i.e. $v_i = v_j$ for all $(i,j)$ that are edges of a connected component, although the values can be different among different connected components. Therefore, any such $v$ can be described as a linear combination of $e^1, \ldots, e^r$. &#9633;

---

**Example B.3** *Rank of the Laplacian of a three-node graph.* Consider again the graph of example B.1. The eigenvalues of the Laplacian of the graph are $\lambda_1 = 0$, $\lambda_2 = 3$, and $\lambda_3 = 3$. Since the graph is connected, it has a single eigenvalue equal to zero. The corresponding eigenvector of $\lambda_1$ is $e = (1,1,1)$.

---

## B.2    Circuits

Electric power systems are large circuits that consist of both **passive elements** (transmission lines, transformers, loads represented as impedances) as well as **active elements** that generate or consume electric power (generators, or loads represented as active and reactive power sinks).

The state of an electric circuit can be described by the voltage difference between each node of the circuit and a reference point, called **neutral**. Once voltages are determined, the amount of current flowing over all passive elements and the amount of power flowing along each line of the circuit can be determined.

In alternating current (AC) electric power systems, voltage and current fluctuate in a sinusoidal pattern.[2] In Europe the nominal frequency of power systems is 50 Hz, whereas in the United States the nominal frequency is 60 Hz. Sinusoidal signals can be described as two-dimensional vectors rotating with constant frequency around the origin, as shown in figure B.3. The horizontal projection is the value of the signal as a function of time. The magnitude of the vector represents the **amplitude** of the signal. The angle of the vector with respect to the horizontal axis at time 0 is the **phase angle**. The sinusoidal signal can be represented equivalently through its **phasor**, which is a complex number that collapses the minimum information needed to describe the signal: its phase angle, as well as its **root mean square (RMS)** value, which can be shown to be equal to the amplitude divided by $\sqrt{2}$.

A passive electric element is characterized by a complex number, its **impedance** $Z$, or equivalently its **admittance**, $Y = Z^{-1}$, which is determined by its electrical characteristics. What characterizes a passive element is the fact that the current

---

[2] This chapter focuses on AC systems. The treatment of direct current (DC) systems is simpler as far as the analysis of power flow is concerned. Moreover, AC systems represent the vast majority of high-voltage transmission systems, although there is an increasing resurgence of DC networks, which are appealing due to the improved controllability and other technical advantages that they offer.

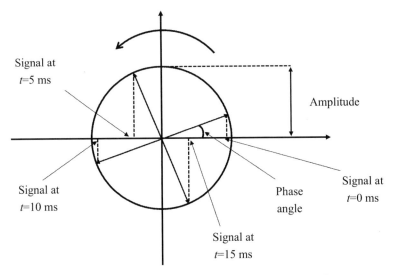

**Figure B.3** A sinusoidal signal is the projection of a vector rotating at constant angular speed on the horizontal axis.

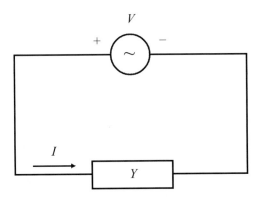

$$I = YV, S = VI^*, = P + Qi$$

**Figure B.4** For a passive element with admittance $Y$, the current flowing over the element is given by $I = YV$. The complex power consumed by the element is $S = P + Qi = VI^*$.

induced by the application of a given voltage at the terminals of the passive element is given by the following expression,[3] as shown in figure B.4:

$$I = YV.$$

---

[3] A special case of this is **Ohm's law**, which applies to purely resistive passive elements.

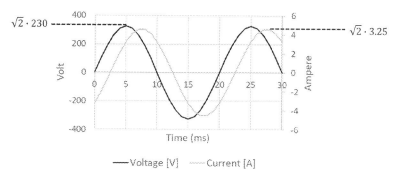

Figure B.5 The evolution of voltage and current over time in example B.4.

**Example B.4** *Current flowing through a passive element.* Consider a passive element with the following admittance, $Y = 0.01 - 0.01i\,\Omega^{-1}$, and suppose that a voltage with an RMS of 230 V is applied across the element. Using the definition of impedance, the current flowing through the passive element is given by

$$I = VY = 230 \cdot (0.01 - 0.01i) = 2.3 - 2.3i.$$

The RMS of the current is $\sqrt{2.3^2 + 2.3^2} = 3.25$ Ampere and the phase difference between the current and the voltage is $\arctan\left(\frac{-2.3}{2.3}\right) = -45°$. This implies that the current lags voltage by $\frac{45}{360}$ of a full cycle (where a full cycle lasts 20 ms in systems with a frequency of 50 Hz), hence the current peaks 2.5 ms after the voltage. This is presented in figure B.5.

An electric power system can be represented as a graph, where the nodes of the graph correspond to buses and the arcs of the graph correspond to passive network elements. Buses are locations in the power system where active network elements are connected, or junctions where multiple elements meet. If $m, n$ are indices for two buses, then $(m, n)$ indexes a line that connects bus $m$ with bus $n$. Power consumption along a branch $(m, n)$ of an electric circuit is computed as the product of the voltage along the branch and the conjugate of the current along the branch (see figure B.4):

$$S_{mn} = P_{mn} + Q_{mn}i = V_{mn}I_{mn}^{\star}.$$

Here, $S_{mn}$ is the **apparent power** that is consumed on the line, $P_{mn}$ is the **real power** consumed on the line, and $Q_{mn}$ is the **reactive power** consumption along the line. This is indicated in figure B.4.

Passive electrical elements can be classified among resistors, inductors, capacitors, and their combinations. This classification is based on the admittance of the passive element, $Y = G + Bi$, where $G$ is the **conductance**, and $B$ is the **susceptance**. Resistors, such as wires, have positive conductance ($G > 0, B = 0$), and consume real power ($P_{mn} > 0$). Inductors, such as coils, have negative susceptance ($B < 0, G = 0$), and consume reactive power $Q_{mn} > 0$. Capacitors have positive susceptance ($G = 0, B > 0$) and produce reactive power. Typically, transmission lines and transformers in power systems are reactive (i.e. $B < 0$) and slightly resistive (i.e. $G > 0$ but $G << |B|$).

## B.3   The power flow equations

The definition of impedance describes how current determines voltage across passive elements, and vice versa. This relation between currents and voltages, along with Kirchhoff's laws, can be used to derive the voltages and currents of an electrical circuit. **Kirchhoff's current law** states that the total amount of current flowing into a node equals the total amount of current flowing out of a node, i.e. current cannot accumulate. **Kirchhoff's voltage law** states that the accumulated voltage change across any loop of an electrical circuit equals zero. These laws create a set of equations among currents and voltages on the network that imply that we cannot independently ship power from point $m$ to point $n$, since this will cause power to also flow on other lines of the network. This feature differentiates electricity transmission systems from transmission systems of other goods (e.g. railroads and highways) and has significant economic implications.

Consider a power network with $N + 1$ buses, numbered $0, \dots, N$, as in figure B.6. Choose bus 0 as the reference bus (ground bus). Each bus and the ground bus constitute a port. The inputs of the ports can have active elements, whereas the outputs of the ports, characterized by their admittance, correspond to passive elements. We therefore have $N$ ports in the network, and this topology can represent any circuit with active and passive elements. Let the reference directions for bus

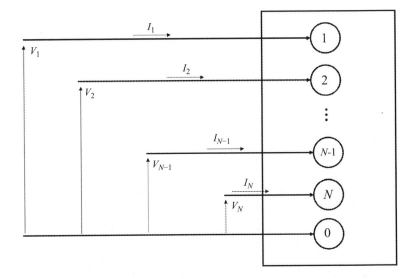

$$S_m = V_m I_m^* = P_m + Q_m i, \; m = 1, \dots, N$$

**Figure B.6** $N$-port transmission network and the *production* of apparent power at a node $m$.

currents be chosen as in figure B.6. Denote the bus currents as $I_{bus} = (I_1, I_2, \ldots, I_N)^T$, bus voltages as $V_{bus} = (V_1, V_2, \ldots, V_N)^T$, and $Y_{bus} \in \mathbb{C}^{N \times N}$ as the **admittance matrix** of the network. Each entry of the matrix is computed as follows: (i) off-diagonal elements $Y_{mn}$ are equal to the negative of the admittance between bus $m$ and bus $n$ for $m \neq n$ and (ii) the diagonal entries $Y_{mm}$ are equal to the sum of the admittance between node $m$ and the ground plus the admittance between node $m$ and all of its adjacent nodes. Kirchhoff's current law can be used to derive the **performance equations** of the circuit in admittance form:

$$I_{bus} = Y_{bus} V_{bus}.$$

These equations describe the amount of current that will flow into each port when the voltages $V_{bus}$ are applied at the terminals of each port. At the $m$th bus we have

$$I_m = \sum_{n=1}^{N} Y_{mn} V_n.$$

Conjugating this equation, we get

$$S_m = V_m I_m^\star = V_m \sum_{n=1}^{N} Y_{mn}^\star V_n^\star.$$

Separating into real and imaginary parts, we get

$$P_m = Re(V_m I_m^\star) = Re\left( V_m \sum_{n=1}^{N} Y_{mn}^\star V_n^\star \right), \tag{B.1}$$

$$Q_m = Im(V_m I_m^\star) = Im\left( V_m \sum_{n=1}^{N} Y_{mn}^\star V_n^\star \right). \tag{B.2}$$

In terms of polar coordinates, these equations can be expressed as

$$P_m = |V_m| \sum_{n=1}^{N} |V_n|(G_{mn} \cos(\theta_{mn}) + B_{mn} \sin(\theta_{mn})), \tag{B.3}$$

$$Q_m = |V_m| \sum_{n=1}^{N} |V_n|(G_{mn} \sin(\theta_{mn}) - B_{mn} \cos(\theta_{mn})), \tag{B.4}$$

where $Y_{mn} = G_{mn} + B_{mn}i$ and $\theta_{mn}$ is the phase angle difference of voltages $V_m$ and $V_n$. These equations are called the **power flow equations**.

There are $N$ complex equations, thus $2 \cdot N$ real power flow equations for a system with $N$ buses. There are $2 \cdot N$ complex variables (complex power and voltage at each node), which translate to $4 \cdot N$ real variables. In a power flow problem, $2 \cdot N$ of these variables are typically fixed according to the following rule:

- There is a unique **swing bus** or **slack bus**, in which the voltage magnitude and voltage phase are fixed (with the phase commonly being set to zero).
- There are $M$ **load buses** or **P-Q buses**, for which the real and reactive power withdrawal are fixed.

**Table B.1** The power flow problem can be decomposed to a system of $N+M-1$ strongly coupled equations (indicated in dark gray) and $N-M+1$ equations (indicated in light gray) that imply the remaining variables by substitution, once the strongly coupled system is solved.

| | $P$ | $Q$ | $|V|$ | $\theta$ | How many |
|---|---|---|---|---|---|
| Slack bus (generator) | | | √ | √ | 1 |
| PV buses (generators) | √ | | √ | | $N-M-1$ |
| PQ buses (loads) | √ | √ | | | $M$ |

- There are $N-M-1$ **production buses** or **P-V buses**, for which the real power and the voltage magnitude are fixed.

This classification of buses allows us to partition the power flow problem into a part that is strongly coupled, and amounts to a system of $N+M-1$ real variables in $N+M-1$ real constraints,[4] and a remaining system that is trivial to solve, by substitution, once the strongly coupled system is solved. This structure is indicated in table B.1. The rows of the table indicate the three different categories of buses in a power flow problem. Columns $P$, $Q$, $V$, and $\theta$ are marked with a tick for each row if the corresponding variable of that bus type is already known. For instance, in a slack bus the absolute value of the bus voltage and the angle of the bus are given. The last column indicates the number of power flow equations corresponding to that specific bus type. The dark gray boxes indicate a system of $N+M-1$ equations that constitute the coupled part of the problem. The light gray part of the table indicates equations that can be easily resolved once the dark gray system has been solved, and essentially amount to substitutions. For instance, once the dark gray system is solved, it is easy to compute the real and reactive power of the slack bus by substituting the variables that are already known into the power flow equations (equations (B.3) and (B.4)) of the slack bus. Thus, the nontrivial part of the power flow equations essentially amounts to a system of $N+M-1$ real nonlinear equations in $N+M-1$ real unknowns.

**Example B.5** *A two-node AC system.* Consider the system of figure B.7. All values in the figure are **per unit**.[5] Node 2 is a PQ bus, while node 1 is the slack bus of the system. The admittance matrix of the system can be expressed as follows:

---

[4] This system is obtained by expressing equation (B.3) or (B.4) of a given bus type, as long as the corresponding real or reactive power, respectively, is given. Specifically, the strongly coupled system consists of (i) $N-M-1$ equations (B.3) since only real power is known in the $N-M-1$ PV buses, (ii) $M$ equations (B.3), since real power is known in PQ buses, and (iii) $M$ equations (B.4), since reactive power is known in PQ buses.

[5] Per unit values are normalized and unitless values for power, impedance, conductance, voltage, and current. They are used in power systems analysis in order to cope with the fact that transformers imply a different impedance value on each different side of their winding, and the fact that different physical quantities are measured in units with different orders of magnitude. The normalization that is used in per unit conversions results in impedance values that are invariant and converts all values to comparable

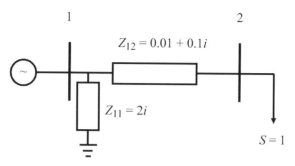

**Figure B.7** The two-node system of example B.5. All values in the figure are given per unit.

$$Y_{\text{bus}} = \begin{bmatrix} \frac{1}{2i} + \frac{1}{0.01+0.1i} & -\frac{1}{0.01+0.1i} \\ -\frac{1}{0.01+0.1i} & \frac{1}{0.01+0.1i} \end{bmatrix} = \begin{bmatrix} 0.99 - 10.4i & -0.99 + 9.9i \\ -0.99 + 9.9i & 0.99 - 9.9i \end{bmatrix}.$$

The power flow equations are as follows:

$$P_1 = G_{11} + |V_2|(G_{12}\cos(-\theta_2) + B_{12}\sin(-\theta_2))$$
$$= 0.99 + |V_2|(-0.99\cos(-\theta_2) + 9.9\sin(-\theta_2)),$$
$$Q_1 = -B_{11} + |V_2|(G_{12}\sin(-\theta_2) - B_{12}\cos(-\theta_2))$$
$$= 10.4 + |V_2|(-0.99\sin(-\theta_2) - 9.9\cos(-\theta_2)),$$
$$P_2 = |V_2|^2 G_{22} + |V_2|(G_{21}\cos(\theta_2) + B_{21}\sin(\theta_2))$$
$$\Rightarrow -1 = 0.99|V_2|^2 + |V_2|(-0.99\cos(\theta_2) + 9.9\sin(\theta_2)),$$
$$Q_2 = |V_2|(G_{21}\sin(\theta_2) - B_{21}\cos(\theta_2)) - |V_2|^2 B_{22}$$
$$\Rightarrow 0 = |V_2|(-0.99\sin(\theta_2) - 9.9\cos(\theta_2)) + 9.9|V_2|^2.$$

Note that this is a system of four nonlinear equations in four unknowns (the unknowns being $P_1$, $Q_1$, $|V_2|$, and $\theta_2$). The solution of this nonlinear system (see problem B.6) is as follows:

$$P_1 = 1.010,$$
$$Q_1 = 0.603,$$
$$|V_2| = 0.985,$$
$$\theta_2 = -5.83°.$$

The resulting current is computed as:

$$I_{12} = (V_1 - V_2)/Z_{12} = 1.016\angle - 5.83°.$$

Note that the current $I_{12}$ is parallel to the voltage at bus 2, $V_2$, which is anticipated since the load in bus 2 is consuming only active power. The voltage and current phasors are depicted graphically in figure B.8. Note that the phase angle of bus 2

orders of magnitude. This results in numerical stability, enables error-checking, and facilitates the interpretation of the results of power system analysis.

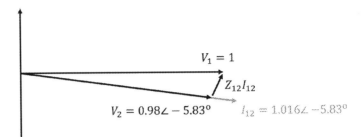

**Figure B.8** The current and voltage phasors of example B.5. Note that the voltage of bus 1 leads the voltage of phase 2, i.e. $\theta_1 - \theta_2 > 0$.

leads the phase angle of bus 1. This is related to the shipment of real power from bus 1 to bus 2, as becomes apparent later when we linearize the power flow equations. Geometrically, it follows from the fact that the angle $Z_{12}$ is relatively close to $90°$, since the resistance $Z_{12}$ is almost purely reactive, thus $Z_{12}I_{12}$ is a vector that is almost vertical to $V_2$, as indicated in figure B.7. Since $V_1$ is the sum of $V_2$ and $Z_{12}I_{12}$, $V_1$ leads $V_2$. The real power losses in our example, which we return to in section B.6, are equal to $P_1 + P_2 = 1.010 - 1 = 0.010$.

## B.4    From power injections to bus angles

In this and the following section, we demonstrate how net injections of active power imply real power flows on each line of the network in the linearized power flow model. This result is used extensively in electricity market models. This amounts to the so-called power transfer distribution factors. We develop the result in two phases:

- In this section we demonstrate that net injections imply voltage phase angles for each bus of the network.
- In the following section we demonstrate that voltage phase angles imply flows on lines.

Combining these two results implies that net injections of real power at each bus of the network imply line flows along each line of the network, and the linear mapping is what amounts to the power transfer distribution factors. The point is illustrated in figure B.1.

Before proceeding with the formal development, we pause briefly to reflect at the result by juxtaposing it against the "opposite" mapping: from line flows to net injections. It is intuitive to see that net injections in a general transportation network are typically lower-dimensional than flows along lines. The most "slim" of connected networks (in the sense of having the fewest lines) with $N$ nodes is a tree (also referred to as a radial network), which is a network with $N - 1$ lines. More general networks (and ones with richer structure) have a number of lines that are at least equal to the number of nodes. Flows thus constitute a higher-dimensional vector than injections at nodes. And if flows are given, one can infer net injections from the conservation constraint of each node: since injections equal the net of outgoing flows minus incoming flows, if flows are given then net injections can be inferred. Thus, flows imply injections, and thus pack more information. The converse is typically not true in general transportation networks: injections do not imply flows in a unique way. But it is true in the case of linearized power flow, and the point of this and the following section is to demonstrate this formally.

The following approximations are often acceptable in the power flow equations:

- Line resistance is negligible: $G_{mn} = 0$.
- The phase angle differences across the branches $\theta_{mn} = \theta_m - \theta_n$ are small, so that $\sin(\theta_{mn}) \simeq \theta_{mn}$ and $\cos(\theta_{mn}) \simeq 1$.
- The voltage magnitude on each bus is approximately equal to nominal: $|V_m| \simeq 1$ (where we assume that the power flow equations are expressed in the per unit system).

This results in the **linearized power flow equations**. These are also referred to as the **direct current power flow equations**, or the lossless DC power flow model. The qualifier lossless is due to the fact that these equations neglect thermal losses on lines, since it is assumed that the resistance of the passive elements is zero.

A consequence of these simplifying assumptions is that reactive power flows are ignored, and real power injection in each port is approximated by

$$P_m = \sum_{n=1}^{N} B_{mn}(\theta_m - \theta_n). \tag{B.5}$$

---

**Example B.6** *Two-node DC power flow.* We return to example B.5 and consider the linearized power flow, which can be expressed as follows:

$$P_1 = B_{12} \cdot (\theta_1 - \theta_2) = 9.9 \cdot (0 - \theta_2),$$
$$P_2 = B_{21}(\theta_2 - \theta_1) \Rightarrow -1 = 9.9(\theta_2 - 0).$$

Note that, compared to the system of example B.5, we now have two equations (the power flow equations related to reactive power flows have been dropped) in two unknowns ($P_1$ and $\theta_2$). This system has the following solution:

$$P_1 = 1,$$
$$\theta_2 = -5.79°.$$

The solution is reasonably close to that of example B.5 ($-5.83°$), and there are no losses in the network: $P_1 + P_2 = 0$.

---

From a physical standpoint, this linearization ignores any shunt passive elements (since the real part of all elements is ignored, and the reactive part of shunt elements is only implicated in the reactive power flow equations, which are ignored in the linearized power flow). Furthermore, the linearization ignores the real part of any series passive element.

For passive elements that are purely reactive, we know that $Y_{mn} = -y_{mn} = -\frac{1}{X_{mn}i} = iX_{mn}^{-1}$, where $X_{mn}$ is the **reactance** of line $mn$. And since $Y_{mn} = B_{mn}i$, we conclude that $B_{mn} = X_{mn}^{-1}$. Thus, the linearized power flow can be approximated in terms of the reactance of each series line in the network as follows:

$$P_m = \sum_{n=1, n \neq m}^{N} \frac{1}{X_{mn}} \theta_m - \sum_{n=1, n \neq m}^{N} \frac{1}{X_{mn}} \theta_n. \tag{B.6}$$

---

**Example B.7** *Direct derivation of the DC power flow equations for the two-node example.* Returning to example B.5, we can approximate the linearized power flow equations directly without computing the admittance matrix $Y_{bus}$ of the system:

$$P_1 = \frac{\theta_1 - \theta_2}{X_{12}} = \frac{0 - \theta_2}{0.1},$$

$$P_2 = \frac{\theta_2 - \theta_1}{X_{12}} \Rightarrow -1 = \frac{\theta_2 - 0}{0.1}.$$

This highlights why the shipment of power from node 1 to node 2 over a reactive line results in voltage at node 2 lagging voltage at node 1, as indicated in figure B.8. The solution of this system is:

$$P_1 = 1,$$
$$\theta_2 = -5.73°.$$

The result for $\theta_2$ is less accurate than in example B.6 (where $\theta_2$ is $-5.79°$), but still reasonably close to the AC solution (where $\theta_2$ is $-5.83°$). As in the case of example B.6, there are no losses in the network: $P_1 + P_2 = 0$.

---

Note that power injections $P_m$ are linearly dependent on bus angles, $\theta_m$, and the dependence is described by a matrix that is the weighted Laplacian of the graph of the electric network, where the weights on the lines are given by $B_{mn}$. Denote $T^F = (T_{mn}), m, n \in \{1, \ldots, N\}$, where $T^F$ is an $N \times N$ matrix the elements of which are

$$T_{mn}^F = \begin{cases} -\frac{1}{X_{mn}}, & (m, n) \in A, m \neq n, \\ \sum_{n'=1, n' \neq m}^{N} \frac{1}{X_{mn'}}, & m = n, \\ 0, & (m, n) \notin A. \end{cases} \tag{B.7}$$

Here, $A$ is the set of arcs in the network. It is easy to see (and it becomes relevant later) that $T^F$ is a symmetric matrix.

The linear mapping from phase angles $\theta^F \in \mathbb{R}^N$ to net injections of real power $P^F \in \mathbb{R}^N$ at every node can be described as:

$$P^F = T^F \theta^F. \tag{B.8}$$

Proposition B.1 guarantees that, if the graph is connected, then the Laplacian $T^F$ has rank $N - 1$. This implies that there exist $N - 1$ constraints that capture the full information of the linear mapping, and thus one constraint can be dropped. It is interesting that any $N-1$ choice will suffice, meaning that keeping *any* $N-1$ equalities of the system maintains the full information of the system of $N$ equations. To see why this is the case, note that energy conservation holds (see problem B.4) for the linear system of equations (B.6):

$$\sum_{m=1}^{N} P_m = 0. \tag{B.9}$$

This means that any row can be dropped, since the $i$th row of the matrix $T^F$ can be obtained as minus the sum of the other rows of the matrix. Since $T^F = \left(T^F\right)^T$, this also implies that the $i$th column can be obtained as minus the sum of the other columns. We can thus pick the $(N-1) \times (N-1)$ submatrix of $T^F$, which we denote as $T$, and which consists of dropping the $i$th row and the $i$th column, and still maintain all the information of the original system of equations (B.8). Since the rank of the remaining matrix is $N - 1$, this remaining matrix $T$ is invertible, and creates a one-to-one mapping from the $N - 1$ bus angles to the $N - 1$ power injections. The remaining node is the **hub node**, which corresponds to the swing bus or slack bus defined in the previous section: its bus angle is set to zero, $\theta_h = 0$, and its net power injection is given by the following equation:

$$P_h = - \sum_{n \in \{1, \dots, N\} - \{h\}} P_n.$$

This implies that the power flow equations can be rewritten by ignoring the phase angle of the hub node (which is fixed at zero) and the power injection of the hub node (since it can be derived from the power injections of all other buses). This results in the following $(N - 1) \times (N - 1)$ system of equations:

$$P = T\theta, \tag{B.10}$$

where $P = (P_m), m \in \{1, \dots, N\} - \{h\}$ and $\theta = (\theta_m), m \in \{1, \dots, N\} - \{h\}$.

## B.5    From bus angles to line flows

Having seen that power injections $P \in \mathbb{R}^{N-1}$ in nodes $\{1, \dots, N\} - \{h\}$ of the network uniquely define bus angles $\theta \in \mathbb{R}^{N-1}$ in nodes $\{1, \dots, N\} - \{h\}$, it is now shown that these in turn uniquely define power flows on lines.

**Proposition B.2**   *Power flow on line $(m,n)$ is given as*

$$P_{mn} = \frac{\theta_m - \theta_n}{X_{mn}}. \tag{B.11}$$

*Proof*   The argument is identical to the arguments that lead to equation (B.6). In fact, equation (B.6) states that the net injection of power in a given node is equal to the sum of outgoing flows to adjacent nodes, with each flow being equal to $B_{mn}(\theta_m - \theta_n) = \frac{1}{X_{mn}}(\theta_m - \theta_n)$.    □

Since power injections determine bus angles, which in turn determine power flows along lines, the first step towards deriving a mapping from power injections to line flows is to define the matrix $M$ as $M = (M_{kn}), k \in A, n \in \{1, \ldots, N\} - \{h\}$, where

$$M_{kn} = \begin{cases} \frac{1}{X_k}, & \text{if } k = (n, \cdot), \ n \neq h, \\ -\frac{1}{X_k}, & \text{if } k = (\cdot, n), \ n \neq h, \\ 0, & \text{otherwise.} \end{cases} \tag{B.12}$$

By the definition of $M$, and from equation (B.11), it follows that

$$P_L = M\theta, \tag{B.13}$$

where $P_L$ is the vector of power flows along the lines of the network. From equation (B.10), it follows that

$$P_L = MT^{-1}P. \tag{B.14}$$

This is the desired mapping from net injections $P$ to line flows $P_L$. Power transfer distribution factors summarize this information: given a line $k$ and a bus $n$, the **power transfer distribution factor (PTDF)** is the amount of power flow induced on line $k$ by a transfer of 1 MW of power from bus $n$ to the hub node. The value of a PTDF is therefore dependent on the choice of hub node. The PTDF of bus $n$ on line $k$ (denoted as $F_{kn}$) is obtained as:

$$F_{kn} = M_k^T \left(T^{-1}\right)_n,$$

where $M_k^T$ is the $k$th row of $M$, and $\left(T^{-1}\right)_n$ is the $n$th column of the matrix $T^{-1}$.

The modeling of DC power flow using PTDFs $(F_{kn})$ is thus equivalent to using the reactance of lines $(X_{mn})$. The representation of the power flow equations through PTDFs is most useful when the goal is to study market interactions because they describe the impact of the actions of market agents on a constrained resource, namely the capacity of a line. This representation is used most frequently throughout the textbook. The representation of the network through reactance can be advantageous when the topology of the network changes (e.g. due to line failures or line switching).

---

**Example B.8** *Computation of PTDFs on a four-node network.* Consider the network of figure B.9, with the indicated reactance $X_{mn}$ and the indicated power injections and withdrawals, where 0 is the hub node. Compute the PTDF from bus 1 to line 2-3.

**Solution**
The matrix $T$ is given by:

$$T = \begin{bmatrix} \frac{1}{1} + \frac{1}{1.5} & -\frac{1}{1} & 0 \\ -\frac{1}{1} & \frac{1}{1} + \frac{1}{2.5} + \frac{1}{2} & -\frac{1}{2} \\ 0 & -\frac{1}{2} & \frac{1}{2} + \frac{1}{3} \end{bmatrix}.$$

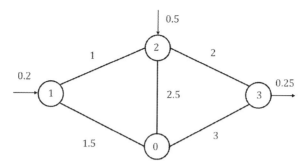

**Figure B.9** The network of example B.8.

Then

$$T^{-1} = \begin{bmatrix} 0.96 & 0.6 & 0.36 \\ 0.6 & 1.0 & 0.6 \\ 0.36 & 0.6 & 1.56 \end{bmatrix}.$$

The power injection vector is $P = (0.2, 0.5, -0.25)^T$. The bus angles are:

$$\theta = T^{-1}P = \begin{bmatrix} 0.96 & 0.6 & 0.36 \\ 0.6 & 1.0 & 0.6 \\ 0.36 & 0.6 & 1.56 \end{bmatrix} \begin{bmatrix} 0.2 \\ 0.5 \\ -0.25 \end{bmatrix} = \begin{bmatrix} 0.402 \\ 0.470 \\ -0.018 \end{bmatrix}.$$

The power flow on each line is

$$P_{12} = \frac{\theta_1 - \theta_2}{X_{12}} = \frac{0.402 - 0.47}{1} = -0.068,$$

$$P_{10} = \frac{\theta_1 - \theta_0}{X_{10}} = \frac{0.402 - 0}{1.5} = 0.268,$$

$$P_{23} = \frac{\theta_2 - \theta_3}{X_{23}} = \frac{0.47 - (-0.018)}{2} = 0.244,$$

$$P_{20} = \frac{\theta_2 - \theta_0}{X_{20}} = \frac{0.47 - 0}{2.5} = 0.188,$$

$$P_{30} = \frac{\theta_3 - \theta_0}{X_{30}} = \frac{-0.018 - 0}{3} = -0.006.$$

Denoting $M_k^T \in \mathbb{R}^3$ as the row of matrix $M$ that corresponds to line $k$, the matrix $M$ that determines line flows as a function of bus angles, $P_L = M\theta$, is

$$M = \begin{bmatrix} M_{01}^T \\ M_{12}^T \\ M_{23}^T \\ M_{02}^T \\ M_{03}^T \end{bmatrix} = \begin{bmatrix} -\frac{1}{1.5} & 0 & 0 \\ \frac{1}{1} & -\frac{1}{1} & 0 \\ 0 & \frac{1}{2} & -\frac{1}{2} \\ 0 & -\frac{1}{2.5} & 0 \\ 0 & 0 & -\frac{1}{3} \end{bmatrix}.$$

The change of the flow on line 2-3 associated with a unit power injection in bus 1 is

$$F_{23,1} = M_{23}^T \left( T^{-1} \right)_1 = (0, 0.5, -0.5)(0.96, 0.6, 0.36)^T = 0.12.$$

**Figure B.10** The symmetric network of example B.9.

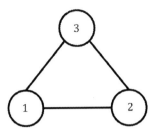

**Example B.9** *PTDFs of a symmetric triangular network.* Consider the triangular symmetric network of figure B.10, where $X$ is the reactance of each line. Compute the PTDF matrix of the network when node 3 is the hub node.

**Solution**

The matrix $T$ is given as:

$$T = \begin{bmatrix} \frac{1}{X} + \frac{1}{X} & -\frac{1}{X} \\ -\frac{1}{X} & \frac{1}{X} + \frac{1}{X} \end{bmatrix} = \frac{1}{X}\begin{bmatrix} 2 & -1 \\ -1 & 2 \end{bmatrix}.$$

The matrix $T^{-1}$ is then

$$T^{-1} = X\begin{bmatrix} 0.667 & 0.333 \\ 0.333 & 0.667 \end{bmatrix},$$

and

$$M = \begin{bmatrix} M_{12}^T \\ M_{23}^T \\ M_{13}^T \end{bmatrix} = \begin{bmatrix} \frac{1}{X} & -\frac{1}{X} \\ 0 & \frac{1}{X} \\ \frac{1}{X} & 0 \end{bmatrix}.$$

Finally,

$$F_{12,1} = \left(\frac{1}{X}, -\frac{1}{X}\right) \cdot (0.667X, 0.333X)^T = 0.333,$$

$$F_{13,1} = \left(\frac{1}{X}, 0\right) \cdot (0.667X, 0.333X)^T = 0.667,$$

$$F_{23,1} = \left(0, \frac{1}{X}\right) \cdot (0.667X, 0.333X)^T = 0.333,$$

$$F_{12,2} = \left(\frac{1}{X}, -\frac{1}{X}\right) \cdot (0.333X, 0.667X)^T = -0.333,$$

$$F_{13,2} = \left(\frac{1}{X}, 0\right) \cdot (0.333X, 0.667X)^T = 0.333,$$

$$F_{23,2} = \left(0, \frac{1}{X}\right) \cdot (0.333X, 0.667X)^T = 0.667.$$

When injecting a unit of power from node 1 to the hub node, node 3, there are two paths to the hub. The first path is the "electrically short" path, 1-3, which has a reactance of $X$. The "electrically long" path goes through 1-2-3 and encounters double the reactance, $2 \cdot X$. The physical intuition is that current splits in a way that is inversely proportional to reactance. Two thirds of the current flow along the "electrically short" path, and one third flows along the "electrically long" path.

---

## B.6    Losses

Losses are commonly represented in electricity market models, and affect the formation of market clearing prices. It is common practice to represent losses as a linear function of net power injections at every node of the system.

Complex power losses along a line $k = (m, n)$ can be computed as:

$$S_k = V_m I_{mn}^\star + V_n I_{nm}^\star = V_m I_{mn}^\star - V_n I_{mn}^\star = (V_m - V_n) I_{mn}^\star$$
$$= (V_m - V_n) y_{mn}^\star (V_m - V_n)^\star = |V_m - V_n|^2 y_{mn}^\star.$$

We now derive $|V_m - V_n|^2$ as a function of voltage magnitudes and phase angles:

$$V_m - V_n = |V_m| \cos(\theta_m) + i |V_m| \sin(\theta_m) - |V_n| \cos(\theta_n) - i |V_n| \sin(\theta_n)$$
$$\Rightarrow |V_m - V_n|^2 = (|V_m| \cos(\theta_m) - |V_n| \cos(\theta_n))^2 + (|V_m| \sin(\theta_m) - |V_n| \sin(\theta_n))^2$$
$$= |V_m|^2 \cos^2(\theta_m) + |V_n|^2 \cos^2(\theta_n) - 2|V_m||V_n| \cos(\theta_m) \cos(\theta_n) +$$
$$|V_m|^2 \sin^2(\theta_m) + |V_n|^2 \sin^2(\theta_n) - 2|V_m||V_n| \sin(\theta_m) \sin(\theta_n)$$
$$= |V_m|^2 + |V_n|^2 - 2|V_m||V_n| \cos(\theta_m) \cos(\theta_n) - 2|V_m||V_n| \sin(\theta_m) \sin(\theta_n)$$
$$= |V_m|^2 + |V_n|^2 - 2|V_m||V_n| \cos(\theta_m - \theta_n).$$

Isolating the real part from $S_k$, we obtain real power losses as:

$$L_k = g_{mn} \left( |V_m|^2 + |V_n|^2 - 2|V_m||V_n| \cos(\theta_m - \theta_n) \right).$$

We can then replace $\cos(\theta_m - \theta_n)$ by its second-order Taylor approximation, $\cos(\theta_m - \theta_n) \simeq 1 - \frac{(\theta_m - \theta_n)^2}{2}$, to arrive to the following approximation of losses:

$$L_k \simeq g_{mn} \left( |V_m|^2 + |V_n|^2 - 2|V_m||V_n| + |V_m||V_n|(\theta_m - \theta_n)^2 \right).$$

Recalling that, in linearized power flow, $|V_m| \simeq |V_n| \simeq 1$, it follows that:

$$L_k \simeq g_{mn} (\theta_m - \theta_n)^2.$$

Further recalling that, according to the assumptions of the linearized power flow, $g_{mn} = \frac{R_{mn}}{R_{mn}^2 + X_{mn}^2} \simeq \frac{R_{mn}}{X_{mn}^2}$ and $(\theta_m - \theta_n)^2 = P_{mn}^2 X_{mn}^2$, we have that:

$$L_k \simeq R_{mn} P_{mn}^2.$$

---

**Example B.10** *Approximation of losses.* Returning to example B.7, we can approximate losses as

$$L_{12} = R_{12} \cdot P_{12}^2 = 0.01 \cdot 1^2 = 0.01,$$

where we use the value of linearized power flow $P_{12}$, as computed in example B.7. Comparing to the exact result of example B.5, where losses are computed to be equal to 0.010, we note that the approximation is exact up to the third digit.

---

Having derived a relation between losses and line flows, we are now interested in deriving a linear relation between losses and net power injection, which can be used in a market clearing model. We use a first-order Taylor expansion to approximate the square of flows on line $k$, $P_k^2$:

$$P_k^2 \simeq \bar{P}_k^2 + 2 \cdot \bar{P}_k \cdot (P_k - \bar{P}_k) = -\bar{P}_k^2 + 2 \cdot \bar{P}_k \cdot P_k,$$

where $\bar{P}_k$ is the base flow at the point where the system is currently dispatched. Noting further that $P_k = \sum_{n \in N} PTDF_{kn} \cdot P_n$, we finally arrive at our target approximation of losses, as a linear function of net injections $P_n$:

$$L = \sum_{k \in K} L_k = \sum_{k \in K} R_k \cdot \left( -\bar{P}_k^2 + 2 \cdot \bar{P}_k \cdot \sum_{n \in N} PTDF_{k,n} \cdot P_n \right).$$

## Problems

**B.1** *Laplacian of a complete graph.* A **complete graph** is a graph where each node is connected to each other node via an edge. Derive the Laplacian matrix of a complete graph.

**B.2** *Rank of the Laplacian of a connected graph.* Prove that the weighted Laplacian of a connected graph $G = (N, E)$ has rank $|N| - 1$.

**B.3** *Power flow equations in polar coordinates.* Derive equations (B.3) and (B.4) from equations (B.1) and (B.2).

**B.4** *Conservation of energy in linearized power flow.* Prove equation (B.9).

**B.5** *Conductance of a highly inductive passive element.* True or false: if a passive element is highly inductive ($R << X$), i.e. its resistance is negligible, then its conductance is very large.

**B.6** *Two-node AC power flow.* Implement a mathematical programming code that solves the power flow of example B.5.

## Bibliography

**Section B.2**   Part of the material in this section is inspired by Glover, Sarma, and Overbye (2012).

**Section B.3**   Part of the material in this section is inspired by Glover, Sarma, & Overbye (2012). Table B.1 is inspired by the discussion of the power flow equations in Vournas and Kontaxis (2010). The B-theta linearization of the power flow equations, as well as a multitude of other linear and other convex approximations and relaxations of the power flow equations, are presented in Taylor (2015).

**Section B.4**    The material is inspired by Singh (2006), as well as the ECE427 course notes of Professor Fernando Alvarado at the University of Wisconsin on PTDFs, linearized power flows, and DC power flows.

**Section B.5**    The material in Section B.3 is inspired by Singh (2006), as well as the ECE427 course notes of Professor Fernando Alvarado at the University of Wisconsin on PTDFs, linearized power flows, and DC power flows.

**Section B.6**    Losses and their interactions with the formation of market prices are discussed in Schweppe et al. (1988) and Hogan (1992). The linearized model presented in this section is based on section 2.3 of Eldridge et al. (2016).

# APPENDIX C

# Solutions to problems

**Chapter 1**

**Problem 1.1** The optimal capacities are as follows:

- DR: 502.4 MW
- Oil: 952.7 MW
- Gas: 1416.5 MW
- Coal: 2853.3 MW
- Nuclear: 7354.7 MW

**Problem 1.2** The code is available on the textbook website. The optimal capacities are as follows:

- DR: 502.5 MW
- Oil: 953.5 MW
- Gas: 1417.3 MW
- Coal: 2852.6 MW
- Nuclear: 7353.8 MW

Deviations from the solution of problem 1.1 can be attributed to rounding effects, since the approach which is based on screening curves is based on rounding frequencies to hours.

**Problem 1.3** The code is available on the textbook website.

**Problem 1.4** The correct answer is option 3.

**Problem 1.5** The screening curve is $L(f) = 1 - G(f)$.

**Problem 1.6** False.

**Chapter 2**

**Problem 2.1** This is proven on p. 244 of Boyd and Vandenberghe (2008).

**Problem 2.2** First question: option 1. Second question: option 1.

**Problem 2.3** The KKT conditions are:

$$y = 1,$$
$$0 \le \lambda_1 \perp x \ge 0,$$
$$0 \le \lambda_2 \perp 2 - x \ge 0,$$
$$-1 + \lambda_1 + \lambda_2 = 0,$$
$$-2 - \mu = 0.$$

The primal-dual optimal pair is $x = 2$, $y = 1$, $\lambda_1 = 0$, $\lambda_2 = 1$, $\mu = -2$.

**Problem 2.4** The idea is to show that

$$g(u_0, v_0) - f(x_0)^T(u - u_0) - h(x_0)^T(v - v_0) \leq g(u, v).$$

This can be done by replacing $g(u_0, v_0)$ by its definition, and noting that $x_0$ maximizes the Lagrangian function for $(u_0, v_0)$ at $x_0$.

**Problem 2.5** The argument is identical to that of the proof of proposition 2.5, where $Ax + By \leq b$ is replaced by $g(x, y) \leq 0$. The generalized conditions are:

$$Cx + Dy - d = 0,$$
$$0 \leq \lambda \perp g(x, y) \leq 0,$$
$$0 \leq x \perp \lambda^T \nabla_x g(x, y) + \mu^T C - \nabla_x f(x, y)^T \geq 0,$$
$$\lambda^T \nabla_y g(x, y) + \mu^T D - \nabla_y f(x, y)^T = 0.$$

**Problem 2.6** The KKT conditions of this problem are as follows:

$$0 \leq q \perp MC + \mu - \pi \geq 0,$$
$$0 \leq x \perp IC - \mu \geq 0.$$

The subgradient of this cost with respect to produced power $p$ is precisely the dual multiplier $\pi$, according to proposition 2.8. As long as $p > 0$, then $q > 0$, which implies that $x > 0$. The fact that $x > 0$ implies, from the second KKT condition, that $\mu = IC$. And, as long as $q > 0$, the first condition implies that $\lambda = \mu + MC = IC + MC$.

## Chapter 3

**Problem 3.1** The first and second injection patterns are feasible, the third one is not.

**Problem 3.2** The king should bid his true valuation; the argument is identical to that of example 3.12.

**Problem 3.3** The code is available on the textbook website. The cost of the three-period lookahead is $1396.8.

**Problem 3.4** The market clearing price is 85 $/MWh. The offers pay/are paid as follows:

- Offer B1: pays $8500
- Offer B2: pays $0
- Offer S1: is paid $1700
- Offer S2: is paid $4250
- Offer S3: is paid $2550
- Offer S4: is paid $0

**Problem 3.5** False, because there are better incentives for truthful bidding and the inframarginal rents contribute towards covering investment costs.

**Problem 3.6** False, because there is a single price, and supply equals demand.

**Problem 3.7** The total cost functions for the two- and three-generator case are expressed as follows:

$$TC^{(2)}(Q) =$$
$$\begin{cases} 2 \cdot 5.8 \cdot 200 + 20 \cdot Q + 0.1 \cdot Q^2 \text{ \$/h}, & 0 \text{ MW} \leq Q < 400 \text{ MW}, \\ 2 \cdot 5.8 \cdot 200 + 2 \cdot 12000 + 500 \cdot (Q - 400) \text{ \$/h}, & 400 \text{ MW} \leq Q \leq 440 \text{ MW}, \\ +\infty \text{ \$/h}, & Q > 440 \text{ MW}, \end{cases}$$

$$TC^{(3)}(Q) =$$
$$\begin{cases} 3 \cdot 5.8 \cdot 200 + 20 \cdot Q + \frac{0.2}{3} \cdot Q^2 \text{ \$/h}, & 0 \text{ MW} \leq Q < 600 \text{ MW}, \\ 3 \cdot 5.8 \cdot 200 + 3 \cdot 12000 + \frac{1000}{3} \cdot (Q - 600) \text{ \$/h}, & 600 \text{ MW} \leq Q \leq 660 \text{ MW}, \\ +\infty \text{ \$/h}, & Q > 660 \text{ MW}, \end{cases}$$

The average cost functions are the total cost functions divided by $Q$.

**Problem 3.8** Denoting the total cost function of the $n$-generator aggregation as $MC^{(n)}$, and using corresponding notation for the other cost functions, we can establish the following:

- $FC^{(n)} = n \cdot FC^{(1)}$
- $MC^{(n)}(n \cdot Q) = MC^{(1)}(Q)$
- $VC^{(n)}(Q) = n \cdot VC^{(1)}(\frac{Q}{n})$

This allows us to prove that, if $Q^\star$ is the point at which $AC^{(n)}$ intersects $MC^{(n)}$, then $Q^\star/n$ is the point at which $AC^{(1)}$ intersects $MC^{(1)}$. And this in turn allows us to show that $AC^{(n)}(Q^\star) = AC^{(1)}(Q^\star/n)$. In order to find $Q^\star$ for the single-generator case, we solve the equation $MC^{(1)}(Q) = AC^{(1)}(Q)$, which is a quadratic equation. Its nonnegative solution is $Q^\star = 76.16$ MW, which we plug back into the expression for marginal cost in order to compute $MC^{(1)}(Q^\star) = 50.46$ \$/h.

**Problem 3.9** The marginal cost range is $[MC, +\infty)$. Thus, the marginal cost range includes the long-run marginal cost $MC + IC$ which is computed in exercise 2.6.

**Problem 3.10** Use the fact that, by definition,

$$OC = \sum_{t=0}^{T} \frac{FC}{(1+r)^t},$$

and standard identities regarding the sum of geometric series:

$$\sum_{t=0}^{T} \frac{1}{(1+r)^t} = \frac{1}{r}\left(1 - \frac{1}{(1+r)^T}\right).$$

## Chapter 4

**Problem 4.1** The proposition generalizes as follows. Given an optimal solution of the economic dispatch problem, there exists a threshold $\lambda$ such that:

- If a generator is operating strictly within its operating limits ($P_g^- < p_g < P_g^+$), then $MC_g(p_g) = \lambda$. If a load is consuming strictly within its dispatch interval ($0 < d_l < D_l$), then $MB_l(d_l) = \lambda$.
- If a generator is producing $P_g^-$, then $MC_g(p_g) \geq \lambda$. If a load is consuming zero, then $MB_l(d_l) \leq \lambda$.
- If a generator is producing at peak capacity, then $MC_g(p_g) \leq \lambda$. If a load is consuming at peak capacity, then $MB_l(d_l) \geq \lambda$.

**Problem 4.2** The code is available on the textbook website.

1. The production cost is \$23600 and the market clearing price is 35 \$/MWh.
2. 1. The production cost is \$23600. The consumer benefit is \$629400. The generator profit is \$24000. The consumer surplus is \$581800. The total welfare is \$605800.
   2. The market clearing price is 250 \$/MWh. There are 1000 MW which are in the money. There are 60 MW which are out of the money. The market clearing price cannot be 249 \$/MWh, nor can it be 251 \$/MWh.

**Problem 4.3** The KKT conditions can be grouped as follows:
Market clearing:

$$\sum_{g \in G} p_g - \sum_{l \in L} d_l = 0.$$

Generator profit maximization conditions (three groups, one for each generator technology):

$$0 \leq p_g \perp -\lambda + MC_g(p_g) + \mu_g \geq 0, g \in G,$$
$$0 \leq \mu_g \perp P_g - p_g \geq 0, g \in G.$$

Load surplus maximization conditions (three groups, one for each utility):

$$0 \leq d_l \perp -MB_l(d_l) + \lambda + v_l \geq 0, l \in L,$$
$$0 \leq v_l \perp D_l - d_l \geq 0, l \in L.$$

The market equilibrium is identical to that computed in problem 4.2, question 1.

**Problem 4.4** The Cournot equilibrium solution for $n$ identical players is

$$p_i = \frac{1}{(n+1) \cdot b}(a - MC_i(p_i)),$$

whereas the perfect competition solution is

$$p_i^{\star} = \frac{1}{n \cdot b}(a - MC_i(p_i)).$$

We thus conclude that

$$\lim_{n \to \infty} \frac{p_i}{p_i^{\star}} = 1.$$

**Problem 4.5**

1. The KKT conditions of the problem are:

$$d_1 - \sum_{g \in G_1} p_g = 0,$$

$$d_2 - \sum_{g \in G_2} p_g = 0,$$

$$0 \le d_1 \perp \lambda_1 - V + \nu \ge 0,$$

$$0 \le d_2 \perp \lambda_2 - V + \nu \ge 0,$$

$$0 \le p_g \perp -\lambda_1 + \mu_g \ge 0, g \in G_1,$$

$$0 \le p_g \perp MC_g - \lambda_2 + \mu_g \ge 0, g \in G_2,$$

$$0 \le \mu_g \perp P_g - p_g \ge 0, g \in G,$$

$$0 \le \nu \perp D - d_1 - d_2 \ge 0.$$

2. If $d_1 > 0$ and $d_2 > 0$, then $\nu = V - \lambda_1 = V - \lambda_2$; thus $\lambda_1 = \lambda_2$. If prices are different, consumers will choose to buy from the market where prices are lower, thus exerting an upward pressure on prices, until prices equalize in both markets.

## Chapter 5

**Problem 5.1** The condition states that the market cannot be in equilibrium unless the price of power in location $n$, $\rho_n$, is equal to the price of power at the hub node, $\phi$, plus the price of transmission rights for shipping the power from the hub node to node $n$, $\sum_{k \in K} F_{kn} \lambda_k^- - \sum_{k \in K} F_{kn} \lambda_k^+$.

**Problem 5.2** The equality is the complementary dual constraint of $f_k$. A conclusion cannot be drawn about the relative value of $\rho_n$ and $\rho_m$ when power flows from $n$ to $m$, because $\gamma_k$ has no sign constraints.

**Problem 5.3** The code is available on the textbook website. A notable aspect of the model is that there is a different LMP for every different location of the network.

**Problem 5.4** The code is available on the textbook website. The main point of this example is to observe that zonal pricing violates the flow limit of line 1–6.

**Problem 5.5** The subdifferential $\partial g(u)$ of a function $g$ at a certain point $u$ is a closed convex set. Moreover, $\rho_3$ is a subgradient of the objective function $g(u)$ of the optimal power flow model with respect to the right-hand side $u$ of the corresponding constraint of the model at $u = 200$.

**Problem 5.6** The KKT conditions of figure 5.3 that implicate the network are equivalent to the following network operator surplus maximization problem:

$$\max_{r,f} \sum_{n \in N} \rho_n \cdot r_n$$

$$(\psi_k): \quad f_k = \sum_{n \in N} F_{kn} \cdot r_n, k \in K$$

$$(\lambda_k^+): \quad f_k \le T_k, k \in K$$

$$(\lambda_k^-): \quad -f_k \le T_k, k \in K.$$

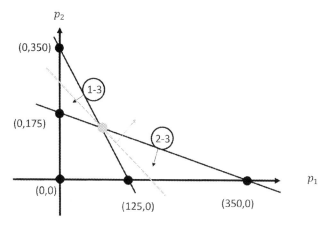

**Figure C.1** The feasible set of problem 5.8. The constraints are indicated in black, while the objective function is indicated in gray. The optimal solution is the gray dot, and is equal to $(50, 150)$.

**Problem 5.7** This is not possible. It can be argued by using the result of problem 5.6, or using the KKT conditions of figure 5.3 to arrive to a contradiction.

**Problem 5.8** The feasible set in the space of variables $(p_1, p_2)$ is presented in figure C.1. The optimal solution is $(p_1, p_2) = (50, 150)$. The increase of demand in node 3 by 1 MW corresponds to the exact same feasible set; the only thing that changes is the constant term in the objective function, which is originally $\min_{p_1, p_2} 40 \cdot p_1 + 80 \cdot p_2 + 140 \cdot (300 - p_1 - p_2)$ and becomes $\min_{p_1, p_2} 40 \cdot p_1 + 80 \cdot p_2 + 140 \cdot (301 - p_1 - p_2)$.

**Problem 5.9** Rewrite the $(ZPT)$ problem as follows:

$$(ZPT): \quad \max_{p,d,r,f} \sum_{l \in L} MB_l \cdot d_l - \sum_{g \in G} MC_g \cdot p_g$$

$$(\rho_z): \quad r_z - \sum_{g \in G_z} p_g + \sum_{l \in L_z} d_l = 0, z \in Z$$

$$(\phi_z): \quad -r_z + \sum_{a=(z,\cdot)} f_a - \sum_{a=(\cdot,z)} f_a = 0, z \in Z$$

$$(\lambda_a^-): \quad -f_a \leq -ATC_a^-, a \in A$$

$$(\lambda_a^+): \quad f_a \leq ATC_a^+, a \in A$$

$$(\mu_g): \quad p_g \leq P_g, g \in G$$

$$(\nu_l): \quad d_l \leq D_l, l \in L$$

$$p_g \geq 0, g \in G$$

$$d_l \geq 0, l \in L.$$

The KKT conditions are given as follows:

$$0 \leq \mu_g \perp P_g - p_g \geq 0, g \in G,$$

$$0 \leq p_g \perp MC_g + \mu_g - \rho_{z(g)} \geq 0, g \in G,$$

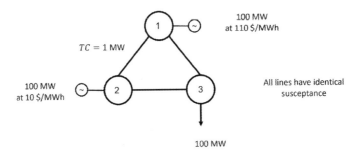

**Figure C.2** The network of problem 5.11 which is used to illustrate that power can flow from a more expensive to a cheaper location.

$$0 \leq \mu_l \perp D_l - d_l \geq 0, l \in L,$$
$$0 \leq d_l \perp -MB_l + v_l + \rho_{z(l)} \geq 0, l \in L,$$
$$0 \leq \lambda_a^- \perp f_a - ATC_a^- \geq 0, a \in A,$$
$$0 \leq \lambda_a^+ \perp ATC_a^+ - f_a \geq 0, a \in A,$$
$$(r_z): \quad \rho_z - \phi_z = 0, z \in Z,$$
$$(f_a): \quad \lambda_a^+ - \lambda_a^- + \phi_{z(F(a))} - \phi_{z(T(a))} = 0, a \in A.$$

Here, $Z(a)$ and $T(a)$ correspond, respectively, to the from and to zone of a link $a$. Suppose that $f_a > 0$, then it must be the case that $\lambda_a^- = 0$ (as long as $ATC_a^- < 0$). And since $\rho_z = \phi_z$, we also know that $\rho_{z(T(a))} - \rho_{z(F(a))} = \lambda_a^+ \geq 0$, which proves that the price at the destination is no less than the price at the origin. This is not always true for $(OPF)$; see problem 5.11.

**Problem 5.10** Using the KKT conditions of the previous exercise, we know that, since $ATC_a^- < f_a < ATC_a^+$, we have that $\lambda_a^+ = \lambda_a^- = 0$. This implies that $\rho_{z(T(a))} = \rho_{z(F(a))}$.

**Problem 5.11** In the network of figure C.2, the power flow from node 2 to node 3 is 49.5 MW, while the price in node 2 is 110 \$/MWh and the price in node 3 is 60 \$/MWh. The code is available on the textbook website.

**Problem 5.12** The code is available on the textbook website.

**Problem 5.13**

1. The price in the north is 10 \$/MWh; the price in the south is 200 \$/MWh. Consumer surplus is \$0.
2. The zonal price is 10 \$/MWh. There is a downward redispatch of 100 MW in the north, and an upward redispatch of 100 MW of demand in the south. The system operator is paid \$1000 from generators in the north, and pays \$20000 to consumers in the south.
3. No.
4. In the zonal design, consumers enjoy a benefit of \$60000. They pay \$4000 in the zonal auction and they are paid \$20000 in redispatch. They thus enjoy a surplus of \$76000 (essentially their entire load is subsidized, even the part that is not served, and they are additionally paying a lower price for the power that they do

end up consuming). The consumer lobby would certainly support the zonal design.

**Problem 5.14**

1. We have 30000 MW of production in the north of country G and 0 MW of production in the south of country G. The zonal price in country G is 20 $/MWh.
2. No, we overload the line from the north to the south of country G.
3. We have 12000 MW in the north and 18000 MW in the south, 20 $/MWh in the north, and 200 $/MWh in the south.
4. Zonal pricing because it results in a price of 20 $/MWh instead of 200 $/MWh.

**Chapter 6**

**Problem 6.1** The KKT conditions of $(ORDC)$ when $R_g$ is unlimited for all $g \in G$ can be written out as follows:

$$d - \sum_{g \in G} p_g = 0,$$

$$dr - \sum_{g \in G} r_g = 0,$$

$$0 \leq \mu_g \perp P_g - p_g - r_g \geq 0, g \in G,$$

$$0 \leq p_g \perp MC_g(p_g) - \lambda + \mu_g \geq 0, g \in G,$$

$$0 \leq r_g \perp \mu_g - \mu \geq 0, g \in G,$$

$$0 \leq d \perp \lambda - MB(d) \geq 0,$$

$$0 \leq dr \perp \mu - MR(dr) \geq 0.$$

For the case of a unit $g_m$ such that $p_{g_m} > 0$, $r_{g_m} > 0$, the fourth KKT condition implies that:

$$\lambda = MC_{g_m}(p_{g_m}) + \mu_{g_m},$$

and the fifth KKT condition implies that $\mu_{g_m} = \mu$. The equality implies that agent $g_m$ receives an equal profit margin in the energy and reserves market.

**Problem 6.2** Write out the KKT conditions of $(EDR)$. Observe that either the entire load and the entire reserve requirement is covered (in which case $\mu^* = 0$), or not (in which case $\mu^* = MB(d^*) - MC_{g_m}(p_{g_m}^*)$). An important point of the exercise is that reserve prices can be highly volatile (either zero or a very high value driven by the valuation of curtailed demand) when reserve requirements are inelastic.

**Problem 6.3** The code is available on the textbook website.

**Problem 6.4** The code is available on the textbook website.

**Problem 6.5** The code is available on the textbook website.

**Problem 6.6** The model can be described as follows:

$$\max_{d,di,p,r} VOLL \cdot di + \sum_{l \in L} V_l \cdot d_l - \sum_{g \in G} MC_g \cdot p_g$$

$$d_l \leq D_l, l \in L$$

$$p_g + r_g \leq P_g^+, g \in G$$

$$di + \sum_{l \in L} d_l - \sum_{g \in G} p_g = 0$$

$$DR - \sum_{g \in G} r_g = 0$$

$$d_l \leq D_l, l \in L$$

$$di \leq DI$$

$$di, d, p, r, dr \geq 0.$$

The code is available on the textbook website.

**Problem 6.7** The introduction of a fixed reserve requirement means that the most expensive part of the merit order is not available. We thus need to consider the following cases:

- The system is shedding inelastic load when $D + R > 10000 \Rightarrow D > 9000$. In this case, the inelastic demand sets the energy price at VOLL.
- The system is shedding price-elastic load when $D + R \leq 10000$ and $D + R + 100 \geq 10000$, i.e. $8900 \leq D \leq 9000$. In this case, the price is determined by the valuation of the price-elastic demand:
  $\lambda^\star = 1000 - 10 \cdot (10000 - D - R) = 1000 - 10 \cdot (9000 - D)$.
- The system is covering the entire price-elastic load when
  $D + R + 100 \leq 10000 \Rightarrow D \leq 8900$. In this case, the price is determined by the supply curve at the level of total served demand, i.e.
  $\lambda^\star = MC_G(D) = 0.015 \cdot (D + 100)$.

Any differences in price with respect to the results of the previous problem are due to the fact that the math programming model is a discrete approximation of the continuous model.

**Problem 6.8** The code is available on the textbook website.

## Chapter 7

**Problem 7.1** Regarding equation (7.2), if $v_{g\tau}$ is equal to 1 for any of the periods $t - UT_g + 1$ through $t$, then $u_{gt}$ is lower bounded by 1. Since this constraint is repeated over time periods, it will force $u_{gt} = 1$ whenever $v_{g\tau} = 1$ for $\tau \leq t - UT_g + 1$.

Regarding equation (7.3), if $z_{g\tau}$ is equal to 1 for any of the periods $t - DT_g + 1$ through $t$, then $1 - u_{gt}$ is lower bounded by 1. Since this constraint is repeated over time periods, it will force $u_{gt} = 0$ whenever $z_{g\tau} = 1$ for $\tau \leq t - DT_g + 1$.

**Problem 7.2** The code is available on the textbook website.

**Problem 7.3** The code is available on the textbook website.

**Problem 7.4** Applying the result of Balas (1998) to the feasible set of order C yields:

$$
conv(F) = \left\{ (p,u) \in \mathbb{R}^2 \middle| \begin{array}{c} (p,u) = (0,0) + \left(p^1, u^1\right) \\ 11u^1 \leq p^1 \leq 12u^1 \\ u^1 \leq \xi_0^1 \\ \xi_0^0 + \xi_0^1 = 1 \\ (p^1, u^1, \xi_0^1, \xi_0^0) \geq 0 \end{array} \right\}.
$$

**Problem 7.5** The code is available on the textbook website.

**Problem 7.6** The code is available on the textbook website.

## Chapter 8

**Problem 8.1** The code is available on the textbook website.

**Problem 8.2** The code is available on the textbook website.

**Problem 8.3** The code is available on the textbook website. We have $SP = 77$, $EEV = 72.5$, and $WS = 110$, thus $VSS = 4.5$ and $EVPI = 33$.

**Problem 8.4** The code is available on the textbook website.

**Problem 8.5** The lattice unfolds into 2 nodes in stage 2 (with transition probabilities 0.5 and 0.5, and values 10 and 0), 6 nodes in stage 3 (with transition probabilities 0.7, 0.2, and 0.1 into the triplet 15, 5, and 0), and 18 nodes in stage 4 (with transition probabilities 0.1, 0.8, and 0.1 into triplets 20, 10, and 0).

**Problem 8.6** The code is available on the textbook website.

**Problem 8.7** The code is available on the textbook website.

**Problem 8.8** All three-stage trees satisfy the Markov property.

## Chapter 9

**Problem 9.1** An important advantage of a CfD, in terms of solvency, is the fact that the cash flow that is required for settling the CfD is lower than that of a forward contract (and, in expectation, should equal zero).

**Problem 9.2** To prove subadditivity, note that $\mathcal{R}(\xi + \zeta) = \max_{\omega \in \Omega}(\xi(\omega) + \zeta(\omega)) \leq \max_{\omega \in \Omega} \xi(\omega) + \max_{\omega \in \Omega} \zeta(\omega) = \mathcal{R}(\xi) + \mathcal{R}(\zeta)$.

For positive homogeneity, note that $\mathcal{R}(\lambda \cdot \xi) = \max_{\omega \in \Omega}(\lambda \cdot \xi(\omega)) = \lambda \cdot \max_{\omega \in \Omega} \xi(\omega) = \lambda \cdot \mathcal{R}(\xi)$.

For monotonicity, we have that if $\xi \preceq \zeta$ then $\mathcal{R}(\xi) = \max_{\omega \in \Omega} \xi(\omega) \leq \max_{\omega \in \Omega} \zeta(\omega) = \mathcal{R}(\zeta)$.

For translation invariance, we have that $\mathcal{R}(\xi + t) = \max_{\omega \in \Omega}(\xi(\omega) + t) = \max_{\omega \in \Omega} \xi(\omega) + t = \mathcal{R}(\xi) + t$.

**Problem 9.3** It is necessary to show that, if $\xi$ and $\zeta$ are two random variables and $\lambda \in (0, 1)$, then $\mathcal{R}(\lambda \cdot \xi + (1 - \lambda) \cdot \zeta) \leq \lambda \cdot \mathcal{R}(\xi) + (1 - \lambda) \cdot \mathcal{R}(\zeta)$.

**Problem 9.4** To prove homogeneity, consider any $\lambda \geq 0$. By the definition of VaR, we have

$$VaR_\alpha(\lambda \cdot \xi) = \min\{t | \mathbb{P}[\lambda\xi \le t] \ge \alpha\}$$
$$= \min\left\{t | \mathbb{P}[\xi \le \frac{t}{\lambda}] \ge \alpha\right\} = \lambda \min\left\{\frac{t}{\lambda} | \mathbb{P}[\xi \le \frac{t}{\lambda}] \ge \alpha\right\}$$
$$= \lambda \cdot VaR_\alpha(\xi).$$

To prove monotonicity, consider $\xi, \zeta$ such that $\xi \preceq \zeta$. Since for any $t$ we have $\mathbb{P}[\xi \le t] \ge \mathbb{P}[\zeta \le t]$, it follows that the minimum $t_\zeta$ for which we achieve $\mathbb{P}[\zeta \le t_\zeta] \ge \alpha$ also achieves $\mathbb{P}[\xi \le t_\zeta] \ge \alpha$, for any $\alpha$. Hence, $VaR_\alpha(\xi) \le VaR_\alpha(\zeta)$.

To prove translation invariance, we resort again to the definition of value at risk:

$$VaR_\alpha(\xi + c) = \min\{t | \mathbb{P}[\xi + c \le t] \ge \alpha\}$$
$$= \min\{t | \mathbb{P}[\xi \le t - c] \ge \alpha\} = \min\{t - c | \mathbb{P}[\xi \le t - c] \ge \alpha\} + c$$
$$= VaR_\alpha(\xi) + c.$$

**Problem 9.5** We first prove homogeneity. Consider any $\lambda \ge 0$:

$$CVaR_\alpha(\lambda \cdot \xi) = \min\left\{t + \frac{1}{1 - \alpha}\mathbb{E}[(\lambda \cdot \xi - t)^+]\right\}$$
$$= \min\left\{t + \lambda\frac{1}{1 - \alpha}\mathbb{E}[(\xi - \frac{t}{\lambda})^+]\right\} = \lambda \cdot \min\left\{\frac{t}{\lambda} + \frac{1}{1 - \alpha}\mathbb{E}[(\xi - \frac{t}{\lambda})^+]\right\}$$
$$= \lambda \cdot CVaR_\alpha(\xi).$$

In order to prove translation invariance, consider any $r \in \mathbb{R}$:

$$CVaR_\alpha(\xi + r) = \min\left\{t + \frac{1}{1 - \alpha}\mathbb{E}[(\xi + r - t)^+]\right\}$$
$$= \min\left\{(t - r) + r + \frac{1}{1 - \alpha}\mathbb{E}[(\xi - (t - r))^+]\right\}$$
$$= r + \min\left\{(t - r) + \frac{1}{1 - \alpha}\mathbb{E}[(\xi - (t - r))^+]\right\}$$
$$= r + CVaR_\alpha(\xi).$$

In order to prove monotonicity, we note that first-order stochastic dominance implies second-order stochastic dominance. An equivalent definition of second-order stochastic dominance is that for every convex function $u(\cdot)$ we have $\mathbb{E}[u(\xi)] \le \mathbb{E}[u(\zeta)]$. Therefore, for any $t \in \mathbb{R}$ we have $t + \frac{1}{1-\alpha}\mathbb{E}[(\xi - t)^+] \le t + \frac{1}{1-\alpha}\mathbb{E}[(\zeta - t)^+]$, which implies that

$$\min_t\left\{t + \frac{1}{1 - \alpha}\mathbb{E}[(\xi - t)^+]\right\} \le \min_t\left\{t + \frac{1}{1 - \alpha}\mathbb{E}[(\zeta - t)^+]\right\}.$$

From this we get that if $\xi \preceq \zeta$ then $CVaR_\alpha(\xi) \le CVaR_\alpha(\zeta)$.

In order to prove subadditivity, we note that for $a, b \in \mathbb{R}$ we have that $(a + b)^+ \le a^+ + b^+$. It follows that

$$t_1 + \frac{1}{1 - \alpha}(\xi_1 - t_1)^+ + t_2 + \frac{1}{1 - \alpha}(\xi_2 - t_2)^+ \ge$$
$$(t_1 + t_2) + \frac{1}{1 - \alpha}(\xi_1 + \xi_2 - (t_1 + t_2))^+.$$

Taking expectation, and minimizing *jointly* with respect to $(t_1, t_2)$ yields $CVaR_\alpha(\xi_1 + \xi_2) \leq CVaR_\alpha(\xi_1) + CVaR_\alpha(\xi_2)$.

**Problem 9.6** The conditional distribution over the three least favorable outcomes is as follows:

- $1000 with a probability of (4/16)
- $0 with a probability of (10/16)
- −$1000 with a probability of (2/16)

The expectation is thus:

$$CVaR_{0.84} = 1000 \cdot (4/16) + (0) \cdot (10/16) + (-1000) \cdot (2/16) = \$125.$$

The code is available on the textbook website.

**Problem 9.7** Consider a lottery with two possible outcomes, each occurring with a probability of 0.5. Suppose that the lottery requires a payment of $-1 + 2 \cdot \epsilon$ for the first outcome, and 1 for the second outcome. Denote the lottery for $\epsilon$ arbitrarily small as $Z(\epsilon)$. Then, for $\beta = 1$, we have that $Z(0.01) \succeq Z(0)$ but $\mathcal{R}(Z(0)) = 1 > \mathcal{R}(Z(0.01)) = 0.09901$.

**Problem 9.8** The expected price of electricity 12 months from now is 120 \$/MWh.

1. Forward contract payments now from the supplier to its counter-party: $360000. Net payments at the expiration of the forward contract from the supplier to the market/its counter-party: $17000. CfD payments now from the supplier to its counter-party: $0. Net payments at the expiration of the CfD from the supplier to the market/its counter-party: $377000.
2. Both instruments result in total payments from the supplier equal to $377000.
3. The forward contract leads to lower cash reserves on average.

## Chapter 10

**Problem 10.1** The function $r(v)$ that maps consumer valuation to the appropriate level of reliability is given as

$$\mathbb{P}[\text{serve all load with valuation} \leq v]$$
$$= \mathbb{P}[\text{cost-effective capacity} \geq D(v)]$$
$$= \mathbb{P}[\hat{p}(\omega) \geq v],$$

where $\hat{p}(\omega)$ is the competitive equilibrium price of the system. The rest of the analysis for computing the optimal price menu follows identically.

**Problem 10.2** The code is available on the textbook website.

**Problem 10.3** The code is available on the textbook website.

**Problem 10.4** The KKT conditions are as follows:

$$0 \leq \gamma \perp \sum_{i \in I} s_i \leq D,$$
$$0 \leq s_i \perp \gamma - r_i \cdot v + p_i \geq 0, i \in I.$$

Suppose that $s_i^* > 0$ and $s_j^* > 0$ for $i \neq j$. We can then conclude that $r_i \cdot v - p_i = r_j \cdot v - p_j = \gamma$. But if this is the case, then the objective function value of choosing $\bar{s}_i = s_i^* + s_j^*$ and $\bar{s}_j = 0$ (or, the other way around, transferring the entire procurement to option $j$) is equal to that of the $s^*$ solution. This argument can be repeated until either $s_i = D$ or $s_i = 0$ for all $i \in I$. The interpretation of this result is that the consumer can limit its selection to one of the offered options, without having to split its load between multiple options.

**Problem 10.5** Two constraints are needed, that correspond to the lower-level choice of consumers. These correspond to the KKT conditions of the previous exercise. The remaining constraints correspond to upper-level constrains of the system operator, and can be summarized as follows:

- The reliability of an option has to be fulfilled.
- Generators cannot produce above their technical maximum.
- Production should equal demand.
- The demand served under a given option should not exceed the amount of capacity subscribed under this option.
- The objective function prioritizes the service of options with higher reliability to the service of options with lower reliability.

The code is available in the textbook website.

## Chapter 11

**Problem 11.1** Both the centralized capacity expansion in vertical slice form and the decentralized market equilibrium can be described by the KKT conditions of the centralized capacity expansion problem.

**Problem 11.2** The load data should be represented in horizontal slice form. The linear program can be described as follows:

$$\min_{x,p} \sum_{j=1}^{m} \sum_{i=1}^{n} T_j \cdot MC_i \cdot p_{ij} + \sum_{i=1}^{n} I_i \cdot x_i$$

$$\sum_{j=1}^{m} p_{ij} \leq x_i, i = 1, \ldots, n$$

$$\sum_{i=1}^{n} p_{ij} = \Delta D_j, j = 1, \ldots, m$$

$$p, x \geq 0.$$

The code is available on the textbook website, and produces identical results to the vertical slice formulation.

The model based on horizontal slices differs to the one based on vertical slices in two ways: (i) the data that represents the load duration curve needs to be changed, and (ii) the power capacity limit constraint also needs to be changed.

**Problem 11.3** The code is available on the textbook website.

1. The optimal capacity mix is as follows:

   - Coal: 1918 MW
   - Gas: 2165 MW
   - Nuclear: 7086 MW
   - Oil: 0 MW

   The equilibrium prices are as follows:

   - Base periods: 12.56 \$/MWh
   - Medium periods: 27.52 \$/MWh
   - Peak periods: 109.20 \$/MWh

2. The short-term market equilbrium prices are

   - Base periods: 6.5 \$/MWh
   - Medium periods: 25 \$/MWh
   - Peak periods: 350 \$/MWh

   Any price in peak periods between 350 \$/MWh and 3000 \$/MWh is a valid peak period price. For a peak period price of 350 \$/MWh, the long-run profits of each technology (i.e. after investment costs are accounted for) are:

   - Coal: 39.65 \$/MWh
   - Gas: −5.00 \$/MWh (exactly equal to the investment cost)
   - Nuclear: 38.43 \$/MWh
   - Oil: 0.00 \$/MWh

3. Optimal capacities are:

   - Coal: 4083 MW
   - Gas: 0 MW
   - Nuclear: 7086 MW
   - Oil: 0 MW

   The new market clearing prices are:

   - Base periods: 12.56 \$/MWh
   - Medium periods: 25.00 \$/MWh
   - Peak periods: 118.44 \$/MWh

**Problem 11.4** Indeed, the design can support an equilibrium mix with 1000 MW of nuclear.

The equilibrium model can be expressed as follows, where $\lambda_{nuc}$ is a variable in the problem, and where we formulate the model with $\Delta T_j$ being expressed in hours:

$$D_j - \sum_{i=1}^{n} p_{it} = 0, j = 1, \ldots, m,$$

$$\lambda_{nuc} \cdot \sum_{j=1}^{m} \Delta T_j \cdot p_{ij} = 8760 \cdot I_i \cdot x_i + \sum_{j=1}^{m} \Delta T_j \cdot MC_i \cdot p_{ij},$$

$$0 \le p_{ij} \perp MC_i - \lambda_j + \mu_{ij} \ge 0, i \in I - \{nuc\}, j = 1, \dots, m,$$

$$0 \le x_i \perp 8760 \cdot I_i - \sum_{j=1}^{m} \Delta T_j \cdot \mu_{ij} \ge 0, i = 1, \dots, n,$$

$$0 \le \mu_{ij} \perp x_i - p_{ij} \ge 0, i = 1, \dots, n, j = 1, \dots, m,$$

$$0 \le p_{nuc,j} \perp MC_{nuc} - \lambda_{nuc} + \mu_{nuc,j}, j = 1, \dots, m.$$

This is a complementarity problem, which needs to be solved with a nonlinear solver, e.g. knitro. In order to detect a solution, we assist the solver by proposing an equilibrium solution ourselves, in which case the solver simply verifies that the conditions of the equilibrium are indeed satisfied. We concretely build a proposed solution by:

- setting the capacity of the non-dispatchable resources to the proposed value of $x_{nuc} = 1000$ MW and $p_{nuc,j} = 1000$ MWh for $j = 1, \dots, m$,
- setting the price of the non-dispatchable resources, $\lambda_{nuc}$, to the investment plus marginal cost of these resources, $\lambda_{nuc} = 38.5$ \$/MWh, and
- solving a standard capacity expansion problem for the original demand minus the 1000 MW covered by the non-dispatchable units, which gives proposed values for $x_i$, $p_{ij}$, and $\mu_{ij}$, for all the non-dispatchable resources.

The code is available on the course website.

**Problem 11.5**

1. The KKT conditions can be expressed as follows:

$$0 \le \mu_{gt} \perp x_g - p_{gt} \ge 0, g \in G, t \in T,$$

$$\Delta T_t \cdot \left( D_t - \sum_{g \in G} p_{gt} \right) = 0, t \in T,$$

$$0 \le p_{gt} \perp MC_g - \lambda_t + \mu_{gt} \ge 0, g \in G, t \in T,$$

$$0 \le x_g \perp 8760 \cdot IC_g - \sum_{t \in T} \Delta T_t \cdot \mu_{gt} \ge 0, g \in G.$$

2. The optimal solution uses units in merit order:

- Base periods: $p_{nuc,1} = 7086$ MW, $p_{coal,1} = 1 = 0$ MW, $p_{ng,1} = 0$ MW
- Medium periods: $p_{nuc,2} = 7086$ MW, $p_{coal,1} = 1922$ MW, $p_{ng,2} = 0$ MW
- Peak periods: $p_{nuc,3} = 7086$ MW, $p_{coal,3} = 1 = 1922$ MW, $p_{ng,3} = 2161$ MW

The dual vector $\mu$ is as follows:

- Base periods: $\mu_{nuc,1} = 1.12$ \$/MWh, $\mu_{coal,1} = 1$ \$/MWh, $\mu_{ng,1} = 0$ \$/MWh
- Medium periods: $\mu_{nuc,2} = 20.81$ \$/MWh, $\mu_{coal,2} = 2.31$ \$/MWh, $\mu_{ng,2} = 0$ \$/MWh
- Peak periods: $\mu_{nuc,3} = 102.70$ \$/MWh, $\mu_{coal,3} = 84.20$ \$/MWh, $\mu_{ng,3} = 29.20$ \$/MWh

3. The subsidy is equivalent to a decrease in marginal cost for the natural gas technology. The objective function becomes:

$$\min_{p,x} \sum_{g \in G, t \in T} F_t \cdot (MC_g - S_g) \cdot p_{gt} + 8760 \cdot \sum_{g \in G} IC_g \cdot x_g.$$

4. We need to set the subsidy equal to the change in the marginal cost of natural gas, i.e. 270 $/MWh.

**Problem 11.6** The KKT conditions of both the market equilibrium and of $(VOLLP)$ are:

$$0 \leq p_{ij} \perp MC_i - \rho_j + \mu_{ij} \geq 0, i = 1, \ldots, n, j = 1, \ldots, m,$$

$$0 \leq x_i \perp I_i - \sum_{j=1}^{m} \Delta T_j \cdot \mu_{ij} \geq 0, i = 1, \ldots, n, j = 1, \ldots, m,$$

$$0 \leq ls_j \perp VOLL - \rho_j \geq 0, j = 1, \ldots, m,$$

$$0 \leq \mu_{ij} \perp x_i - p_{ij} \geq 0, i = 1, \ldots, n, j = 1, \ldots, m,$$

$$D_j - ls_j - \sum_{i=1}^{n} p_{ij} = 0, j = 1, \ldots, m.$$

**Problem 11.7** The code is available on the course website. Note that, for the equilibrium models, we fix capacity investments and prices to the solutions of the optimization model, and we allow prices to vary within $0.1 of the value obtained from the optimization model in order to overcome numerical tolerance issues.

**Problem 11.8** The $(CRM)$ model decomposes to the following market equilibrium conditions:

- Price adjustment in the energy market: $D_j - ls_j - \sum_{i=1}^{n} p_{ij} = 0, j = 1, \ldots, m.$
- Price adjustment in the capacity market: $xd - \sum_{i=1}^{n} x_i = 0.$
- The policy of setting price to VOLL whenever there is involuntary load shedding: $0 \leq ls_j \perp VOLL - \rho_j \geq 0, j = 1, \ldots, m.$
- Profit maximization of generators $i = 1, \ldots, n$:

$$\max_{p,x} \rho C \cdot x_i + \sum_{j=1}^{m} (\rho_j - MC_i) \cdot \Delta T_j \cdot p_{ij}$$

$$p_{ij} \leq x_i, j = 1, \ldots, m$$

$$p, x \geq 0.$$

- Surplus maximization of the system operator:

$$\max_{xd} \int_{v=0}^{xd} VC(v)dv - \rho C \cdot xd$$

$$xd \geq 0.$$

**Problem 11.9** The code is available on the textbook website.

**Problem 11.10** Compare the KKT conditions of $(CRM)$ to those of $(VOLLP)$. Build an optimal solution to $(CRM)$ by observing that the fact that $VC(\sum_{i=1}^{n} x_i^\star) = 0$

implies that you can choose $\rho C = 0$, $xd = \sum_{i=1}^{n} x_i^*$, and you can choose all other primal-dual variables of the $(CRM)$ problem to be equal to the optimal primal-dual solution of $(VOLLP)$.

**Problem 11.11** The $(ORDC)$ model decomposes to the following market equilibrium conditions:

- Price adjustment in the energy market: $D_j - ls_j - \sum_{i=1}^{n} p_{ij} = 0, j = 1, \ldots, m$.
- Price adjustment in the reserve market: $dr_j - \sum_{i=1}^{n} r_{ij} = 0, j = 1, \ldots, m$.
- The policy of setting price to VOLL whenever there is involuntary load shedding:
  $0 \leq ls_j \perp VOLL - \rho_j \geq 0, j = 1, \ldots, m$.
- Profit maximization of generators $i = 1, \ldots, n$:

$$\max_{p,r,x} \sum_{j=1}^{m} (\rho_j - MC_i) \cdot \Delta T_j \cdot p_{ij} + \sum_{j=1}^{m} \rho R_j \cdot \Delta T_j \cdot r_{ij}$$

$$p_{ij} + r_{ij} \leq x_i, j = 1, \ldots, m$$

$$p, r, x \geq 0.$$

- Surplus maximization of the system operator:

$$\max_{dr} \sum_{j=1}^{m} \Delta T_j \cdot \left( \int_{v=0}^{dr_j} VR_j(v)dv - \rho R_j \cdot dr_j \right)$$

$$dr \geq 0.$$

**Problem 11.12** The code is available on the textbook website.

**Problem 11.13** We can proceed as follows:

- Since the gas technology has been constructed, it must break its investment cost even from the peak period profits. From this we can conclude what the price of peak periods is.
- Since the coal technology has been constructed, and since it needs to break even, it must recover its investment cost from peak and medium periods. But we know the price of peak periods from the previous step, so we can conclude the price of medium periods.
- Since the nuclear technology has been constructed, and since it needs to break even, it must recover its investment cost from peak, medium, and base periods. But we know the price of peak and medium periods, and we can conclude what the price of base periods is.

**Problem 11.14** The CRM price, the profit distribution, and the payoff of the call option are computed by the code of problem 11.9. An excel file with the profit of the oil plant owner and the payoff of the call option is available on the textbook website.

1. The CRM price is 1.62 $/MWh.
2. The plant earns a profit of 1104360 $/h for the four peak periods of the year, and is losing 504.50 $/h at every other hour of the year.
3. The value at risk of the investor is $VaR_{0.01} = 504.2$ $/h.

4. The payoff of the reliability option is $\max(\rho_j - k, 0)$. This means that the call option essentially pays 700 \$/MWh over the four peak periods of the system, and 0 throughout the rest of the year. The expected payoff of the reliability option is thus $700 \cdot \frac{4}{8760} = 0.32$ \$/MWh, or 420.42 \$/h for the full oil capacity (1315.3 MW). It is understood that what is traded is an annual call option, i.e. a strip of call options that covers the entire year.

5. The new distribution of losses now involves a profit of 184070 \$/h for every one of the four peak periods and a loss of 84.08 \$/h during every other period of market operation.

6. We have $VaR_{0.01} = 84.08$ \$/h. The investor is better off.

## Chapter 12

**Problem 12.1** The MPEC can be written as follows:

$$\max_{p \geq 0, \lambda, d, p_c} \lambda \cdot p - 10 \cdot p$$
$$0 \leq d \perp \lambda - (1050 - 28.571 \cdot d) \geq 0$$
$$0 \leq p_c \perp -\lambda + (-950 + 43.478 \cdot p_c) \geq 0$$
$$d - p_c - p = 0.$$

We introduce an additional constraint of nonnegative prices in order to prevent unstable behavior from the solver. The code is available on the textbook website. The analytical solution and the solution of the code are equal within rounding errors.

**Problem 12.2** It suffices to observe that the fringe production is 24.9195 billion barrels, thus above 22.126 billion barrels, at the solution of problem 12.1, to argue that we are in the range where the marginal cost of the fringe suppliers is linear.

In order to compute the equilibrium as an MPEC, we can rewrite the MPEC of problem 12.1 by adjusting the profit maximization problem of the fringe competition:

$$\max_{z, p_c \geq 0} \lambda \cdot p_c - z$$
$$(\gamma_1): \quad z \geq 12 \cdot 22.126 + \int_{x=22.126}^{p_c} (-950 + 43.478 \cdot x) dx$$
$$(\gamma_2): \quad z \geq 12 \cdot p_c.$$

The idea here is that we express the cost incurred by the fringe suppliers as $z = \max\left(12 \cdot 22.126 + \int_{x=22.126}^{p_c} (-950 + 43.478 \cdot x) dx, 12 \cdot p_c\right)$, where 22.126 is the output at which the linear function evaluates to a marginal cost of \$12. The code is available on the textbook website.

**Problem 12.3**

1. The implied marginal cost is $MC = 11.18$ \$/mcf.
2. The adjusted output of the monopoly is $p = 3130.39$ bcf. The price becomes 27.86 \$/mcf. The losses in consumer surplus amount to \$72414 million. The tax that is collected from the monopolist amounts to $10 \cdot 3130.39 = \$31303.9$

million. The policy conclusion is now reversed: the losses in consumer surplus from introducing the tax are approximately two times greater than the benefits of the collected tax, and the tax should not be introduced, unless its level is optimized.

**Problem 12.4** If Russia had the option to divert its sales to Asia, it would do so, up to the level where the price paid by the Asian market equals the price paid by the European market (net of the tax). Thus, the assumption is indeed crucial.

**Problem 12.5** Before the introduction of the tax, the equilibrium quantity is $p^\star = d^\star = 0.5$ and the equilibrium price is $\lambda^\star = 0.5$. After the introduction of the tax, the equilibrium quantity is $p^\star = d^\star = 0.25$, the equilibrium buy price is $\lambda_b^\star = 0.75$, and the equilibrium sell price is $\lambda_s^\star = 0.25$.

**Problem 12.6** The code is available on the textbook website. The market model is as follows:

$$\max_{d,p} \int_{x=0}^{d} MB_L(x)dx - \int_{x=0}^{p} MC_G^C(x)dx$$
$$(\lambda): \quad d - p - P_{OPEC} = 0,$$

where $P_{OPEC}$ is the fixed production of OPEC.

1. The short-run aggregate marginal cost function of the competitive suppliers in the market is:

$$MC_G^{SR,C}(p) = -1440 + 53.5714 \cdot p \text{ \$/barrel.}$$

The short-run aggregate marginal benefit function is

$$MB_L^{SR}(d) = 1560 - 39.4737 \cdot d \text{ \$/barrel.}$$

The long-term supply function for the competitive part of the market is given as:

$$MC_G^{LR,C}(p) = -240 + 10.7143 \cdot p \text{ \$/barrel.}$$

The long-term aggregate marginal benefit function is given as:

$$MB_L^{LR}(d) = 360 - 7.89474 \cdot d \text{ \$/barrel.}$$

2. The new short-term equilibrium becomes as follows: price drops (dramatically) to 14.54 \$/barrel, demand increases to 39.1515 billion barrels per year, and competitive supply decreases to 27.1515 billion barrels. The long-term equilibrium is as follows: price drops to 50.91 \$/barrel, demand increases to 39.1515 billion barrels per year, and competitive supply decreases to 27.1515 billion barrels. Note that demand is equal (within numerical precision) to that of the short-term model.
3. We confirm that we get the same price as in the example.
4. We confirm that we get the same price as in the example, and the total demand remains constant within numerical precision.

**Problem 12.7** The code is available in the textbook website.
**Problem 12.8** The code is available in the textbook website.

**Problem 12.9** We have $p_{cf} = 150$, $p_{ce} = 0$, $p_o = 150$, $\lambda_f = 30$ \$/unit, $\lambda_e = 20$ \$/unit, and $\mu_c = \mu_o = 0$.

There is no coupling between food and energy anymore.

The coupling of prices is due to the fact that the corn production factor needs to be indifferent, from a profit point of view, between the energy market and the food market. Once the production factor is no longer split between the two markets, this condition need no longer hold.

**Problem 12.10** False, it no longer holds if the constraint $\sum_{t=1}^{H} p_t \leq S$ is non-binding (thus $\mu = 0$).

**Appendix A**

**Problem A.1** The code is available on the textbook website.

**Problem A.2**

1. The linear programming model can be expressed as follows:

$$\min_{p \geq 0, p^H \geq 0, d^H \geq 0} \sum_{g \in G, t \in T} MC_g \cdot p_{gt}$$

$$(\lambda_t): \quad D_t + d_t^H - \sum_{g \in G} p_{gt} - p_t^H = 0, t \in T$$

$$p_{gt} \leq P_g, g \in G, t \in T$$

$$\sum_{t \in T} d_t^H = \frac{1}{\eta} \sum_{t \in T} p_t^H.$$

2. The code is available on the textbook website. The resulting dispatch for period 1 is $p_{G_1,1} = 60$ MWh, $p_{G_2,1} = 0$ MWh, $p_1^H = 0$ MWh, $d_1^H = 10$ MWh. The dispatch for period 2 is $p_{G_1,2} = 60$ MWh, $p_{G_2,2} = 32$ MWh, $p_2^H = 8$ MWh, $d_2^H = 0$ MWh. The market clearing price is 40 \$/MWh in period 1 and 50 \$/MWh in period 2.

3. The market clearing prices are such that the hydro unit is willing to consume power in period 1 at 80% of the price of period 2, and sell 80% of it in period 2.

4. The heuristic (mis-)pricing method whereby the cheapest unit producing below its nominal capacity is setting the price results in clearing prices equal to 10 \$/MWh in period 1 and 50 \$/MWh in period 2. These are not equilibrium prices, because at these prices the hydro unit would be willing to pump an arbitrary amount of energy in period 1 and release that energy (subject to efficiency losses) in period 2.

**Problem A.3** The decision variable $x_g$ corresponds to the CO2 abatement in tons, the objective function aims at maximizing the profit of firm $g$ from trading emissions permits, the first constraint represents the maximum amount of abatement that a firm can achieve, and the last constraint imposes that a firm cannot increase its emissions.

The first profit maximization model is equivalent to the second one because the term $\lambda \cdot E_g$ is a constant.

For a price of $\lambda = 22$ \$/ton, the optimal decision of firms 1 through 21 is $x_g^\star = X_g$. For firms 23 through 100, $x_g^\star = 0$. Firm 22 is indifferent about any level of abatement. The decentralized solution results in identical abatement decisions as the centralized one.

The profits of the firms are as follows:

$$\Pi_g = \begin{cases} -53 \cdot 22 + 25 \cdot (22 - g) \text{ million \$}, g = 1, \ldots, 21, \\ -53 \cdot 22 \text{ million \$}, g = 22, \ldots, 100. \end{cases}$$

**Problem A.4**

1. The KKT conditions can be expressed as follows:

$$-z_1 + z_2 = 3,$$
$$0 \leq \lambda_1 \perp x_1 - x_2 - 3 \geq 0,$$
$$0 \leq \lambda_2 \perp -z_1 \geq 0,$$
$$(x_1): \quad 5 - \lambda_1 = 0,$$
$$(z_1): \quad -1 + \lambda_2 - \mu = 0,$$
$$(z_2): \quad 0.5 + \mu = 0,$$
$$0 \leq x_2 \perp 1 + \lambda_1 \geq 0.$$

2. The primal-dual optimal pair is $x_1^\star = 3$, $x_2^\star = 0$, $z_1^\star = 0$, $z_2^\star = 3$, $\lambda_1^\star = 5$, $\lambda_2^\star = 0.5$, $\mu^\star = -0.5$.

**Problem A.5** If $V$ is arbitrarily high, then it is optimal to fully cover demand, $d = D$.

**Problem A.6** False. The optimal solution of the primal model leads to an objective function value of 10, which is also the optimal objective function value of the dual, due to strong duality.

**Appendix B**

**Problem B.1** The Laplacian is expressed as

$$L = D - A = nI_{n \times n} - 1_{n \times n},$$

where $1_{n \times n}$ is an $n \times n$ matrix with all ones as entries.

**Problem B.2** Proposition B.1 establishes that the Laplacian has a single eigenvalue equal to zero. Since the Laplacian is symmetric, the number of nonzero eigenvalues is equal to the rank of the matrix.

**Problem B.3** We have

$$V_m Y_{mn}^\star V_n^\star = |V_m||V_n|e^{i\theta_{mn}}(G_{mn} - iB_{mn})$$
$$= |V_m||V_n|(\cos(\theta_{mn}) + i\sin(\theta_{mn}))(G_{mn} - iB_{mn})$$
$$= |V_m||V_n|[(G_{mn}\cos(\theta_{mn}) + B_{mn}\sin(\theta_{mn})) + i(G_{mn}\sin(\theta_{mn}) - B_{mn}\cos(\theta_{mn}))].$$

Taking real and imaginary parts, and summing over nodes, we arrive to equations (B.1), (B.2).

**Problem B.4** If we sum equation (B.6) over all nodes, we have each term in the first sum of node $m_1$ which links to node $m_2$ being canceled out by the second term of node $m_2$, which links to $m_1$.

**Problem B.5** False, since $G = \frac{R}{R^2+X^2} \simeq 0$.

**Problem B.6** The code is available on the textbook website.

# References

Abada, I., de Maere d'Aertrycke, G.,
Ehrenmann, A. & Smeers, Y. (2019), "What
models tell us about long-term contracts in
times of energy transition," *Economics of
Energy and Environmental Policy* **8**(1),
163–182.

Alaywan, Z., Wu, T. & Papalexopoulos, A.
(2004), "Transitioning the California market
from a zonal to a nodal framework: An
operational perspective," *IEEE PES Power
Systems Conference and Exposition*, IEEE,
862–867.

Allaz, B. & Vila, J.-L. (1993), "Cournot
competition, forward markets and efficiency,"
*Journal of Economic Theory* **59**(1), 1–16.

Andrianesis, P., Bertsimas, D., Caramanis, M. C.
& Hogan, W. W. (2021), "Computation of
convex hull prices in electricity markets with
non-convexities using Dantzig-Wolfe
decomposition," *IEEE Transactions on Power
Systems* **37**(4), 2578–2589.

Aravena, I., Lété, Q., Papavasiliou, A. & Smeers,
Y. (2021), "Transmission capacity allocation
in zonal electricity markets," *Operations
Research* **69**(4), 1240–1255.

Aravena, I. & Papavasiliou, A. (2017),
"Renewable energy integration in zonal
markets," *IEEE Transactions on Power
Systems* **32**(2), 1334–1349.

Arrow, K. J. & Hahn, F. (1971), *General
Competitive Analysis*, Holden-Day.

Artzner, P., Delbaen, F., Eber, J.-M. & Heath, D.
(1999), "Coherent measures of risk,"
*Mathematical Finance* **9**(3), 203–228.

Ausubel, L. M. & Milgrom, P. (2006), "The
lovely but lonely Vickrey auction,"
*Combinatorial Auctions* **17**, 22–26.

Ávila, D., Papavasiliou, A. & Löhndorf, N. (in
press), "Batch learning SDDP for long-term
hydrothermal planning," *IEEE Transactions
on Power Systems*.

Baland, A. (2014), Co-optimization of gas
forward contracts and unit commitment, MSc
thesis, UCLouvain.

Balas, E. (1998), "Disjunctive programming:
Properties of the convex hull of feasible
points," *Discrete Applied Mathematics*
**89**(1–3), 3–44.

Baldick, R. (1995), "The generalized unit
commitment problem," *IEEE Transactions on
Power Systems* **10**(1), 465–475.

Baldick, R. (2006), *Applied Optimization:
Formulation and Algorithms for Engineering
Systems*, Cambridge University Press.

Bertsekas, D. P., Lauer, G. S., Sandell, N. R. &
Posbergh, T. A. (1983), "Optimal short-term
scheduling of large-scale power systems,"
*IEEE Transactions on Automatic Control*
**AC-28**(1), 1–11.

Bertsimas, D., Litvinov, E., Sun, X. A., Zhao, J.
& Zheng, T. (2013), "Adaptive robust
optimization for the security constrained unit
commitment problem," *IEEE Transactions on
Power Systems* **28**(1), 52–63.

Bertsimas, D. & Tsitsiklis, J. N. (1997),
*Introduction to Linear Optimization*, Athena
Scientific.

Biggar, D. R. & Hesamzadeh, M. R. (2022), "Do
we need to implement multi-interval real-time
markets?," *The Energy Journal* **43**(2), 111–131.

Binato, S., Pereira, M. V. F. & Granville, S.
(2001), "A new Benders decomposition
approach to solve power transmission
network design problems," *IEEE Transactions
on Power Systems* **16**(2), 235–240.

Birge, J. R. & Louveaux, F. (2010), *Introduction
to Stochastic Programming*, Springer Series in
Operations Research and Financial
Engineering, Springer.

Bjorndal, M. & Jornsten, K. (2001), "Zonal
pricing in a deregulated electricity market,"
*The Energy Journal* **22**(1), 51–73.

Boiteux, M. (1960), "Peak-load pricing," *The Journal of Business* **33**(2), 157–179.

Borenstein, S., Jaske, M. & Rosenfeld, A. (2002), Dynamic pricing, advanced metering and demand response in electricity markets, Technical Report 105, University of California Energy Institute.

Boucher, J. & Smeers, Y. (2001), "Alternative models of restructured electricity systems, part I: No market power," *Operations Research* **49**(6), 821–838.

Boyd, S. P. & Vandenberghe, L. (2008), *Convex Optimization*, Cambridge University Press.

Bushnell, J., Flagg, M. & Mansur, E. (2017), Capacity markets at a crossroads, Technical Report Working Paper 278, Energy Institute at Haas.

CAISO (2013), "Business practice manual for market operations, version 31," California ISO.

CAISO (2015), Flexible ramping product, revised draft final proposal, Technical report, California ISO.

Callaway, D. S. & Hiskens, I. A. (2011), "Achieving controllability of electric loads," *Proceedings of the IEEE* **99**(1), 184–199.

Camelo, S., Papavasiliou, A., de Castro, L., Riascos, Á. & Oren, S. (2018), "A structural model to evaluate the transition from self-commitment to centralized unit commitment," *Energy Economics* **75**, 560–572.

Caramanis, M., Tabors R., & Stevenson, R. (1982), Utility spot pricing study: Wisconsin, Technical report, MIT energy laboratory.

Carpentier, P., Cohen, G., Culioli, J.-C. & Renaud, A. (1996), "Stochastic optimization of unit commitment: A new decomposition framework," *IEEE Transactions on Power Systems* **11**(2), 1067–1073.

Carrion, M. & Arroyo, J. M. (2006), "A computationally efficient mixed-integer linear formulation for the thermal unit commitment problem," *IEEE Transactions on Power Systems* **21**(3), 1371–1378.

Cartuyvels, J. & Papavasiliou, A. (2022), "Calibration of operating reserve demand curves using a system operation simulator," *IEEE Transactions on Power Systems* **38**(4), 3043–3055.

Chao, H.-P. (2019), "Incentives for efficient pricing mechanism in markets with non-convexities," *Journal of Regulatory Economics* **56**(1), 33–58.

Chao, H.-P., Oren, S. S., Smith, S. A. & Wilson, R. B. (1986), "Multilevel demand subscription pricing for electric power," *Energy Economics* **8**(4), 199–217.

Chao, H.-P., Peck, S. P., Oren, S. S. & Wilson, R. B. (2000), "Flow-based transmission rights and congestion management," *The Electricity Journal* **13**(8), 38–58.

Chao, H.-P. & Wilson, R. (1987), "Priority service: Pricing, investment and market organization," *The American Economic Review* **77**(5), 899–916.

Chen, Y., Gribik, P. & Gardner, J. (2014), "Incorporating post zonal reserve deployment transmission constraints into energy and ancillary service co-optimization," *IEEE Transactions on Power Systems* **29**(2), 537–549.

Cho, J. & Papavasiliou, A. (2022), "Pricing under uncertainty in multi-interval real-time markets," *Operations Research: Articles in Advance*, 1–15.

Cohen, A. I. & Ostrowski, G. (1995), "Scheduling units with multiple operating modes in unit commitment," *Proceedings of Power Industry Computer Applications Conference*, IEEE, 494–500.

Cramton, P. (2017), "Electricity market design," *Oxford Review of Economic Policy* **33**(4), 589–612.

Cramton, P., Ockenfels, A. & Stoft, S. (2013), "Capacity market fundamentals," *Economics of Energy & Environmental Policy* **2**(2), 27–46.

Cramton, P. & Stoft, S. (2005), "A capacity market that makes sense," *The Electricity Journal* **18**(7), 43–54.

Cramton, P. & Stoft, S. (2006), The convergence of market designs for adequate generating capacity with special attention to the CAISO's resource adequacy problem, Technical report, MIT Center for Energy and Environmental Policy Research.

Cramton, P. & Stoft, S. (2008), "Forward reliability markets: Less risk, less market power, more efficiency," *Utilities Policy* **16**(3), 194–201.

Damcı-Kurt, P., Küçükyavuz, S., Rajan, D. & Atamtürk, A. (2016), "A polyhedral study of production ramping," *Mathematical Programming* **158**, 175–205.

de Maere d'Aertrycke, G., Ehrenmann, A. & Smeers, Y. (2017), "Investment with incomplete markets for risk: The need for long-term contracts," *Energy Policy* **105**, 571–583.

De Matos, V. L., Philpott, A. B. & Finardi, E. C. (2015), "Improving the performance of stochastic dual dynamic programming," *Journal of Computational and Applied Mathematics* **290**, 196–208.

De Matos, V., Philpott, A. B., Finardi, E. C. & Guan, Z. (2010), Solving long-term hydro-thermal scheduling problems, Technical report, Electric Power Optimization Centre, University of Auckland.

Dijk, J. & Willems, B. (2011), "The effect of counter-trading on competition in electricity markets," *Energy Policy* **39**(3), 1764–1773.

Dupacova, J., Gröwe-Kuska, N. & Römisch, W. (2003), "Scenario reduction in stochastic programming: An approach using probability metrics," *Math Programming* **95**(3), 493–511.

E-Bridge (2014), Potential cross-border balancing cooperation between the Belgian, Dutch and German electricity transmission system operators, Technical report, Institute of Power Systems and Power Economics and E-Bridge Consulting GMBH.

Ehrenmann, A. (2004), Equilibrium problems with equilibrium constraints and their application to electricity markets, PhD thesis, Cambridge University.

Ehrenmann, A. & Smeers, Y. (2005), "Inefficiencies in European congestion management proposals," *Utilities Policy* **13**, 135–152.

Ehrenmann, A. & Smeers, Y. (2011a), "Generation capacity expansion in a risky environment: A stochastic equilibrium analysis," *Operations Research* **59**(6), 1332–1346.

Ehrenmann, A. & Smeers, Y. (2011b), *Stochastic Equilibrium Models for Generation Capacity Expansion*, Vol. 163 of *Stochastic Optimization Methods in Finance and Energy, International Series in Operations Research and Management Science, Part 2*, Springer, pp. 273–310.

Eldridge, B., O'Neill, R. P. & Castillo, A. R. (2016), Marginal loss calculations for the DCOPF, Technical report, Sandia National Lab. (SNL-NM), Albuquerque, NM (United States).

ERCOT (2015), "ERCOT market training: Purpose of ORDC, methodology for implementing ORDC, settlement impacts for ORDC," ERCOT.

European Commission (2017a), "Commission regulation (EU) 2017/2195 of 23 November 2017 establishing a guideline on electricity balancing," *Official Journal of the European Union*, L312/6.

European Commission (2017b), State aid no. SA. 44464 (2017/N) – Ireland: Irish capacity mechanism, Technical report.

Fabra, N. (2018), "A primer on capacity mechanisms," *Energy Economics* **75**, 323–335.

Fabra, N., von der Fehr, N.-H. & Harbord, D. (2006), "Designing electricity auctions," *The RAND Journal of Economics* **37**(1), 23–46.

Fernandez-Blanco, R., Arroyo, J. M. & Alguacil, N. (2011), "A unified bilevel programming framework for price-based market clearing under marginal pricing," *IEEE Transactions on Power Systems* **27**(1), 517–525.

Fisher, E. B., O'Neill, R. P. & Ferris, M. C. (2008), "Optimal transmission switching," *IEEE Transactions on Power Systems* **23**(3), 1346–1355.

Flach, B., Barroso, L. A. & Pereira, M. V. F. (2010), "Long-term optimal allocation of hydro generation for a price-maker company in a competitive market: Latest developments and a stochastic dual dynamic programming approach," *IET Generation, Transmission and Distribution* **4**(2), 299–314.

Follmer, H. & Schied, A. (2002), "Convex measures of risk and trading constraints," *Finance and Stochastics* **6**(4), 429–447.

Fuller, J. D., Ramasra, R. & Cha, A. (2012), "Fast heuristics for transmission-line switching," *IEEE Transactions on Power Systems* 27(3), 1377–1386.

Gabriel, S. A., Conejo, A. J., Fuller, J. D., Hobbs, B. F. & Ruiz, C. (2012), *Complementarity Modeling in Energy Markets*, Vol. 180, Springer Science & Business Media.

Garcia, M., Nagarajan, H. & Baldick, R. (2020), "Generalized convex hull pricing for the AC optimal power flow problem," *IEEE Transactions on Control of Network Systems* 7(3), 1500–1510.

Gates, B. (2023), "The surprising key to a clean energy future," GatesNotes. www.gatesnotes.com/Transmission.

Gedra, T. W. & Varaiya, P. P. (1993), "Markets and pricing for interruptible electric power," *IEEE Transactions on Power Systems* 8(1), 122–128.

Gentile, C., Morales-Espana, G. & Ramos, A. (2017), "A tight MIP formulation of the unit commitment problem with start-up and shut-down constraints," *EURO Journal on Computational Optimization* 5(1–2), 177–201.

Gérard, C., Ávila, D., Mou, Y., Papavasiliou, A. & Chevalier, P. (2022), "Comparison of priority service with multilevel demand subscription," *IEEE Transactions on Smart Grid* 13(3), 2026–2037.

Gérard, C. & Papavasiliou, A. (2022), "The role of service charges in the application of priority service pricing," *Energy Systems* 13(4), 1099–1128.

Gjelsvik, A., Belsnes, M. M. & Haugstad, A. (1999), "An algorithm for stochastic medium-term hydrothermal scheduling under spot price uncertainty," *Power Systems Computation Conference*, pp. 1079–1085.

Glover, D. J., Sarma, M. S. & Overbye, T. J. (2012), *Power system analysis and design*, Cengage Learning.

Greek delegation, EU Council (2022), "Proposal for a power market design in order to decouple electricity prices from soaring gas prices – information from the Greek delegation," Information Note 11398/22.

Green, R. (2000), "Competition in generation: The economic foundations," *Proceedings of the IEEE* 88(2), 128–139.

Gribik, P. R., Hogan, W. W. & Pope, S. L. (2007), Market-clearing electricity prices and energy uplift, Technical report, John F. Kennedy School of Government, Harvard University.

Gros, D. (2022), Optimal tariff versus optimal sanction: The case of European gas imports from Russia, Technical report, CEPS policy insights.

Gröwe-Kuska, N., Kiwiel, K. C., Nowak, M. P., Römisch, W. & Wegner, I. (2002), "Power management in a hydro-thermal system under uncertainty by Lagrangian relaxation." In Greengard, C. & Ruszczynski, A. (eds.), *Decision Making Under Uncertainty: Energy and Power*, Vol. 128, IMA Volumes in Mathematics and Its Applications, Springer, pp. 39–70.

Guo, Y., Chen, C. & Tong, L. (2021), "Pricing multi-interval dispatch under uncertainty part I: Dispatch-following incentives," *IEEE Transactions on Power Systems* 36(5), 3865–3877.

Han, J. & Papavasiliou, A. (2015), "Congestion management through topological corrections: A case study of central western Europe," *Energy Policy* 86, 470–482.

Han, J. & Papavasiliou, A. (2016), "The impacts of transmission topology control on the European electricity network," *IEEE Transactions on Power Systems* 31(1), 496–507.

Hao, S., Angelidis, G. A., Singh, H. & Papalexopoulos, A. D. (1998), "Consumer payment minimization in power pool auctions," *IEEE Transactions on Power Systems* 13(3), 986–991.

Harker, P. T. (1993), *Lectures on computation of equilibria with equation-based methods*, Center for Operations Research and Econometrics, UCLouvain.

Harvey, S. M. & Hogan, W. W. (2000), "Nodal and zonal congestion management and the exercise of market power," *Harvard University* 21.

Hassler, J. & Sinn, H.-W. (2016), "The fossil episode," *Journal of Monetary Economics* 83, 14–26.

Hedman, K. W., Ferris, M. C., O'Neill, R. P., Fisher, E. B. & Oren, S. S. (2010),

"Co-optimization of generation unit commitment and transmission switching with N-1 reliability," *IEEE Transactions on Power Systems* **25**(2), 1052–1063.

Hedman, K. W., O'Neill, R. P., Fisher, E. B. & Oren, S. S. (2009), "Optimal transmission switching with contingency analysis," *IEEE Transactions on Power Systems* **24**(3), 1577–1586.

Hedman, K. W., O'Neill, R. P., Fisher, E. B. & Oren, S. S. (2010), "Smart flexible just-in-time transmission and flowgate bidding," *IEEE Transactions on Power Systems* **26**(1), 93–102.

Heitsch, H. & Römisch, W. (2003), "Scenario reduction algorithms in stochastic programming," *Computational Optimization and Applications* **24**(2–3), 187–206.

Hellenic Republic of Environment and Energy (2022), A power market design to decouple electricity prices from soaring natural gas prices – analysis of the Greek proposal, Technical Report, Cabinet of the Minister.

Hirst, E. & Kirby, B. (1996), Electric power ancillary services, Technical Report ORNL/CON-426, Oak Ridge National Laboratory.

Hirth, L. & Schlecht, I. (2018), Market-based redispatch in zonal electricity markets, USAEE Working Paper No. 18–369.

Hobbs, B. F. & Oren, S. S. (2019), "Three waves of US reforms: Following the path of wholesale electricity market restructuring," *IEEE Power and Energy Magazine* **17**(1), 73–81.

Hogan, W. W. (1992), "Contract networks for electric power transmission," *Journal of Regulatory Economics* **4**, 211–242.

Hogan, W. W. (2005), On an "energy only" electricity market design for resource adequacy, Technical Report, Center for Business and Government, JFK School of Government, Harvard University.

Hogan, W. W. (2013), "Electricity scarcity pricing through operating reserves," *Economics of Energy and Environmental Policy* **2**(2), 65–86.

Hogan, W. W. (2016a), Electricity market design: Optimization and market equilibrium, *in* "Workshop on Optimization and Equilibrium in Energy Economics, Institute for Pure and Applied Mathematics, Los Angeles, CA."

Hogan, W. W. (2016b), "Virtual bidding and electricity market design," *The Electricity Journal* **29**(5), 33–47.

Hogan, W. W. (2020), "Electricity market design: Multi-interval pricing models," https://scholar.harvard.edu/files/whogan/files/hogan_hepg_multi_period_062220.pdf.

Hogan, W. W. & Pope, S. L. (2019), PJM reserve markets: Operating reserve demand curve enhancements, Technical Report, Harvard University.

Hogan, W. W. & Ring, B. J. (2003), On minimum-uplift pricing for electricity markets, Technical Report, John F. Kennedy School of Government, Harvard University.

Hotelling, H. (1931), "The economics of exhaustible resources," *Journal of Political Economy* **39**(2), 137–175.

Hua, B., Schiro, D. A., Zheng, T., Baldick, R. & Litvinov, E. (2019), "Pricing in multi-interval real-time markets," *IEEE Transactions on Power Systems* **34**(4), 2696–2705.

Hubbert, M. K. (1956), Nuclear energy and the fossil fuels, Shell Development Company, Publication No. 95.

Jevons, W. S. (1865), *The Coal Question: An Inquiry Concerning the Progress of the Nation, and the Probable Exhaustion of Our Coal-Mines*, Macmillan and Co.

Jevons, W. S. (1879), *The Theory of Political Economy*, Macmillan and Co.

Joskow, P. & Tirole, J. (2007), "Reliability and competitive electricity markets," *RAND Journal of Economics* **38**(1), 60–84.

Kaye, R. J., Outhred, H. R. & Bannister, C. H. (1990), "Forward contracts for the operation of an electricity industry under spot pricing," *IEEE Transactions on Power Systems* **5**(1), 46–52.

Keay, M. & Robinson, D. (2017a), The decarbonized electricity system of the future: The two market approach, Technical Report, The Oxford Institute for Energy Studies.

Keay, M. & Robinson, D. (2017b), "Market Design for a Decarbonized Electricity System: The Two Market Approach." *Proceedings from the Eurelectric-Florence School of*

*Regulation Conference, 7 June 2017*, European University Institute, 61–64.

Kunz, F. (2013), "Improving congestion management: How to facilitate the integration of renewable generation in Germany," *The Energy Journal* **34**(4), 55–78.

Lété, Q. (2022), Models and algorithms for quantifying and mitigating the inefficiency of zonal pricing: Short and long-term measures, PhD thesis, UCLouvain.

Lété, Q. & Papavasiliou, A. (2020a), "Impacts of transmission switching in zonal electricity markets – Part I," *IEEE Transactions on Power Systems* **36**(2), 902–913.

Lété, Q. & Papavasiliou, A. (2020b), "Impacts of transmission switching in zonal electricity markets – Part II," *IEEE Transactions on Power Systems* **36**(2), 914–922.

Lété, Q., Smeers, Y. & Papavasiliou, A. (2022), "An analysis of zonal electricity pricing from a long-term perspective," *Energy Economics* **107**, 105853.

Litvinov, E., Zhao, F. & Zheng, T. (2009), "Alternative auction objectives and pricing schemes in short-term electricity markets," 2009 IEEE Power & Energy Society General Meeting.

Löhndorf, N. & Shapiro, A. (2019), "Modeling time-dependent randomness in stochastic dual dynamic programming," *European Journal of Operational Research* **273**(2), 650–661.

Löhndorf, N. & Wozabal, D. (2021), "Gas storage valuation in incomplete markets," *European Journal of Operational Research* **288**(1), 318–330.

Luh, P. B., Blankson, W. E., Chen, Y., Yan, J. H., Stern, G. A., Chang, S.-C. & Zhao, F. (2006), "Payment cost minimization auction for deregulated electricity markets using surrogate optimization," *IEEE Transactions on Power systems* **21**(2), 568–578.

Lyons, K., Fraser, H. & Parmesano, H. (2000), "An introduction to financial transmission rights," *The Electricity Journal* **13**(10), 31–37.

Machado, F. D., Diniz, A. L., Borges, C. L. & Brandao, L. C. (2021), "Asynchronous parallel stochastic dual dynamic programming applied to hydrothermal generation planning," *Electric Power Systems Research* **191**, 106907.

Madani, M., Ruiz, C., Siddiqui, S. & Van Vyve, M. (2018), "Convex hull, IP and European electricity pricing in a European power exchanges setting with efficient computation of convex hull prices," *arXiv:1804.00048*.

Madani, M. & Van Vyve, M. (2015), "Computationally efficient MIP formulation and algorithms for European day-ahead electricity market auctions," *European Journal of Operational Research* **242**, 580–593.

Matsumoto, T., Bunn, D. & Yamada, Y. (2021), "Mitigation of the inefficiency in imbalance settlement designs using day-ahead prices," *IEEE Transactions on Power Systems* **37**(5), 3333–3345.

Meadows, D. H., Meadows, D. L., Randers, J. & Behrens III, W. W. (1972), *The Limits to Growth: A Report to the Club of Rome*, Potomac Associates.

Meibom, P., Barth, R., Hasche, B., Brand, H., Weber, C. & O'Malley, M. (2010), "Stochastic optimization model to study the operational impacts of high wind penetrations in Ireland," *IEEE Transactions on Power Systems* **26**(3), 1367–1379.

Mickey, J. (2015), "Multi-interval real-time market overview," Board of Directors Meeting, ERCOT, October 13, 2015.

Monticelli, A., Pereira, M. V. F. & Granville, S. (1987), "Security-constrained optimal power flow with post-contingency corrective rescheduling," *IEEE Transactions on Power Systems* **2**(1), 175–180.

Morales, J. M., Pineda, S., Conejo, A. J. & Carrion, M. (2009), "Scenario reduction for futures trading in electricity markets," *IEEE Transactions on Power Engineering* **24**(2), 878–888.

Morales-España, G., Gentile, C. & Ramos, A. (2015), "Tight MIP formulations of the power-based unit commitment problem," *OR Spectrum* **37**(4), 929–950.

Morales-España, G., Latorre, J. M. & Ramos, A. (2013), "Tight and compact MILP formulation for the thermal unit commitment problem," *IEEE Transactions on Power Systems* **28**(4), 4897–4908.

Mou, Y., Papavasiliou, A. & Chevalier, P. (2019), "A bi-level optimization formulation of priority service pricing," *IEEE Transactions on Power Systems* **35**(4), 2493–2505.

Muckstadt, J. A. & Koenig, S. A. (1977), "An application of Lagrangian relaxation to scheduling in power-generation systems," *Operations Research* **25**(3), 387–403.

New York ISO (2019), Ancillary services shortage pricing, Technical Report, New York Independent System Operator.

Nordhaus, W. & Sztorc, P. (2013), DICE 2013R: Introduction and users manual, Program on Coupled Human and Earth Systems, www.econ.yale.edu/~nordhaus/homepage/homepage/documents/DICE_Manual_100413r1.pdf.

Nowak, M. P. & Römisch, W. (2000), "Stochastic Lagrangian relaxation applied to power scheduling in a hydro-thermal system under uncertainty," *Annals of Operations Research* **100**(1–4), 251–272.

O'Neill, R. P., Sotkiewicz, P. M., Hobbs, B. F., Rothkopf, M. H. & Stewart, W. R. (2005), "Efficient market-clearing prices in markets with nonconvexities," *European Journal of Operational Research* **164**(1), 269–285.

Oren, S. S. (1987), "Product differentiation and product line pricing in service industries," *First Annual Conference on Pricing*, New York.

Oren, S. S. (2001), "Design of ancillary services markets," *Proceedings of the 34th Hawaii International Conference on System Sciences*.

Oren, S. S. (2004), When is a pay-as-bid preferable to uniform price in electricity markets, Technical Report, University of California Energy Institute.

Oren, S. S. (2005a), Market design and gaming in competitive electricity markets, Technical Report, UC Berkeley.

Oren, S. S. (2005b), "Generation adequacy via call options obligations: Safe passage to the promised land," *The Electricity Journal* **18**(9), 28–42.

Oren, S. S., Spiller, P. T., Varaiya, P. & Wu, F. (1995), "Nodal prices and transmission rights: A critical appraisal," *The Electricity Journal* **8**(3), 24–35.

Özdemir, O. (2013), Simulation modeling and optimization of competitive electricity

markets and stochastic fluid systems, PhD thesis, Tilburg University.

Papalexopoulos, A., Beal, J. & Florek, S. (2013), "Precise mass-market energy demand management through stochastic distributed computing," *IEEE Transactions on Smart Grid* **4**(4), 2017–2027.

Papavasiliou, A. (2020), "Scarcity pricing and the missing European market for real-time reserve capacity," *The Electricity Journal* **33**(10), 106863.

Papavasiliou, A. (2021a), 'Investigation of the possibility of implementing a scarcity pricing mechanism in the context of the operation of the Greek electricity market under the Target Model'.

Papavasiliou, A. (2021b), Overview of EU capacity remuneration mechanisms, Technical Report. Available at www.rae.gr/wp-content/uploads/2021/05/Report-I-CRM-final.pdf.

Papavasiliou, A. & Bertrand, G. (2021), "Market design options for scarcity pricing in European balancing markets," *IEEE Transactions on Power Systems* **36**(5), 4410–4419.

Papavasiliou, A., Cartuyvels, J., Bertrand, G. & Marien, A. (2023), "Implementation of scarcity pricing without co-optimization in European energy-only balancing markets," *Utilities Policy* **81**, 101488.

Papavasiliou, A., Doorman, G., Bjørndal, M., Langer, Y., Leclercq, G. & Crucifix, P. (2022), "Interconnection of Norway to European balancing platforms using hierarchical balancing," *18th International Conference on the European Energy Market (EEM)*, IEEE.

Papavasiliou, A., He, Y. & Svoboda, A. (2015), "Self-commitment of combined cycle units under electricity price uncertainty," *IEEE Transactions on Power Systems* **30**(4), 1690–1701.

Papavasiliou, A. & Oren, S. S. (2013), "Multi-area stochastic unit commitment for high wind penetration in a transmission constrained network," *Operations Research* **61**(3), 578–592.

Papavasiliou, A., Oren, S. S. & O'Neill, R. P. (2011), "Reserve requirements for wind power integration: A scenario-based stochastic programming framework," *IEEE Transactions on Power Systems* **26**(4), 2197–2206.

Papavasiliou, A., Oren, S. S. & Rountree, B. (2015), "Applying high performance computing to transmission-constrained stochastic unit commitment for renewable penetration," *IEEE Transactions on Power Systems* **30**(3), 1690–1701.

Papavasiliou, A. & Smeers, Y. (2017), "Remuneration of flexibility using operating reserve demand curves: A case study of Belgium," *The Energy Journal* **38**(6), 105–135.

Papavasiliou, A., Smeers, Y. & de Maere-d'Aertrycke, G. (2021), "Market design considerations for scarcity pricing: A stochastic equilibrium framework," *The Energy Journal* **42**(5), 195–220.

Parsons, J., Colbert, C., Larrieu, J., Martin, T. & Mastrangelo, E. (2015), Financial arbitrage and efficient dispatch in wholesale electricity markets, Technical Report, MIT Center for Energy and Environmental Policy Research.

Pereira, M. V. F. & Pinto, L. M. V. G. (1991), "Multi-stage stochastic optimization applied to energy planning," *Mathematical Programming* **52**, 359–375.

Philpott, A., Ferris, M. & Wets, R. (2016), "Equilibrium, uncertainty, and risk in hydrothermal electricity systems," *Mathematical Programming* **157**, 483–513.

Pindyck, Robert S. & Rubinfeld, D. L. (2013), *Microeconomics*, 8th ed, Pearson.

PJM Interconnection (2017), Proposed enhancements to energy price formation, Technical report, PJM.

Powell, W. B. (2007), *Approximate Dynamic Programming: Solving the Curses of Dimensionality*, Vol. 703, John Wiley & Sons.

Queyranne, M. & Wolsey, L. A. (2017), "Tight MIP formulations for bounded up/down times and interval-dependent start-ups," *Mathematical Programming* **164**(1–2), 129–155.

Rajan, D. & Takriti, S. (2005), Mininum up/down polytopes of the unit commitment problem with startup costs, Technical Report, IBM Research.

Ralph, D. & Smeers, Y. (2011), Pricing risk under risk measures: An introduction to stochastic-endogenous equilibra, Technical Report, Cambridge University and UCLouvain.

Ralph, D. & Smeers, Y. (2015), "Risk trading and endogenous probabilities in investment equilibria," *SIAM Journal on Optimization* **25**(4), 2589–2611.

Ring, B. J. (1995), Dispatch based pricing in decentralized power systems, PhD thesis, University of Canterbury.

Rockafellar, R. T. & Uryasev, S. (2002), "Conditional value-at-risk for general loss distributions," *Journal of Banking and Finance* **26**, 1443–1471.

Ruiz, P. A., Philbrick, R. C., Zack, E., Cheung, K. W. & Sauer, P. W. (2009), "Uncertainty management in the unit commitment problem," *IEEE Transactions on Power Systems* **24**(2), 642–651.

Samuelson, P. A. (1952), "Spatial price equilibrium and linear programming," *The American Economic Review* **42**(3), 283–303.

Schiro, D. A. (2017), "Flexibility procurement and reimbursement: A multiperiod pricing approach," *FERC Technical Conference*.

Schiro, D. A., Zheng, T., Zhao, F. & Litvinov, E. (2015), "Convex hull pricing in electricity markets: Formulation, analysis, and implementation challenges," *IEEE Transactions on Power Systems* **31**(5), 4068–4075.

Schweppe, F. C., Caramanis, M. C., Tabors, R. D. & Bohn, R. E. (1988), *Spot Pricing of Electricity*, Kluwer Academic Publishers.

Shanker, R. (2003), "Comments on standard market design: Resource adequacy requirements," Federal Energy Regulatory Commission, Docket RM01-12-000.

Shapiro, A., Dentcheva, D. & Ruszczynski, A. (2009), *Lectures on stochastic programming: Modeling and theory*, Society for Industrial and Applied Mathematics.

Shapiro, A., Tekaya, W., da Costa, J. P. & Soares, M. P. (2013), "Risk neutral and risk averse stochastic dual dynamic programming method," *European Journal of Operational Research* **224**, 375–391.

Silbernagl, M., Huber, M. & Brandenberg, R. (2015), "Improving accuracy and efficiency of start-up cost formulations in MIP unit commitment by modeling power plant temperatures," *IEEE Transactions on Power Systems* **31**(4), 2578–2586.

Simoglou, C. K., Biskas, P. N. & Bakirtzis, A. G. (2010), "Optimal self-scheduling of a thermal producer in short-term electricity markets by MILP," *IEEE Transactions on Power Systems* **25**(4), 1965–1977.

Singh, L. P. (2006), *Advanced Power System Analysis and Dynamics*, New Age International.

Sioshansi, R. & Nicholson, E. (2011), "Towards equilibrium offers in unit commitment auctions with nonconvex costs," *Journal of Regulatory Economics* **40**(1), 41–61.

Smeers, Y. (1997), "Computable equilibrium models and the restructuring of the European electricity and gas markets," *Energy Journal* **18**(4), 1–31.

Starr, R. M. (1969), "Quasi-equilibria in markets with non-convex preferences," *Econometrica: Journal of the Econometric Society* **37**(1), 25–38.

Stevens, N. (2016), Models and algorithms for pricing electricity in unit commitment, MSc thesis, UCLouvain.

Stevens, N. & Papavasiliou, A. (2022), "Application of the level method for computing locational convex hull prices," *IEEE Transactions on Power Systems* **37**(5), 3958–3968.

Stoft, S. (2002), *Power System Economics*, IEEE Press and Wiley Interscience.

Taha, H. A. (2017), *Operations Research: An Introduction*, 10th ed, Pearson Education Limited.

Takriti, S., Birge, J. R. & Long, E. (1996), "A stochastic model for the unit commitment problem," *IEEE Transactions on Power Systems* **11**(3), 1497–1508.

Taylor, J. A. (2015), *Convex Optimization of Power Systems*, Cambridge University Press.

Tuohy, A., Meibom, P., Denny, E. & O'Malley, M. (2009), "Unit commitment for systems with high wind penetration," *IEEE Transactions on Power Systems* **24**(2), 592–601.

UK Government (2013), Electricity market reform – contract for difference: Contract and allocation overview, Technical Report 13D/187.

Varian, H. R. (2014), *Intermediate Microeconomics: A Modern Approach*, 9th ed, WW Norton & Company.

Vazquez, C., Rivier, M. & Perez-Arriaga, I. J. (2002), "A market approach to long-term security of supply," *IEEE Transactions on Power Systems* **17**(2), 349–357.

Vickrey, W. (1961), "Counterspeculation, auctions and competitive sealed tenders," *The Journal of Finance* **16**(1), 8–37.

Vournas, C. & Kontaxis, G. (2010), *Introduction to Electric Power Systems*, Symmetria.

Wang, B. & Hobbs, B. F. (2016), "Real-time markets for flexiramp: A stochastic unit commitment-based analysis," *IEEE Transactions on Power Systems* **31**(2), 846–860.

Watson, J.-P. & Woodruff, D. L. (2011), "Progressive hedging innovations for a class of stochastic mixed-integer resource allocation problems," *Computational Management Science* **8**, 355–370.

Wilson, R. (1993), *Nonlinear Pricing*, Oxford University Press.

Wolsey, L. A. & Nemhauser, G. L. (1999), *Integer and Combinatorial Optimization*, Vol. 55, John Wiley & Sons.

Wood, A. J., Wollenberg, B. F. & Sheblé, G. B. (2013), *Power Generation, Operation, and Control*, John Wiley & Sons.

Wu, F., Varaiya, P. P., Spiller, P. & Oren, S. S. (1996), "Folk theorems on transmission access: Proofs and counterexamples," *Journal of Regulatory Economics* **10**(1), 5–23.

Wu, T., Rothleder, M., Alaywan, Z. & Papalexopoulos, A. (2004), "Pricing energy and ancillary services in integrated market systems by an optimal power flow," *IEEE Transactions on Power Systems* **19**(1), 339–347.

Zarnikau, J., Zhu, S., Woo, C. K. & Tsai, C. (2020), "Texas's operating reserve demand curve's generation investment incentive," *Energy Policy* **137**, 111–143.

Zhao, F., Luh, P. B., Yan, J. H., Stern, G. A. & Chang, S.-C. (2008), "Payment cost minimization auction for deregulated electricity markets with transmission capacity constraints," *IEEE Transactions on Power Systems* **23**(2), 532–544.

Zhao, F., Luh, P. B., Yan, J. H., Stern, G. A. & Chang, S.-C. (2010), "Bid cost minimization versus payment cost minimization: A game theoretic study of electricity auctions," *IEEE Transactions on Power Systems* **25**(1), 181–194.

Zhao, J., Zheng, T. & Litvinov, E. (2019), "A multi-period market design for markets with intertemporal constraints," *IEEE Transactions on Power Systems* **35**(4), 3015–3025.

Zheng, T. & Litvinov, E. (2008), "Contingency-based zonal reserve modeling and pricing in a co-optimized energy and reserve market," *IEEE Transactions on Power Systems* **23**(2), 277–286.

Zhou, Z. & Botterud, A. (2014), "Dynamic scheduling of operating reserves in co-optimized electricity markets with wind power," *IEEE Transactions on Power Systems* **29**(1), 160–171.

# Index

Printed in the United States
by Baker & Taylor Publisher Services